Industrial CFD and Fluid Modelling in Engineering

Industrial CFD and Fluid Modelling in Engineering

Guest Editors

Ernesto Benini
Francesco De Vanna

MDPI

Basel • Beijing • Wuhan • Barcelona • Belgrade • Novi Sad • Cluj • Manchester

Guest Editors

Ernesto Benini
Dipartimento di Ingegneria Industriale
Università degli studi di Padova
Padova
Italy

Francesco De Vanna
Dipartimento di Ingegneria Industriale
Università degli studi di Padova
Padova
Italy

Editorial Office
MDPI AG
Grosspeteranlage 5
4052 Basel, Switzerland

This is a reprint of the Special Issue, published open access by the journal *Fluids* (ISSN 2311-5521), freely accessible at: https://www.mdpi.com/journal/fluids/special_issues/CFD_engineering.

For citation purposes, cite each article independently as indicated on the article page online and as indicated below:

Lastname, A.A.; Lastname, B.B. Article Title. *Journal Name* **Year**, *Volume Number*, Page Range.

ISBN 978-3-7258-3576-8 (Hbk)
ISBN 978-3-7258-3575-1 (PDF)
https://doi.org/10.3390/books978-3-7258-3575-1

Contents

About the Editors

Ernesto Benini

Prof. Ernesto Benini is a renowned expert in machine flows and aircraft propulsion. He is a full professor at the University of Padova, specializing in turbomachinery. His research focuses on gas turbines, propulsion efficiency, and computational fluid dynamics. He has authored numerous scientific papers and contributed to advancements in aerospace engineering. He also plays a key role in academic coordination and industrial collaborations.

Francesco De Vanna

Dr. Francesco De Vanna is an assistant professor at the University of Padova, specializing in machine fluid dynamics. His research focuses on computational fluid dynamics and turbulence modeling in compressible flows. He has contributed significantly to several analyses and problems in CFD using Wall-Modeled LES and Immersed Boundary Methods.

fluids

MDPI

Editorial

Industrial CFD and Fluid Modelling in Engineering

Francesco De Vanna

Dipartimento di Ingegneria Industriale, Università degli Studi di Padova, Via Venezia, 1, 35121 Padova, Italy; francesco.devanna@unipd.it

Fluids is proud to present the Special Issue "Industrial CFD and Fluid Modelling in Engineering", a carefully curated collection of pioneering research that underscores the transformative role of Computational Fluid Dynamics (CFD) in addressing the challenges of industrial fluid mechanics. This Special Issue bridges the gap between advanced CFD methodologies and their practical applications, showcasing innovative strategies that tackle the complexities of real-world engineering systems. Industrial systems are often characterized by complex geometries, multifaceted flow dynamics, and demanding computational requirements, and this collection highlights how modern modelling paradigms and simulation techniques are extending the capabilities of applied CFD to meet these challenges. The aim of this editorial is to provide a concise overview of the papers included in this Special Issue, along with relevant references to contextualize their contributions.

It is worth noting that, given the widespread nature of the topic, the articles featured in this Special Issue underscore the diverse applications of CFD, demonstrating its remarkable versatility in tackling a broad range of engineering challenges, starting with ref. [1], which offers a comprehensive review of CFD modelling for Darrieus turbines, comparing 2D, 2.5D, and 3D approaches. The study synthesizes insights from over 250 articles, providing detailed guidance on turbulence models and meshing techniques. Particular emphasis is placed on understanding wake vortex structures and power coefficient predictions, which are critical for accurately simulating turbine performance. Several investigations have, in fact, addressed this topic in the last decade, further advancing CFD modelling approaches for Darrieus Vertical Axis Wind Turbines (VAWTs). For instance, Alaimo et al. [2] employed QBlade, a specialized simulation tool for wind turbine blade design and aerodynamic analysis, to perform 3D CFD simulations of VAWTs, while Lanzafame et al. [3] focused on 2D CFD modelling for micro H-Darrieus wind turbines, leveraging an unsteady Delayed Detached Eddy Simulation (DDES) approach for turbulence modelling.

Environmental control applications represent another critical area of focus in this Special Issue, as demonstrated in ref. [4], where a 3D thermofluidic CFD model for a climatic chamber is developed and validated. This study explores strategies to optimize chamber efficiency through targeted design improvements, such as adjusting fan rotation and incorporating deflectors, ensuring precise control of airflow, temperature, and humidity distributions. Similarly, ref. [5] addresses the modelling of gas dispersion during liquefied natural gas leaks in bunkering operations, comparing the Flame Acceleration Simulator (FLACS) and BASiL models. The research highlights the importance of high-resolution simulations to refine safety zone assessments and mitigate gas dispersion risks, particularly under variable wind conditions. The challenges of enhancing environmental control applications and advancing gas dispersion modelling are pivotal topics within the realm of industrial CFD.

In addition to these topics, aerodynamic performance and structural integrity are explored in ref. [6], which investigates the effects of leading-edge erosion on wind turbine blades. By comparing 2D and 3D models, the study highlights how computationally

Received: 13 January 2025
Accepted: 15 January 2025
Published: 17 January 2025

Citation: De Vanna, F. Industrial CFD and Fluid Modelling in Engineering. *Fluids* **2025**, *10*, 15. https://doi.org/10.3390/fluids10010015

efficient 2D approaches can effectively capture key aerodynamic features while identifying their limitations, particularly in predicting drag forces. This balance between computational efficiency and accuracy is crucial for predictive maintenance and digital twin development. Fluid–structure interaction problems are further examined in the Special Issue in ref. [7], where coupled simulations investigate the interplay between fluid flow and vibro-acoustic interactions. The study provides valuable insights into how flow-induced vibrations affect acoustic wave propagation and structural behaviour, offering design strategies for systems subjected to such forces. The broader topic of fluid–structure interaction is widely addressed in the recent literature. For instance, Ehsan Khalili et al. [8] developed a coupled CFD–FEA approach to study sound generation in a stenosed artery, elucidating how flow-induced vibrations propagate through surrounding tissues. Similarly, De Vanna et al. [9] introduced a robust numerical approach for simulating compressible flows around moving objects using the Ghost-Point-Forcing Method (GPFM). This methodology, as further detailed in ref. [10], is particularly effective in resolving challenges related to moving boundaries in compressible environments, including turbulent flows. Together, all of these contributions underscore the potential of advanced numerical methods to enhance the design and analysis of aerospace, energy, and mechanical systems' aerodynamic performance.

Innovative surrogate modelling techniques are also collected in the present Special Issue, as reported in ref. [11], where heat transfer in oscillating flows is analysed using an efficient method that treats the fluid as a stationary solid with orthotropic thermal conductivity. This approach significantly reduces computational costs while maintaining accuracy, making it highly relevant for thermal design optimization. Similarly, ref. [12] focuses on supersonic flows, investigating the performance of binary fluid ejectors through computational and experimental methods. Their work provides critical insights into flow dynamics, entrainment ratios, and design optimization for applications in refrigeration and gas mixing.

In the context of multiphase flows, this Special Issue collects ref. [13], which explores the dynamic rearrangement of particulate matter in wall-flow filters during regeneration processes. By modelling particle detachment, transport, and deposition patterns, this study offers predictive insights that are instrumental in enhancing engine performance and extending the service life of filtration systems. Similarly, ref. [14] investigates droplet–wall interactions in Selective Catalytic Reduction (SCR) systems, introducing a new impingement model that effectively captures critical phenomena such as film formation, droplet breakup, and heat transfer. This research underscores the pivotal role of impingement behaviour in improving ammonia distribution and optimizing thermal performance. Further advancements in these areas have been achieved through recent studies. For example, Hafen et al. [15] employed lattice Boltzmann methods to analyse particulate detachment and transport within wall-flow filters during regeneration while, in the context of SCR systems, Bai et al. [16] developed a numerical model to examine the heat transfer characteristics and wall film formation caused by spray impingement, highlighting critical factors that influence ammonia distribution and system performance.

Finally, mixing efficiency in industrial systems is a critical area of investigation, as demonstrated in ref. [17], which evaluates the hydrodynamics of stirred reactors for lead recycling. This study identifies optimal tracer injection points, and highlights the significant influence of tank geometry and impeller design on operational performance. The model's accuracy is validated through physical experiments, offering practical recommendations for improving industrial processes. In a related domain, innovative approaches to cryogenic applications are explored in ref. [18], which introduces a cryogenic duplex pressure-swirl atomizer for seafood preservation. This system combines liquid nitrogen and water sprays to achieve rapid cooling and efficient cold preservation, with its performance validated using a robust CFD framework.

To conclude, this Special Issue exemplifies the diverse and impactful applications of CFD in industrial contexts. By advancing the accuracy, efficiency, and applicability of simulation techniques, the collected studies not only push the boundaries of what CFD can achieve, but also pave the way for future developments in industrial fluid mechanics. We hope this collection inspires further research and innovation in this dynamic and essential field.

Conflicts of Interest: The author declares no conflicts of interest.

References

1. Fertahi, S.E.D.; Belhadad, T.; Kanna, A.; Samaouali, A.; Kadiri, I.; Benini, E. A Critical Review of CFD Modeling Approaches for Darrieus Turbines: Assessing Discrepancies in Power Coefficient Estimation and Wake Vortex Development. *Fluids* **2023**, *8*, 242. [CrossRef]
2. Alaimo, A.; Esposito, A.; Messineo, A.; Orlando, C.; Tumino, D. 3D CFD Analysis of a Vertical Axis Wind Turbine. *Energies* **2015**, *8*, 3013–3033. [CrossRef]
3. Lanzafame, R.; Mauro, S.; Messina, M.; Brusca, S. Development and Validation of CFD 2D Models for the Simulation of Micro H-Darrieus Turbines Subjected to High Boundary Layer Instabilities. *Energies* **2020**, *13*, 5564. [CrossRef]
4. Ramezani, B.; Tadeu, A.; Jesus, T.; Brett, M.; Mendes, J. Evaluation of the Thermofluidic Performance of Climatic Chambers: Numerical and Experimental Studies. *Fluids* **2021**, *6*, 433. [CrossRef]
5. Lim, B.H.; Ng, E.Y.K. Comparison of FLACS and BASiL Model for Ro-Pax Ferry LNG Bunkering Leak Analysis. *Fluids* **2022**, *7*, 272. [CrossRef]
6. Carraro, M.; De Vanna, F.; Zweiri, F.; Benini, E.; Heidari, A.; Hadavinia, H. CFD Modeling of Wind Turbine Blades with Eroded Leading Edge. *Fluids* **2022**, *7*, 302. [CrossRef]
7. Volkov, K. Coupled Simulation of Fluid Flow and Vibro-Acoustic Processes in the Channel with a Circular Cylinder. *Fluids* **2022**, *7*, 382. [CrossRef]
8. Ehsan Khalili, M.; Larsson, M.; Müller, B. Immersed boundary method for viscous compressible flows around moving bodies. *Comput. Fluids* **2018**, *170*, 77–92. [CrossRef]
9. De Vanna, F.; Picano, F.; Benini, E. A sharp-interface immersed boundary method for moving objects in compressible viscous flows. *Comput. Fluids* **2020**, *201*, 104415. [CrossRef]
10. De Vanna, F.; Baldan, G.; Picano, F.; Benini, E. On the coupling between wall-modeled LES and immersed boundary method towards applicative compressible flow simulations. *Comput. Fluids* **2023**, *266*, 106058. [CrossRef]
11. Knecht, S.; Zdravkov, D.; Albers, A. Surrogate Models for Heat Transfer in Oscillating Flow with a Local Heat Source. *Fluids* **2023**, *8*, 80. [CrossRef]
12. Bukharin, N.; El Hassan, M. Numerical and Experimental Investigation of Supersonic Binary Fluid Ejector Performance. *Fluids* **2023**, *8*, 197. [CrossRef]
13. Hafen, N.; Marquardt, J.E.; Dittler, A.; Krause, M.J. Simulation of Dynamic Rearrangement Events in Wall-Flow Filters Applying Lattice Boltzmann Methods. *Fluids* **2023**, *8*, 213. [CrossRef]
14. Quissek, M.; Budziankou, U.; Pollak, S.; Lauer, T. CFD Simulation of SCR Systems Using a Mass-Fraction-Based Impingement Model. *Fluids* **2023**, *8*, 213. [CrossRef]
15. Hafen, N.; Thieringer, J.R.; Meyer, J.; Krause, M.J.; Dittler, A. Numerical investigation of detachment and transport of particulate structures in wall-flow filters using lattice Boltzmann methods. *J. Fluid Mech.* **2023**, *956*, A30. [CrossRef]
16. Bai, C.; Liu, K.; Zhao, T.; Liu, J. Numerical Investigation of the Heat Transfer Characteristics and Wall Film Formation of Spray Impingement in SCR Systems. *Appl. Sci.* **2023**, *13*, 5353. [CrossRef]
17. Ramirez-Lopez, A. Analysis of the Hydrodynamics Behavior Inside a Stirred Reactor for Lead Recycling. *Fluids* **2023**, *8*, 268. [CrossRef]
18. Ayala, E.; Rivera, D.; Ronceros, J.; Vinces, N.; Ronceros, G. Design of a Cryogenic Duplex Pressure-Swirl Atomizer through CFDs for the Cold Conservation of Marine Products. *Fluids* **2023**, *8*, 271. [CrossRef]

fluids

MDPI

Article

Design of a Cryogenic Duplex Pressure-Swirl Atomizer through CFDs for the Cold Conservation of Marine Products

Eduardo Ayala [1], Diego Rivera [1,*], Julio Ronceros [1,*], Nikolai Vinces [1] and Gustavo Ronceros [2]

[1] Mechatronics Department, Universidad Peruana de Ciencias Aplicadas, Santiago de Surco, Lima 15023, Peru; u201411982@upc.edu.pe (E.A.); leonardo.vinces@upc.edu.pe (N.V.)

[2] Instituto Latino-Americano de Tecnologia e Infraestrutura, Universidade Federal de Integração Latino—Americana (UNILA), Foz do Iguaçu 85870-650, PR, Brazil; gustavo_ronceros@hotmail.com

* Correspondence: u201414022@upc.edu.pe (D.R.); julio.ronceros@upc.pe (J.R.)

Abstract: The following article proposes the design of a bi-centrifugal atomizer that allows the interaction of sprays from two fluids (water and liquid nitrogen). The liquid nitrogen (LN_2) is below −195.8 °C, a temperature low enough for the nitrogen, upon contact with the atomized water, to cause heat loss and bring it to its freezing point. The objective is to convert the water droplets present in the spray into ice. Upon falling, the ice particles can be dispersed, covering the largest possible area of the seafood products intended for cold preservation. All these phenomena related to the interaction of two fluids and heat exchange are due to the bi-centrifugal atomizer, which positions the two centrifugal atomizers concentrically, resulting in the inevitable collision of the two sprays. Each of these atomizers will be designed using a mathematical model and CFDs tools. The latter will provide a better study of the flow behavior of both fluids inside and outside the bi-centrifugal atomizer. Hence, the objective revolves around confirming the validity of the mathematical model through a comparison with numerical simulation data. This comparison establishes a strong correlation (with a maximum variance of 1.94% for the water atomizer and 10% for the LN_2 atomizer), thereby ensuring precise manufacturing specifications for the atomizers. It is important to highlight that, in order to achieve the enhanced resolution and comprehension of the fluid both inside and outside the duplex atomizer, two types of meshes were utilized, ensuring the utilization of the optimal option. Similarly, the aforementioned meshes were generated using two distinct software platforms, namely ANSYS Meshing (tetrahedral mesh) and ANSYS ICEM (hexahedral mesh), to facilitate a comparative analysis of the mesh quality obtained. This comprehension facilitated the observation of water temperature during its interaction with liquid nitrogen, ultimately ensuring the freezing of water droplets at the atomizer's outlet. This objective aligns seamlessly with the primary goal of this study, which revolves around the preservation of seafood products through cold techniques. This particular attribute holds potential for various applications, including cooling processes for food products.

Keywords: Ansys Fluent; CFD; Abramovich theory; Kliachko theory; multiphase VOF; cryogenic atomizer; liquid nitrogen; open-end pressure-swirl atomizer; closed pressure-swirl atomizer

Citation: Ayala, E.; Rivera, D.; Ronceros, J.; Vinces, N.; Ronceros, G. Design of a Cryogenic Duplex Pressure-Swirl Atomizer through CFDs for the Cold Conservation of Marine Products. *Fluids* **2023**, *8*, 271. https://doi.org/10.3390/fluids8100271

Academic Editors: D. Andrew S. Rees and Giuliano De Stefano

Received: 6 July 2023
Revised: 11 August 2023
Accepted: 22 September 2023
Published: 1 October 2023

1. Introduction

This research endeavors to delve into the meticulous examination and enhancement of preservation methodologies within the realm of Peru's fishing industry. An industry that has adeptly harnessed the creation of legitimate employment, financial gains for the Peruvian government, and proficient export practices, thereby engendering a pivotal contribution to the decentralized growth of the nation's economy. This sector, which occupies a paramount standing within the economic framework, stands amongst the four preeminent contributors to the country's foreign exchange earnings. The Central Reserve Bank of Peru substantiates that the fishing sector contributes a noteworthy 7% to the overall national export figures [1]. Despite these commendable achievements, the sector encounters intricate challenges arising from the intrinsic variability of vessels and fishing volumes. These challenges necessitate a diverse array of refrigeration techniques, an initial investment that can prove financially burdensome, and a subsequent long-term maintenance commitment, further compounded by their comparatively elevated energy consumption. This confluence of factors often steers many vessels towards adopting less efficient and budget-conscious systems. For instance, the prevalent utilization of ice machines to sustain the integrity of catches during maritime sojourns is a common practice. When faced with these multifaceted challenges, an innovative approach emerged: the potential integration of a pioneering cold preservation method, harnessing the distinctive attributes of cryogenic fluids, prominently liquid nitrogen (LN_2), characterized by a remarkably low boiling point of -195.8 °C [2]. The exploration undertaken in this research extends to the meticulous evaluation and viability assessment of this novel preservation technique.

This investigation capitalizes on the efficacy of dual or bi-centrifugal atomization mechanisms, denoted as duplex atomizers, employing centrifugal or pressure-swirl atomizers [3]. The discernible conical geometry exhibited by the resultant spray from these atomizers facilitates immediate interaction between the atomized substances [4,5]. Consequently, this dynamic interaction triggers the dissipation of heat from water droplets, facilitated by the presence of liquid nitrogen particles [6–8]. In its essence, this research converges upon the central objective of addressing the multifaceted preservation challenges encountered within the sphere of the fishing industry. This endeavor inherently unfurls avenues for novel perspectives, particularly pertaining to the proficient cooling of comestible products and the innovation of novel industrial applications.

2. Mathematical Model

In this project, a bi-centrifugal atomizer model with tangential channels [9,10] will be used. In order to achieve this, input parameters are defined, which, when used in the equations of the mathematical models, will provide us with the main dimensions of the centrifugal atomizers. The mathematical model to be employed consists of two stages: in the first stage, the fluid is considered inviscid, and then in the final stage, viscosity losses are taken into account [11]. These stages are show in Figure 1.

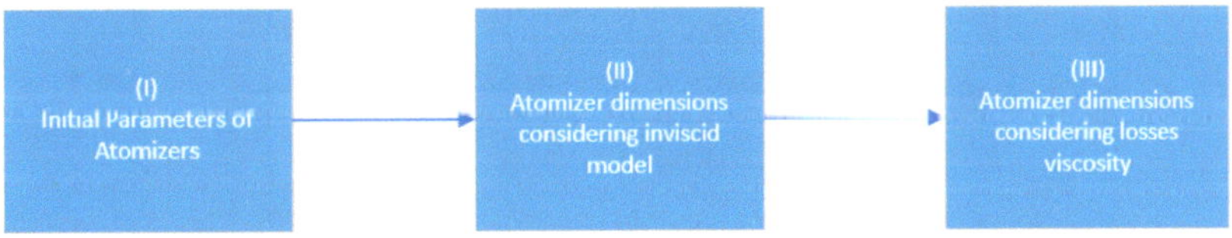

Figure 1. Methodology block diagram.

This design method was selected due to the nonmeasurable and challenging-to-model nature of the air core radius. Therefore, an approach is sought that transforms everything into a function of the annular section coefficient (φ) [12]. As a result, the decision was made to utilize a conical model as a foundation, as depicted in the article "Development of a mathematical model and 3D numerical simulation of the internal flow in a conical swirl atomizer" [3]. In this context, Equation (2) is a derivation of the conical geometric parameter equation (A_c) (Equation (1)) [3].

$$A_c = \frac{\pi r_o R_{inj}}{n f_p} \cos \psi \sin \beta \ldots\ldots \quad (1)$$

Given its tangential model nature, it is considered that ($\psi = 0°$ y $\beta = 90°$), which results in a method with which to determine the geometric parameter (A) with measurable variables (Equation (2)). This equation is then transformed into a function of the annular section coefficient (Equation (4)):

$$A = \frac{\pi r_o R_{inj}}{n f_p} \ldots\ldots \tag{2}$$

Firstly, the initial design parameters of the atomizer need to be defined. These parameters are as follows: they are the design parameters for the required operation of the atomizer and are required for the injection system that supports the pressure of n channels with a spray angle; these parameters are defined at the beginning and by using reverse engineering for the atomizer geometry.

Additionally, the physical properties of each fluid must be obtained, with the main properties as follows:

Once the initial parameters (Table 1) and fluid properties (Table 2) are defined, the next step is to determine the parameters by assuming an ideal liquid. We begin by calculating the annular section coefficient "φ" by using Equation (3), where the annular section coefficient is derived from the spray semi-angle.

$$\sin\alpha \cong \frac{2\sqrt{2}(1-\varphi)}{(1+\sqrt{1-\varphi})\sqrt{2-\varphi}} \ldots\ldots \tag{3}$$

Table 1. Initials parameters.

Initial Parameters	Water	LN_2
Spray Angle, 2α (°)	130	135
Mass Flow Rate, m (kg. s^{-1})	0.245	0.035
Pressure differential, ΔP (kPa)	350	350
Number of channels, "n"	4	4

Table 2. Physical properties.

Physics Properties of Fluid	Water	LN_2
Density, ρ, (kg.m^{-3})	1000	806.08
Kinematic viscosity, ν (m^2.s^{-1})	10^{-6}	19×10^{-8}
Absolut viscosity, μ (kg. $(m.\text{-}s)^{-1}$)	1.003×10^{-3}	160.65×10^{-6}
Surface tension, σ, (N. m^{-1})	0.072	0.0263

Once the annular section coefficient is determined, the constructive geometric parameter of the injector "A" and its corresponding discharge coefficient "Cd" can be calculated using Equations (4) and (5), respectively.

$$A = \frac{\sqrt{2}(1-\varphi)}{\varphi\sqrt{\varphi}} \ldots\ldots \tag{4}$$

$$C_d = \sqrt{\frac{\varphi^3}{2-\varphi}} \ldots\ldots \tag{5}$$

With these obtained data, the radius of the injector discharge orifice "r_o" can be calculated using Equation (6):

$$C_d = \frac{\dot{m}}{\pi r_o^2 \sqrt{2\rho\Delta P}} \ldots \tag{6}$$

Next, Equation (7) relates the swirl chamber radius "Rs" to the radius of the tangential channel "r_{inj}" (Figure 2). Additionally, Equation (8) relates the main diameters of the atomizer using the constructive geometric parameter "A", considering that the atomizers are open (C = 1) [13], which means that "Rs = r_o".

$$\mathrm{R_{inj}} = R_S - \mathrm{r_{inj}} \ldots\ldots\ldots\ldots \quad (7)$$

$$\mathrm{A} = \frac{r_o R_{inj}}{n.r_{inj}^2} \ldots\ldots\ldots\ldots\ldots\ldots \quad (8)$$

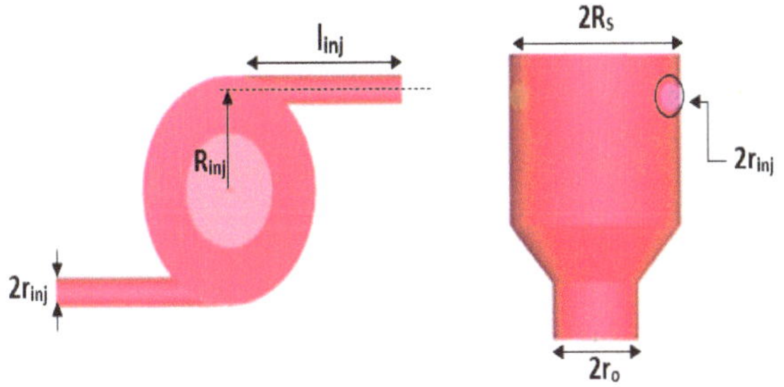

Figure 2. Main dimensions of the centrifugal atomizer.

Finally, by inserting Equation (7) into Equation (8), Equation (9) is obtained, which allows for the calculation of the radius of the tangential channel "r_{inj}".

$$\mathrm{r_{inj}^2} + \frac{\mathrm{R_s}}{\Lambda.\mathrm{n}}.\mathrm{r_{inj}} - \frac{\mathrm{R_s^2}}{\Lambda.\mathrm{n}} = 0 \ldots\ldots \quad (9)$$

With the obtained parameters, the length of each tangential channel can be defined, taking into consideration the following:

$$\mathrm{l_{ent}} = 15\mathrm{r_{inj}} \ldots\ldots\ldots\ldots\ldots \quad (10)$$

Next, the effects of viscosity are taken into account, where, in the first instance, Equation (11) is used to determine the inlet velocity in each channel of the atomizer, which is related to the Reynolds number (Re) and the Blasius coefficient ($\lambda = 0.3164\mathrm{Re}^{-1/8}$) [14,15].

$$U_{ent} = \frac{\dot{\mathrm{m}}}{\rho \mathrm{n}\pi \mathrm{r_{inj}^2}} \ldots \quad (11)$$

For an open atomizer, the equivalent geometric parameter "A_{eq}" is equal to "A" since, in this case, K = 1 (loss coefficient due to the decrease in angular momentum). Then, by applying Equation (12), the annular section coefficient "φ_{eq}" is obtained, which is important for calculating the equivalent discharge coefficient "Cd_{eq}" (Equation (13)) and the equivalent spray semi-angle "α_{eq}" (Equation (14)).

$$\mathrm{A_{eq}} = \mathrm{AK} = \frac{\mathrm{R_{inj} r_o}}{\mathrm{n r_{inj}^2} + \frac{\lambda}{2}\mathrm{R_{inj}}(\mathrm{R_s} - \mathrm{r_o})} = \frac{\sqrt{2}(1-\varphi_{eq})}{\varphi_{eq}\sqrt{\varphi_{eq}}} \ldots \quad (12)$$

$$\mathrm{Cd_{eq}} = \frac{1}{\sqrt{\frac{2-\varphi_{eq}}{\varphi_{eq}^3} + \xi_{tot}\left(\frac{Ar_o}{R_{inj}}\right)^2}} \ldots \quad (13)$$

$$\sin\alpha_{eq} = \frac{2\mathrm{Cd_{eq}A_{eq}}}{\left(1+\sqrt{1-\varphi_{eq}}\right)\sqrt{1-\xi_{tot}\mathrm{Cd_{eq}^2}\left(\frac{\mathrm{Ar_o}}{\mathrm{R_{inj}}}\right)^2}} \ldots \quad (14)$$

Finally, by applying Equation (6) again in terms of the equivalent discharge coefficient "Cd_{eq}", the corrected radius of the outlet orifice "r'_o" (Equation (15)) is obtained. Then, by applying the aforementioned equations, the corrected dimensions of the atomizer are obtained. The tables below show the dimensions obtained using the inviscid model and the viscous model [16] for both the LN_2 atomizer (Table 3) and the water atomizer (Table 4).

$$r'_o = \sqrt{\frac{\dot{m}}{\pi . Cd_{eq}\sqrt{2\rho\Delta P}}} \ldots\ldots \quad (15)$$

Table 3. Parameters of the LN_2 atomizer.

Parameters	Ideal Liquid	Losses Viscosity
φ	0.125	0.125
A	27.85	27.85
cd	0.0324	0.0237
R_s	3.802	4.45
r_o	3.802	4.45
r_{inj}	0.477	0.552
l_{ent}	4.765	5.574

Table 4. Parameters of the water (H_2O) atomizer.

Parameters	Ideal Liquid	Losses Viscosity
φ	0.173	0.173
A	16.258	16.258
cd	0.05323	0.0392
R_s	7.445	8.672
r_o	7.445	8.672
r_{inj}	1.196	1.393
l_{ent}	11.962	13.932

3. Numerical Simulation

By using the dimensions obtained from the mathematical model, the bi-centrifugal atomizer was designed using the ANSYS SPACECLAIM software (Figure 3). In the upper part, the nitrogen atomizer with its four tangential channels can be visualized, and in the lower part, with a larger diameter and located concentrically, the water atomizer with its respective inlet channels can be seen.

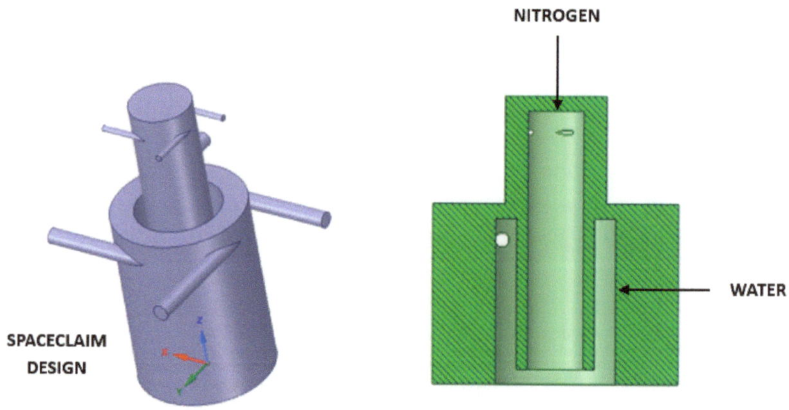

Figure 3. Left side: isometric view of the atomizer; **right side**: section view of the atomizer.

Within the domain of computational fluid dynamics (CFD), the orchestration of meshing emerges as a paramount endeavor in the numerical scrutiny of fluid dynamics. Of noteworthy significance are the two principal classifications of meshes: the tetrahedral and hexahedral variants. Each configuration presents a distinctive array of merits and demerits, thereby underscoring the profound significance of the choice between the two. This decision is contingent upon the intricate inclinations of the simulation at hand.

Subsequently, the mesh for the bi-centrifugal atomizer was fashioned utilizing the Ansys Meshing software. It is imperative to underscore that both atomizers were concentrically amalgamated, leading to the generation of a tetrahedral mesh [17]. Moreover, to refine the mesh design, a hexahedral mesh was engendered for the purpose of juxtaposing and scrutinizing simulation outcomes. (Figures 4–6).

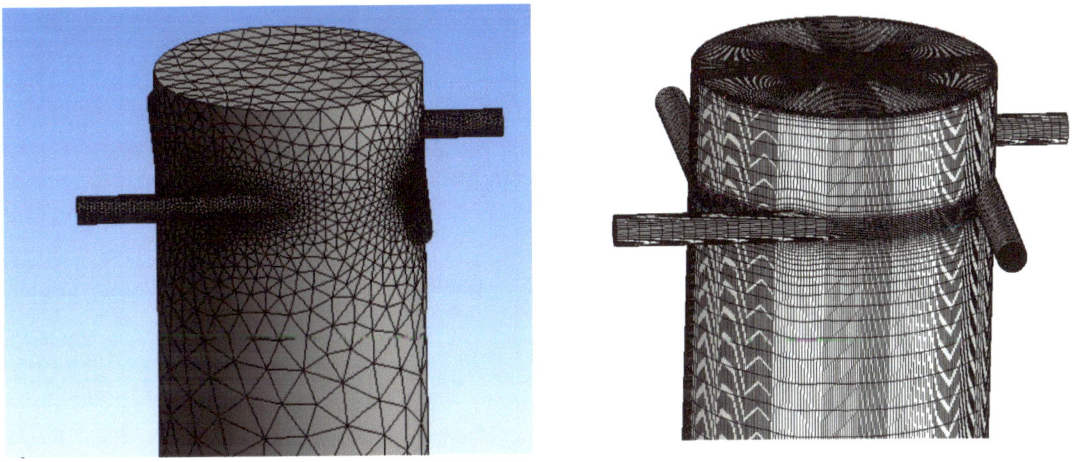

Figure 4. Mesh of nitrogen atomizer—**left side**: tetrahedral; **right side**: hexahedral.

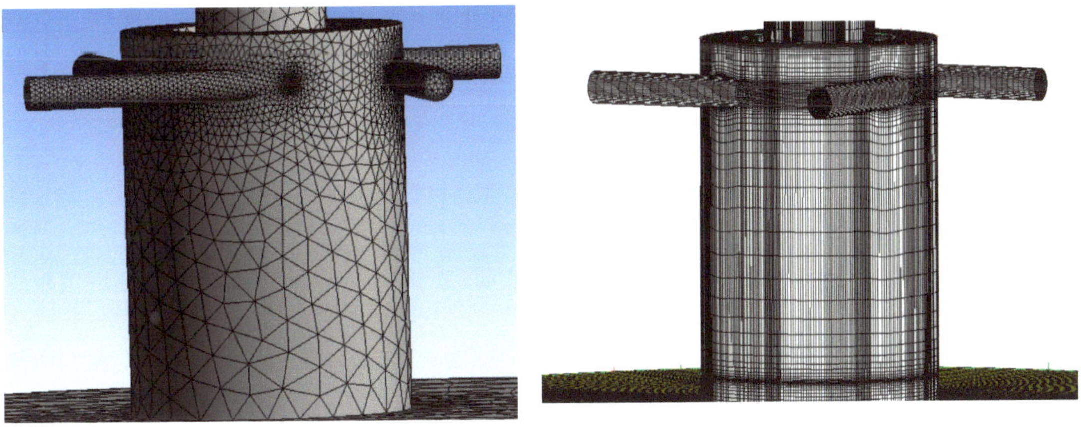

Figure 5. Mesh of water atomizer—**left side**: tetrahedral; **right side**: hexahedral.

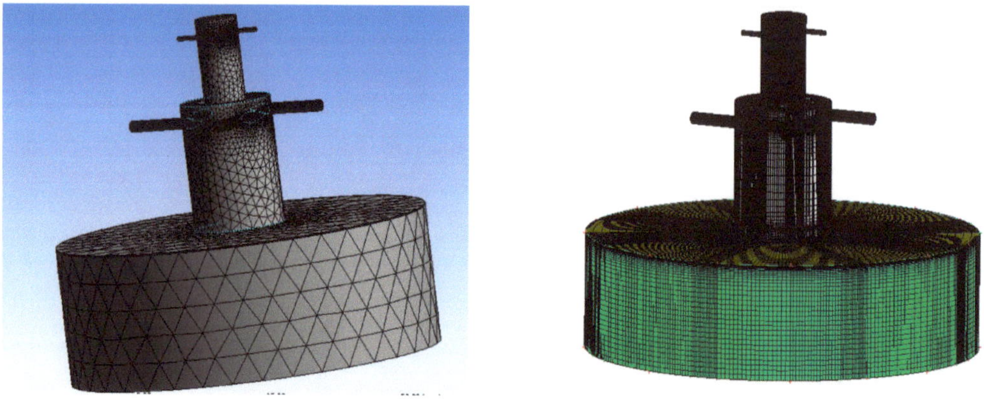

Figure 6. Mesh of duplex atomizer—**left side**: tetrahedral; **right side**: hexahedral.

In the numerical simulation, a tetrahedral and hexahedral mesh were used to control the internal flow of both atomizers, taking into account the inlet boundary conditions with a specific pressure, as well as a specific outlet pressure and no slip on the walls [18]. Subsequently, in order to achieve mesh independence in the simulation, it is necessary to consider the dimensionless distance from the wall, which is known as (**y+**). Based on this dimensionless parameter, it is possible to identify the appropriate region to address turbulence-related phenomena, as indicated by Equation (16).

$$y^+ = \frac{\rho U_\tau y}{\mu} \ldots \tag{16}$$

Then, by considering a range of speeds of the free flow (U∞) and a hydraulic diameter for the inlet channels (Dh = 1 mm), in the meshes made, a skin coefficient "Cf" can be obtained, which is used in Equation (17) [19]. By calculating the shear stress of the wall (τ_W), the friction velocity (U_τ), and Equations (18) and (19), respectively, by considering the closest distance between the cell and the wall, it is possible to replace Equation (16) and obtain a range of values.

$$C_f = 0.078 Re^{-0.25} \ldots \tag{17}$$

$$\tau_w = 0.5 C_f \rho U_\infty^2 \ldots \tag{18}$$

$$U_\tau = \sqrt{\frac{\tau_w}{\rho}} \ldots \tag{19}$$

Finally, there is a range y+, which belongs to the viscous sublayer region (blue circle of Figure 7), that is used. The wall option "ENHANCED WALL TREATMENT" function is recommended for ($\mathbf{y^+}$ < 5) [20], and for the convergence criterion, it uses the k-epsilon RNG turbulence model [21]; this gives the energy equation and the multiphase "Volume of Fluid" (VOF) model [22].

In Table 5, the features of each generated mesh are depicted, indicating their y+, node count, cell count, and the time required to complete 10,000 iterations. The numerical simulations were conducted using an Intel Core i7 processor on an HP Envy Leap 17.

Multiphase is mentioned according to the VOF approach, which may exist in liquid-gas or liquid-liquid interactions, with our case being a type of separate system; in other words, it is a continuous flow, such as in the case of water with oil. It is called a phase if it is in a state or if it is the place it occupies (dispersed or continuous), as in the experiments carried out using VOF simulation for single rising drops in three liquid-liquid extraction systems using csf and css interfacial force models [23]; this included water and n-butanol, water—n.butyl acetate, and water—toluene, where water constituted the continuous phase in all cases, and the other components constituted the dispersed phase [24].

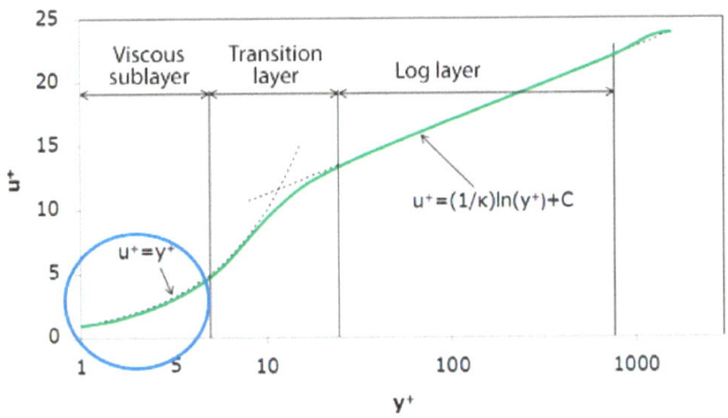

Figure 7. Velocity distribution [16].

Table 5. Characteristics of the mesh.

Characteristics	Tetrahedral	Hexahedral
y (mm)	3.2×10^{-3}	4×10^{-3}
y+	3.002	3.655
Cells	266,119	6,048,104
Faces	617,024	18,265,032
Nodes	99,037	6,169,490
10,000 Iterations Time (hours)	20	42

It is important to note that the phases involved in the simulation are air, nitrogen, and water (Equation (20)) [25]. Initially, the numerical simulation was conducted for each atomizer independently. Each atomizer was subjected to five different inlet conditions: ΔP = 150, 250, 350, 450, and 550 kPa. Additionally, a numerical simulation was performed with both injectors in operation, using ΔP = 350 kPa for the nitrogen atomizer and ΔP = 350 kPa for the water atomizer. The initial temperatures were set to 4.85 °C and −190.15 °C for the water and nitrogen atomizers, respectively.

$$\begin{aligned}&\frac{\partial}{\partial t}(\eta\rho\Phi)_m+\frac{\partial}{\partial x_i}(\eta\rho u_i\Phi)_m=\frac{\partial}{\partial x_i}\left(\Gamma_\Phi\frac{\partial\Phi}{\partial x_i}\right)_m+(\eta S_\Phi)_m+\sum_{n=1}^{N_p}C_{\Phi,\mathrm{mn}}(\Phi_n-\Phi_m)\\&+m\sum_{n=1}^{N_p}C_{\Phi,\mathrm{mn}}(\dot{m}_{mn}\Phi_n-\dot{m}_{mn}\Phi_m)\ \ldots\end{aligned} \quad (20)$$

where

η: Volumetric fraction of fluid.
Φ: Scalar.
ρ: Density.
u_i: Velocity component in the i direction.
Γ_Φ: Diffusion coefficient for a scalar Φ.
S_Φ: Source term for a scalar Φ.
N_p: Nú Total number of phases.
$C_{\Phi,mn}$: Mass transfer coefficient between phases m and n.
$\dot{m}_{mn}$: Mass variation per unit volume of phase m to phase n.

In the context of computational fluid dynamics (CFDs), ANSYS ICEM CFD emerges as a sophisticated and specialized mesh generation tool. Meticulously crafted to address the intricacies of complex geometries and nuanced simulations, ICEM CFD empowers users with the capacity to forge high-quality meshes. The inherent flexibility of this tool enables a seamless adaptation to intricate geometrical configurations, while its advanced control mechanisms facilitate precise adjustments, thereby ensuring meticulous alignment between simulation objectives and mesh refinement. In contrast, within the intricate tapestry of ANSYS simulation solutions, ANSYS Workbench Meshing assumes a pivotal role as an integral component, presenting an accessible pathway to effective mesh generation. This tool streamlines the process through the provision of automation and integration within the ANSYS Workbench environment. It caters to users aiming for a smooth transition from geometry to analysis while harnessing automation to expedite mesh creation. The simplified approach of Workbench Meshing caters to users who prioritize efficiency and a user-friendly interface in consonance with the pursuit of prompt and dependable simulation outcomes. In the context of this project, both options were considered, ultimately leading to the selection of ANSYS ICEM CFD as the preferred choice [26].

For this project, we used RANS models (Reynolds-Averaged Navier-Stokes), which constitute a category of mathematical models employed in the numerical simulation of turbulent flows within fluid mechanics. These models are predicated on decomposing flow variables into a mean component (temporally averaged) and a fluctuating component. These models find extensive utilization in addressing turbulent flow challenges across engineering and scientific research domains. Through numerically solving time-averaged Navier-Stokes equations, RANS models facilitate the estimation of statistical flow features, such as mean velocity and concentration profiles [27,28]. A clear example of the Rans Method is the journal A Study of RANS Turbulence Models in Fully Turbulent Jets: A Perspective for CFD-DEM Simulations, in which the authors investigate RANS turbulence models in fully turbulent jets, and their relevance for CFD-DEM simulations focus on understanding computational fluid dynamics (CFDs) with simulations of discrete particle mechanics. These methods allow for the prediction of the behavior of turbulent flow in specific conditions. Where CFDs is used to analyze and predict the behavior of the fluid flow in a predetermined domain [29].

Computational fluid dynamics (CFDs) simulations can prove exceptionally valuable in the design and optimization of water/nitrogen atomizers or other spray systems. Here are several ways in which CFDs can contribute to the design process [30,31]:

- Geometry design: CFDs enables the evaluation of different atomizer designs prior to physical fabrication.
- Fluid-structure interaction: CFDs can be used to assess how the flow of a liquid and gas affects the atomizer's structure and vice versa.
- Virtual experiment design: CFDs enables virtual experiments under different operating conditions before physical implementation.
- Interaction analysis with the environment: CFDs can simulate how atomized droplets interact with surfaces, air currents, or other fluids in the environment.

Figure 8 shows a flowchart in which the summarized steps are indicated in order to reach the expected results of the Project, for which, as explained above, in the creation of geometries, it is necessary to define the initial parameters and use reverse engineering, and with the ideal method and viscosity losses, it is possible to obtain the geometry that satisfies the operation of the atomizer; then, we proceed to the mesh elaboration, and once in Fluent, the boundary conditions and interactions are chosen for the fluids; when this is simulated if it does not converge, the mesh must be improved, and the same process must be carried out; if it does converge, the final results are observed.

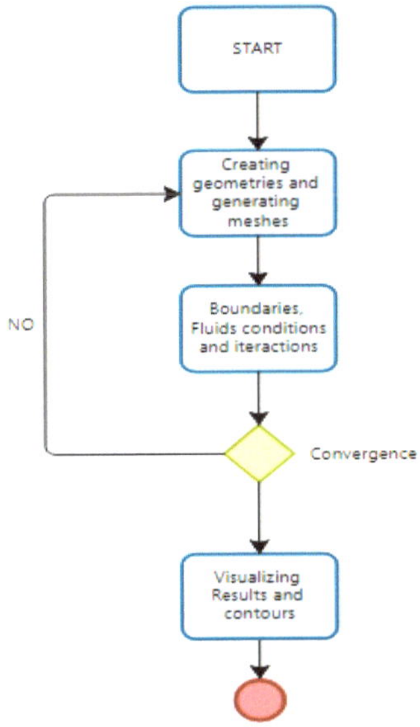

Figure 8. Flowchart.

4. Results

During this project phase, the outcomes attained through the Fluent software are subject to analysis. In both mesh types (tetrahedral and hexahedral), the findings are presented in Tables 6 and 7, encompassing the parameters, such as the mass flow attained under varying inlet pressures, for both nitrogen and water.

Table 6. Mass flow as a function of a drop in pressure from LN_2.

LIQUID NITROGEN							
	Mass Flow, (kg/s)						
Differential Pressure, (ΔP/kPa)	Mathematical Model	Simulation Tetrahedral	Deviation Mathematical Model Respect to Mesh Tetrahedral (%)	Simulation Hexahedral	Deviation Mathematical Model Respect to Mesh Hexahedral (%)	Heat Transfer Rate (KW) Tetrahedral	Heat Transfer Rate (KW) Hexahedral
150	0.0229	0.0252	−10.0437	0.0259	−13.1004	−11.076	−11.373
250	0.0296	0.0324	−9.4595	0.0332	−12.1622	−14.241	−14.578
350	0.035	0.0382	−9.1429	0.0392	−12.0000	−16.791	−17.213
450	0.0397	0.0432	−8.8161	0.0443	−11.5869	−18.988	−19.453
550	0.0439	0.0478	−8.8838	0.0491	−11.8451	−21.011	−21.560

Table 7. Mass flow as a function of a drop in pressure from H_2O.

WATER							
	Mass Flow, (kg/s)						
Differential Pressure, (ΔP/kPa)	Mathematical Model	Simulation Tetrahedral	Deviation Mathematical Model Respect to Mesh Tetrahedral (%)	Simulation Hexahedral	Deviation Mathematical Model Respect to Mesh Hexahedral (%)	Heat Transfer Rate (KW) Tetrahedral	Heat Transfer Rate (KW) Hexahedral
150	0.1600	0.1631	−1.9375	0.1721	−7.5625	−13.557	−14.395
250	0.2070	0.2089	−0.9179	0.2205	−6.5217	−17.364	−18.444
350	0.2450	0.2482	−1.3061	0.262	−6.9388	−20.631	−21.915
450	0.2780	0.2789	−0.3237	0.2944	−5.8993	−23.183	−24.625
550	0.3070	0.3079	−0.2932	0.325	−5.8632	−25.594	−27.184

In Figure 9a,b, we can appreciate the curve formed by the results obtained in the simulation at different pressures and different mesh models for the mass flows. From the figure, we can conclude that higher pressure leads to an increase in the mass flow. These results can be corroborated with the research by Liu X et al. [32], in which they state that higher pressure leads to an increase in mass flow based on their experimental study of pressure swirl atomizers with water and liquid nitrogen.

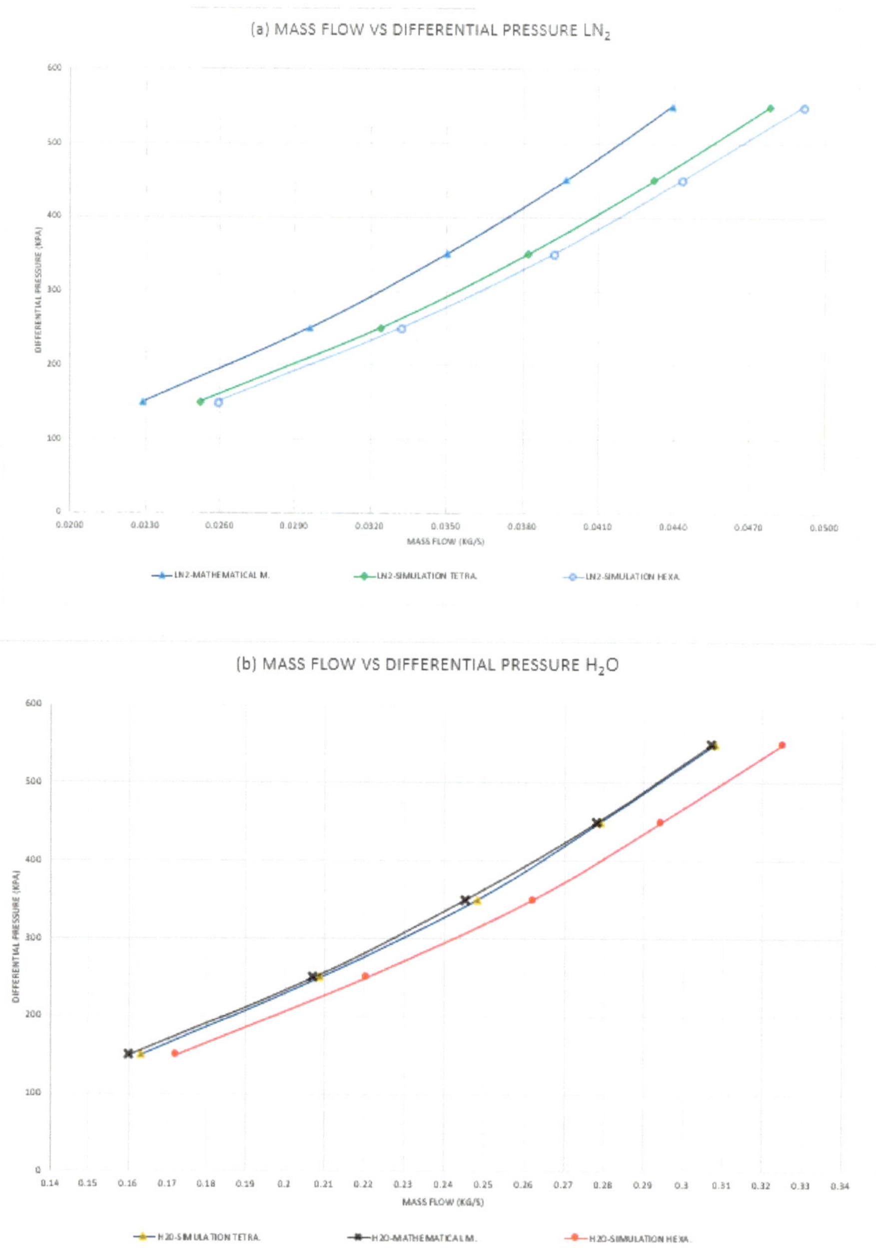

Figure 9. (**a**) Mass flow rate vs. pressure: liquid nitrogen; (**b**) mass flow rate vs. pressure: water.

In Figure 10, we can observe the density contours of the involved fluids, such as water (red color), liquid nitrogen (yellow color), and air (blue color). We can see the phase change of the liquid nitrogen when it interacts with air and water at the atomizer outlet, as well as the conical shape of the resulting sprays in the hexahedral model.

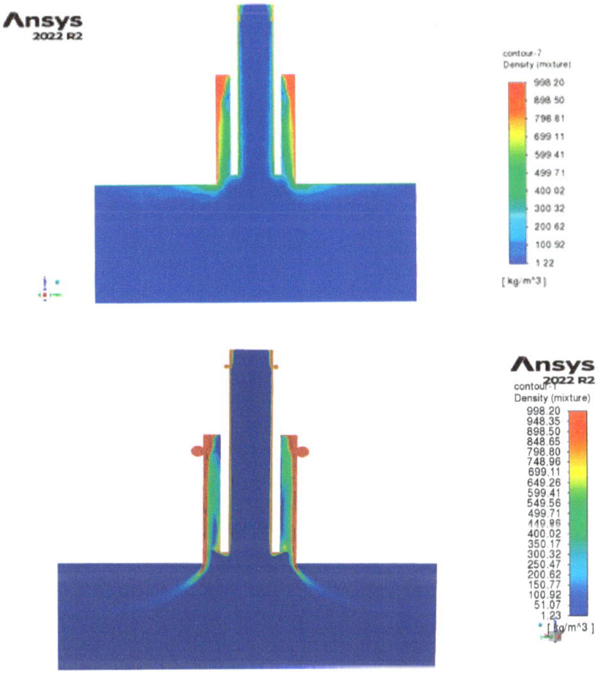

Figure 10. Density contour of duplex pressure-swirl atomizer (nitrogen atomizer: 350 Kpa; water atomizer: 350 Kpa) with tetrahedral mesh (**upper side**) and hexahedral mesh (**lower side**).

In Figure 11, we can observe the low pressures recorded in the "air core" of both the nitrogen and water atomizers, which results in high-speed air re-entry into the swirl chambers (approximately 10 to 17 m/s), as can be seen in the velocity vectors in Figure 12.

Figure 11. *Cont.*

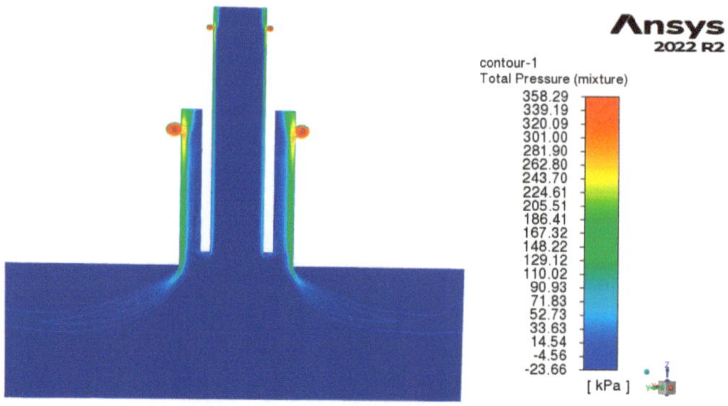

Figure 11. Total pressure contour of duplex pressure swirl atomizer (nitrogen atomizer: 350 Kpa; water atomizer: 350 Kpa) with tetrahedral mesh (**upper side**) and hexahedral mesh (**lower side**).

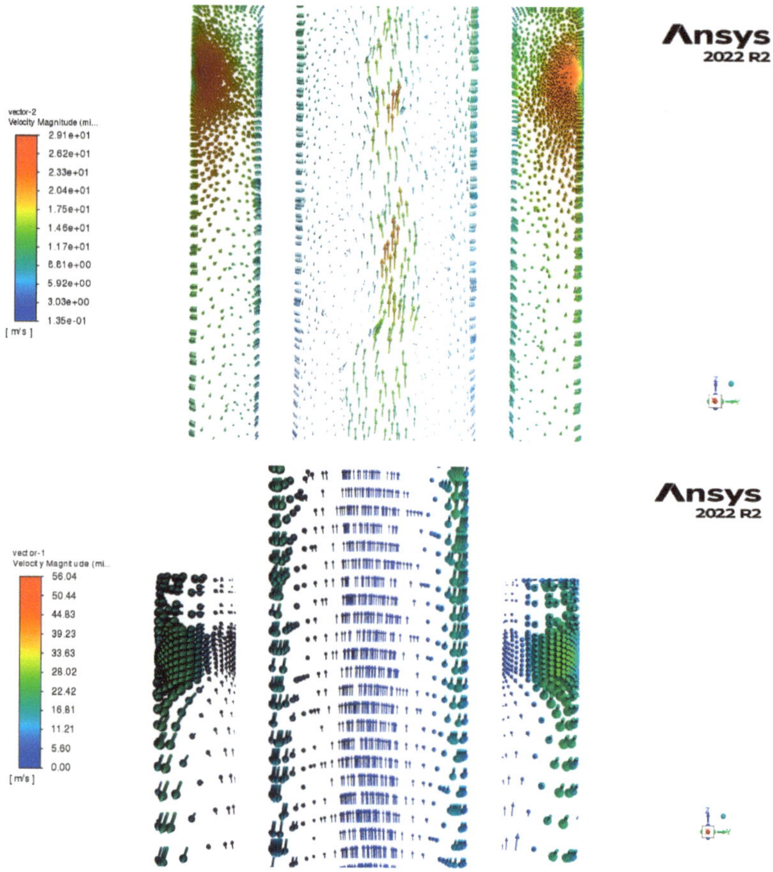

Figure 12. Velocity vectors of duplex pressure-swirl atomizer (nitrogen atomizer: 350 Kpa; water atomizer: 350 Kpa) with tetrahedral mesh (**upper side**) and hexahedral mesh (**lower side**).

Finally, we will analyze Figure 13, which represents the temperature contour. It should be mentioned that the initial temperatures were 4.85 °C and −190 °C for the water and nitrogen atomizers, respectively, through the tangential channels. In Figure 11, we can observe three important regions: A, B, and C. Region A is within a temperature range of [−16.56 °C; 12.38 °C], where we can see that the water film is still in a liquid state (approximately 12.38 °C) and its temperature rapidly decreases upon collision with nitrogen at the atomizer outlet, reaching a temperature of −16.56 °C. This guarantees that the phase change of water from liquid to solid will occur soon (water freezing point: 0 °C), which is the objective of this work. On the other hand, in region B, we can observe a temperature range of [−190.18 °C; −146.78 °C], indicating that this region will always remain at cryogenic temperatures, ensuring the rapid cooling of any nearby fluid. Finally, in region C, we can see the temperature behavior of both fluids after being expelled from the atomizer, within a range of [−16.56 °C; 2.09 °C], demonstrating that the cold environment generated by liquid nitrogen has a relevant effectiveness, achieving the objective of lowering the water temperature and (experimentally) expecting hail. In addition, the development of the hexahedral mesh helps to observe higher quality and better interaction of the phases.

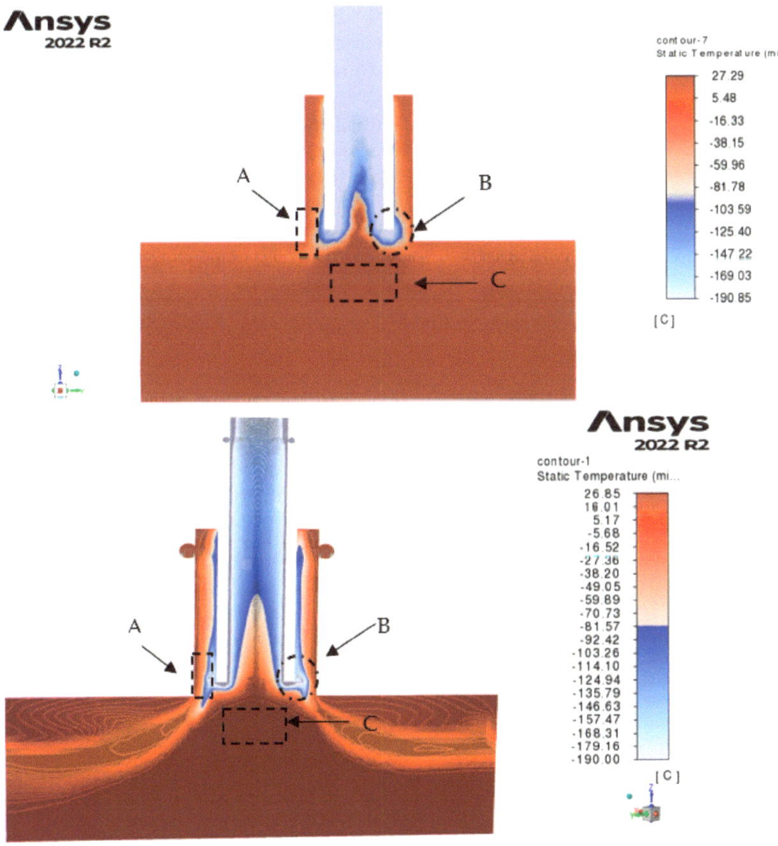

Figure 13. Temperature contour of duplex pressure-swirl atomizer (nitrogen atomizer: 350 Kpa; water atomizer: 350 Kpa) with tetrahedral mesh (**upper side**) and hexahedral mesh (**lower side**).

5. Conclusions

The mass flow rate as a function of the pressure differential (ΔP) is important as a starting point for the design of atomizers. Therefore, we sought to verify if the values from the mathematical model could be validated with the values from the numerical simulation, as shown in the curves of Figure 9a,b, demonstrating such approximation (with a maximum error of 1.94% for the water atomizer and 10% for the LN_2 atomizer), which ensures that the final dimensions of the atomizers are correct for manufacturing. It is worth noting that the models used in the numerical simulation, such as the RNG k-ε turbulence model and the Volume of Fluid (VoF) multiphase model, were crucial in understanding the flow behavior both inside and outside these atomizers. This led to the recording of the temperature of water when interacting with liquid nitrogen, ensuring the solidification of water droplets at the atomizer outlet, which is the purpose of this work related to the cold preservation of seafood products. This is corroborated by the research "Cryogenic fluid dynamics of pressure swirl injectors at supercritical conditions" [33], where it can be observed that at the outlet of these types of atomizers, the cryogenic fluid atomization has a large volume range at low temperatures, which can be used for the cooling of food products and metallurgical treatments. The implementation of the hexahedral mesh reveals an enhanced simulation interface and phase interaction. Furthermore, upon comparing both mesh models, in the case of the tetrahedral model, it incurs lower computational consumption and exhibits a reduced interface quality; however, the mass flow values closely approximate those of the mathematical model. Conversely, within the hexahedral model, higher computational resources are utilized due to the greater number of elements. Nevertheless, the mass flow values also approximate the mathematical model, and the interface quality and result visualization attain an excellent standard. Consequently, the hexahedral mesh model emerges as the optimal choice and is recommended for projects of this nature.

Author Contributions: Conceptualization, E.A. and D.R.; methodology, E.A. and D.R.; software, D.R.; validation, J.R.; writing—review and editing, J.R. and N.V.; visualization, N.V.; supervision J.R., N.V. and G.R. All authors have read and agreed to the published version of the manuscript.

Funding: This research was funded by direction of the Peruvian University of Applied Sciences grant number EXPOST-2023-2 And The APC was funded by 1600.

Data Availability Statement: Data available upon request due to privacy and ethics restrictions. The data presented in this study are available upon request from the corresponding author. The data are not publicly available due to the sensitivity of the data and information created and analyzed in this study, the models it can also be found of bibliographic sources.

Acknowledgments: The authors of this research wish to express their gratitude to the UPC (Universidad Peruana de Ciencias Aplicadas) and its research department for their guidance and support during the duration of this study and its financing through the UPC-EXPOST-2023-2 incentive. which led to the development of this project.

Conflicts of Interest: The authors declare no conflict of interest.

Nomenclature

A	geometrical characteristics parameter of pressure swirl atomizer of tangential inlets
A_E	equivalent geometrical characteristics parameter due to the viscosity of swirl atomizers
A_c	geometrical characteristics parameter of conical pressure swirl atomizer
Cd	discharge coefficient
f_p	cross-sectional area of inlet port
K	coefficient of loss due to liquid viscosity
n	number of inlet channels
$\dot{m}$	mass flow rate
ΔP	differential pressure
Re	Reynolds number
R_s	swirl chamber radius
R_{inj}	radius to axis inlet channel
r_a	air core radius
r_o	outlet orifice radius
u	vectorial velocity
U_{in}	inlet entrance velocity
U,W	velocities

Greek Symbols

α	half-spray angle
ξ	losses Coefficient
η	volume fraction of fluid
φ	film flow area coefficient
Φ	scalar
λ	resistance coefficient of Blasius
μ	liquid absolute viscosity
ρ	liquid density
σ	liquid surface tension
v	liquid kinemátic viscosity

Subscripts

a	air core
eq	equivalent parameter due to viscosity
inj	parameters related to inlet channels.
liq	liquid
r	radial component
s	swirl chamber.
tot	total
θ	tangential component
z	axial component

References

1. Sociedad Nacional de Pesquería. Industria Pesquera: Contribución a la Economía Peruana. 2018. Available online: https://www.snp.org.pe/relevancia-economica/ (accessed on 1 May 2023).
2. Rodriguez, H. Propiedades del Nitrógeno (N). 2022. Available online: https://www.nationalgeographic.com.es/ciencia/propiedades-nitrogeno-n_18217 (accessed on 2 May 2023).
3. Rivas, J.; Pimienta, A.; Rivas, G. Development of a mathematical model and 3D numerical simulation of the internal flow in a conical swirl atomizer. *J. At. Sprays* **2014**, *24*, 97–114. [CrossRef]
4. Rivas, J.R.; Pimenta, A.P.; Salcedo, S.G.; Rivas, G.R.; Suazo, M.G. Study of internal flow of a bipropellant swirl injector of a rocket engine. *J. Braz. Soc. Mech. Sci. Eng.* **2018**, *40*, 289. [CrossRef]
5. Çengel, Y.A.; Cimbala, J.M. *Fluid Mechanics: Fundamentals and Applications*; McGraw-Hill Education: New York, NY, USA, 2014.
6. Lefebvre, A.H.; McDonell, V.G. *Atomization and Sprays*; CRC Press: Boca Raton, FL, USA, 2017.
7. Elger, D.F.; Williams, D.C.; Crowe, C.T.; Roberson, J.A. *Engineering Fluid Mechanics*; John Wiley & Sons: Hoboken, NJ, USA, 2015.
8. Fox, R.W.; McDonald, A.T.; Pritchard, P.J. *Introduction to Fluid Mechanics*; John Wiley & Sons: Hoboken, NJ, USA, 2011.
9. Aminjan, K.K.; Kundu, B.; Ganji, D.D. Study of pressure swirl atomizer with tangential input at design point and outside of design point. *Phys. Fluids* **2020**, *32*, 127113. [CrossRef]
10. Bayvel, L.; Orzechowski, T. *Liquid Atomization*; CRC Press: Boca Raton, FL, USA, 2006.
11. Rivas, J.R.R. Modelo Matemático e Simulação Numérica do Spray em Injetores de Uso Aeroespacial. Ph.D. Dissertation, Instituto Tecnológico de Aeronáutica, São José dos Campos, SP, Brazil, 2015.
12. Pimenta, A.P.; Lima, A.R. Analytical model for conical injector flows. In Proceedings of the International Congress Mechanical Engineering ABCM, Brasilia, Brazil, 5–9 November 2007.
13. Fu, Q.-F.; Yang, L.-J.; Zhang, W.; Cui, K.-D. Spray Characteristics of an Open-End Swirl Injector. *At. Sprays* **2012**, *22*, 431–445. [CrossRef]
14. Ronceros, J.; Porto, A.; Sumara, J.; Ronceros, G. An improved theoretical formulation for sauter mean diameter of pressure-swirl atomizers using geometrical parameter of atomization. *Propuls. Power Res.* **2022**, *11*, 240–252. [CrossRef]
15. Reddy, K.U.; Mishra, D.P. Studies on spray behavior of pressure swirl atomizer in transition regime. *J. Propul. Power* **2008**, *24*, 74–80. [CrossRef]
16. Savonov, R.I. Análise Numérica e Teórica de Injetores Tipo "Swirl" Empregados em Motores-Foguete a Propelente Líquido. Ph.D. Dissertation, Instituto Tecnológico de Aeronáutica, São José Dos Campos, SP, Brazil, 2011.
17. Vehi Josep, M. *Mallado de Geometrías Complejas Mediante CFD*; Escuela Tecnica Superior de Ingenieria Industrial del Barcelona: Barcelona, Spain, 2016.
18. Ochowiak, M.; Janecki, D.; Krupińska, A.; Włodarczak, S.; Wilk, T.; Olszewski, R. Conical Two-Phase Swirl Flow Atomizers—Numerical and Experimental Study. *Energies* **2021**, *14*, 1745. [CrossRef]
19. Ronceros, G.A.R. Simulação Numérica da Convecção Forçada Turbulenta Acoplada à Condução de Calor em Dutos Retangulares. Ph.D. Dissertation, ITA—Instituto Tecnológico de Aeronáutica, São Jose dos Campos, SP, Brazil, 2010.

20. *ANSYS FLUENT Theory Guide Release 15.0*; ANSYS Inc.: Canonsburg, PA, USA, 2013; Available online: http://www.pmt.usp.br/academic/martoran/notasmodelosgrad/ANSYS%20Fluent%20Theory%20Guide%2015.pdf (accessed on 20 May 2023).
21. Hirt, C.W.; Nichols, B.D. Volume of fluid (VOF) Method for the Dynamics of Free Boundarys. *J. Comput. Phys.* **1981**, *39*, 201–225. [CrossRef]
22. Natarajan, V.; Unnikrishnan, U.; Hwang, W.-S.; Choi, J.-Y.; Yang, V. Numerical study of two-phase flow dynamics and atomization in an open-type liquid swirl injector. *Int. J. Multiph. Flow* **2021**, *143*, 103702. [CrossRef]
23. Roshdi, S.; Kasiri, N.; Rahbar-Kelishami, A. VOF simulation of single rising drops in three liquid-liquid extraction systems using csf and css interfacial force models. *Braz. J. Chem. Eng.* **2018**, *35*, 1315–1331. [CrossRef]
24. Rajesh, V.M.; Vivek, B. Volume Of Fluid Simulations of Gas-Liquid—Liquid flow in minichannels. *Chem. Eng. J.* **2018**, *345*, 688–705. [CrossRef]
25. Basic Course on Turbulence and Turbulent Flow Modeling 10: 10.1 Wall Condition of RANS Model, 10.2 Wall Function, 10.3 Turbulence Models and Wall Condition. 2019. Available online: https://www.cradle-cfd.com/media/column/a163 (accessed on 25 May 2023).
26. ANSYS Inc. *ANSYS Meshing User's Guide*; Release 13.0; ANSYS, Inc.: Canonsburg, PA, USA.
27. Turbulence Modeling for CFD. pp. 87–90. Available online: http://ci.nii.ac.jp/ncid/BA21991687 (accessed on 30 May 2023).
28. White, F.M. *Fluid Mechanics*; McGraw-Hill Education: New York, NY, USA, 2011.
29. Weaver, D.S.; Miskovic, S. A study of RANS turbulence models in fully Turbulent jets: A perspective for CFD-DEM Simulations. *Fluids* **2021**, *6*, 271. [CrossRef]
30. Blazek, J. *Computational Fluid Dynamics: Principles and Applications*, 2nd ed.; Elsevier Science: Amsterdam, The Netherlands, 2006.
31. Ghaib, K. *Introduction to Computational Fluid Dynamics*; Springer: Berlin/Heidelberg, Germany, 2023.
32. Liu, X.; Xue, R.; Ruan, Y.; Chen, L.; Zhang, X.; Hou, Y. Flow characteristics of liquid nitrogen through solid-cone pressure swirl nozzles. *Appl. Therm. Eng.* **2017**, *110*, 290297. [CrossRef]
33. Zong, N.; Yang, V. Cryogenic fluid dynamics of pressure swirl injectors at supercritical conditions. *Phys. Fluids* **2008**, *20*, 056103. [CrossRef]

fluids

Article

Analysis of the Hydrodynamics Behavior Inside a Stirred Reactor for Lead Recycling

Adan Ramirez-Lopez

Department of Industrial Engineering, Technological and Autonomous Institute of México (ITAM), Rio Hondo #1 col, Progreso Tizapan, Mexico City 01080, Mexico; adan.ramirez@itam.mx

Abstract: This work focuses on an analysis of hydrodynamics to improve the efficiency in a batch reactor for lead recycling. The study is based on computational fluid dynamics (CFD) methods, which are used to solve Navier–Stokes and Fick's equations (continuity and momentum equations for understanding hydrodynamics and concentration for understanding distribution). The reactor analyzed is a tank with a dual geometry with a cylindrical body and a hemisphere for the bottom. This reactor is symmetrical vertically, and a shaft with four blades is used as an impeller for providing motion to the resident fluid. The initial resident fluid is static, and a tracer is defined in a volume inside to measure mixing efficiency, as is conducted in laboratory and industrial practices. Then, an evaluation of the mixing is performed by studying the tracer concentration curves at different evolution times. In order to understand the fluid flow hydrodynamics behavior with the purpose of identifying zones with rich and poor tracer concentrations, the tracer's concentration was measured at monitoring points placed all around in a defined control plane of the tank. Moreover, this study is repeated independently to evaluate different injection points to determine the best one. Finally, it is proved that the selection of an appropriate injection point can reduce working times for mixing, which is an economically attractive motivation to provide proposals for improving industrial practices.

Keywords: hydrodynamics behavior; stirring of liquid; computational fluid dynamics; lead recycling

Citation: Ramirez-Lopez, A. Analysis of the Hydrodynamics Behavior Inside a Stirred Reactor for Lead Recycling. *Fluids* **2023**, *8*, 268. https://doi.org/10.3390/fluids8100268

Academic Editors: D. Andrew S. Rees and Francesco De Vanna

Received: 28 July 2023
Revised: 20 September 2023
Accepted: 21 September 2023
Published: 28 September 2023

1. Introduction

Lead acid batteries are referred to as secondary batteries or accumulators because they are rechargeable. They are divided into starter and industrial batteries [1–4]. Both are used to provide large quantities of energy. Batteries are used to start cars, operate electric vehicles, as energy storage mediums for industrial applications, as short-term emergency power sources, etc.; nevertheless, acid and lead are very dangerous materials [2–7]. These can cause fires, explosions, or undesirable chemical reactions. Lead can result in serious damage to natural life and human health, especially to children; thus, recycling lead is very important in order to reduce pollution and the potential risks to nature [3,5,7–12]. Moreover, recycling requires only a minimum of effort in comparison to the process required to produce lead from ore, making it an attractive and economical option. This is an important economic motivation for developing new recycling procedures and improving existing industrial methods. Here, a real industrial reactor with two different sections stirred by an impeller for lead recycling was analyzed. Using this method, computational fluid dynamics (CFD) has become a very popular and powerful tool to analyze hydrodynamics behavior in chemical reactors [4,7,9–14]. The evaluation of the mixing performance is very important in order to reduce working times and increase the stirring efficiency; for this reason, this work can be considered a first approach at establishing reliable criteria to evaluate mixing performance [6–10,12–14]. Tracer injection is a useful technique to evaluate dispersion in fluids. Here, tracer injection was conducted by replacing a defined lead volume at the original volume, and the tracer distribution was evaluated by measuring the concentration at the different monitoring points in a defined control plane of the chemical reactor. Some authors have used CFD models to evaluate mixing in reactors [7–9,14–23]. Nevertheless, mixing is a complex problem that is a function of each particular condition according to the liquid's properties and the geometrical disposition of the reactor [7–12]. An understanding of hydrodynamics behavior is essential to establishing parameters to evaluate the mixing performance and then to improve industrial practices in order to reduce working times [1–14].

Lead recycling technology was developed due to the necessity of re-using lead in many metallurgical and chemical products, and special interest has been focused on the automotive industry because the batteries used contain large quantities of lead. Lead recycling involves a process whereby a heavy liquid is confined in a tank reactor after melting and then a stirring is applied [24–26]. Many authors have studied the pyro-metallurgical process in order to develop sustainable green recycling methods [19,27–29]; thus, many have focused their efforts on analyzing the problem of lead recycling and the hydrodynamics behavior inside industrial reactors [20–23]. Other authors have worked on the management of waste, which is also a problem due to pollutants and the toxicity of the components in batteries [19,20,27–29]. Moreover, some researchers have studied pyro and hydro-metallurgical procedures making scale reactors according to physical modeling [19,20,29]. Some of have analyzed hydrodynamics behavior in situ in industrial chemical reactors with application to other chemical or biological industrial processes [26,30]. Thus, there similarities among these works as many of these fluids have similar behaviors. Other authors have studied chemical reactivity in order to find the most appropriate conditions for lead recycling [23–26,30–33].

An understanding of hydrodynamics behavior is very important in the chemical and metallurgical industries, where the melting of metal is conducted. Metals are heated until becoming liquid and, frequently, agents for avoiding oxidation, control of impurities, and others to improve the quality are added using different physical or mechanical methods. Many authors have worked on forecasting the path of impurities on ladles, tundishes, molds, and other metallurgical equipment, with the main goal of improving the cleaning and refining process, especially focusing on the steelmaking process, because it is the most used metal [34–38]. Therefore, this work explores hydrodynamics behavior as a function of the tracer distribution to improve actual industrial practices, reducing mixing times.

2. Mixing Analysis

Mixing in chemical processes is often evaluated as a function of the quality in the homogeneity of one phase distribution in another. The mixing in of liquids is very important for improving industrial practices. Therefore, the diffusion of liquids is critical to evaluate the mixing efficiency. Hydrodynamics studies based on an analysis of concentration and residence time distribution curves (RTDCs) have been used by many authors to understand the mixing evolution [4–8,12–17,39–43]. Figure 1 shows three typical curves of an analysis of mixing as a function of time. Here, points (A) and (B) are the initial and final steady states. Curve (a) shows a very quick diffusion at the beginning. As a consequence, the efficiency of the mixing is high [15–18,24,25,30,39–43]. Nevertheless, this efficiency decreases over time. The second curve (b) shows a constant evolution of the mixing; this kind of curve is very difficult to find in real chemical processes because it is a constant evolutive ideal case; this means that the evolution of the chemical reaction will always be at the same velocity [16,19,26–30,41,43]. The third curve (c) shows a slow mixing evolution at the beginning, but the efficiency increased over time. During real practice, often these three behaviors are present during the mixing evolution inside chemical reactors. A knowledge of the hydrodynamics in chemical reactors where liquids are being mixed is very important in order to modify the stirring conditions and improve industrial practices.

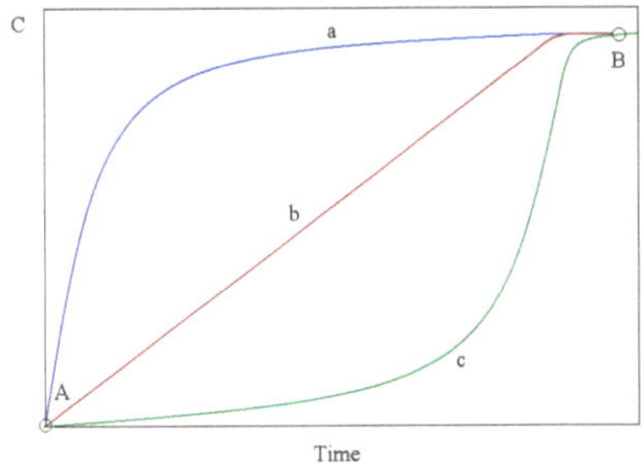

Figure 1. Typical concentration curves of the Residence Time Distribution Curves RTDCs showing the evolution of fluid dispersions during the mixing of liquids in chemical reactors.

Different kinds of stirring systems are used in order to promote mixing such as the rolling of chemical reactors. Nevertheless, this practice is expensive and is only used in some cases. Another way is to inject the chemical components using a pressure system [10–18]. These procedures are frequently used to inject gases using tubes and inlets [19–22,24–27,30]. But the most popular methods are those in which mechanical devices are used. Mechanical stirring is promoted using engines and impellers to provide kinetic energy to the resident fluid inside the reactor [32,33,41–43]. The fluid movement in the analyzed tank in this work was promoted using an impeller with four flat blades; then, the analysis was performed by evaluating the tracer concentration everywhere inside the reactor in order to find the best injection point. Thus, it is also important to obtain a criterion for determining when a good mix has been obtained.

3. Geometrical Description of the Tank and Stirrer

The reactor was a symmetrical tank vertically. It was formed by two sections: an upper cylindrical form and a hemispherical bottom, as shown in Figure 2a. It was built computationally, defining only the reactor wall as solid. The impeller was formed by a symmetrical vertical shaft with four flat blades in the lower part, each disposed at 90°. Both elements were discretized using nonstructured mesh formed with nodes, which could be analyzed with two-dimensional triangular cells or three-dimensional tetragonal cells. The largest cells were placed on the wall and the smallest cells were placed on the impeller 1blades.

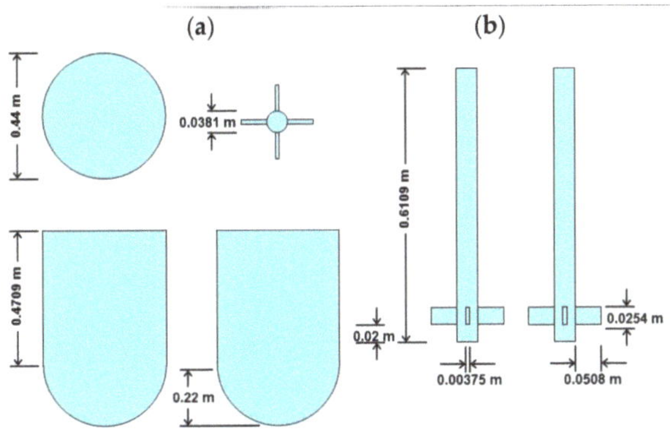

Figure 2. Sketch of the chemical reactor for lead recycling: (**a**) tank reactor; (**b**) impeller.

A control plane was defined vertically for reference. Then, this plane was divided into two subplanes, as shown in Figure 3a,b. Here, each of the injection points was placed, and the tracer was replaced the original resident fluid, while in the second semi-plane, the monitoring points were placed. These are the points where the tracer's concentration was measured. Both planes were placed at 180° to each other, dividing the reactor into 2 sections. There were 22 injection points distributed in the tank to evaluate the influence of the injected place on the mixing time; then, every injected point was evaluated independently. The impeller speed was the same for every simulation, and the influence of the walls on the mixing was considered at the points where the fluid impacted and changed the path. Although, there were no injection points below the impeller, because during real mixing operations the tracer was injected using a tubing inlet system submerged in the bath making this geometrical disposition complicated; nevertheless, there were some monitoring points in the reactor bottom because this zone is critical for mixing as a consequence of frequent problems with tracer diffusion in this region [1–6,20–27,30,31].

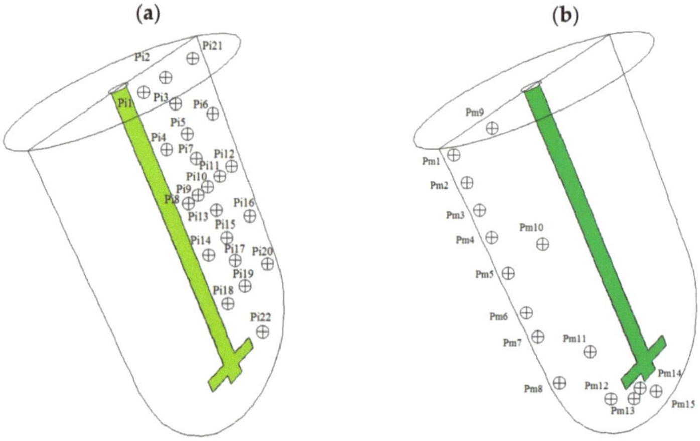

Figure 3. Control plane inside the lead recycling reactor: (**a**) semi-plane used for placing the injection points: (**b**) semi-plane used for placing the monitoring points.

The nearest injection points for the impeller were pi_{18}, pi_{19}, pi_{17}, and pi_{22}, and the nearest injection points for the wall were pi_6, pi_{12}, pi_{16}, and pi_{21}. These points were placed on different positions in the injection plane to evaluate the influence of the reactor geometry and the impeller during rotation in order to identify the death zones and zones with tracer excess because both zones had evidenced poor mixing.

The walls of the reactor were joined to form a single empty volume and meshed using 1062 nonstructured, 2D triangular cells of regular size, which formed a continuous flat surface surrounding the entire lead volume. The cylindrical and hemispherical sections were measured as 0.66922 m^2 and 0.28224 m^2, respectively.

The impeller was a cylindrical shaft that was 0.610 m in length, with 4 blades in the lowest position that were 0.0508 m × 0.0254 m and 0.00375 thick. A finer 2D triangular mesh with 16,890 cells for discretizing the shaft and 657 cells for every blade was used. The impeller speed was 32 RPMs, and the moving rotating frame (MRF) condition was established in concordance. The computational model used for solving the domain was k-ε in the software ANSYS-Fluent (17.1).

The tank and the impeller were made with stainless-steel in order to resist the impacts of the liquid lead against the cylindrical walls, and the blades could promote the required stirring.

4. Initial Assumptions

The following assumptions and boundary conditions were established for every simulation run:

(a) The influence of the shaft is not significant because it is placed in the middle of the reactor to achieve symmetrical conditions, and no vibration is assumed during rotation. Thus, the shaft is a rotatory wall in the middle of the tank without influence over the fluid dynamics;

(b) The reactor walls are free of defects; as a consequence, there is no friction or drag effect; nevertheless, the path of the fluid and speed vectors can be modified due to the liquid particles impacting the walls at different angles;

(c) The impeller rotatory speed is constant during the simulation. It is constant at 200 radians per minute. The impeller is submerged inside the reactor at a central position;

(d) The surface of the liquid lead is flat; this means that no turbulence condition is assumed over the surface or that the tank is a closed volume;

(e) The lead temperature of the liquid lead is assumed at 327 °C during the simulation, and the system is assumed as isothermal;

(f) The tracer is injected in only one injection point at the start time (t = 0) for each simulation case. Thus, the tracer's concentration at all monitoring points is equal to zero ($C_{m1} = C_{m2}, \ldots C_{m15} = 0$). Moreover, the resident liquid is only lead with constant properties, and the tracer is also assumed to have the same properties as lead;

(g) The velocity vectors at the beginning of the simulation are shown in Figure 4a,b, all around the tank and near the blades of the impeller, respectively. Then, a certain volume of lead is replaced with the tracer, which has the same properties as lead. Then, the fluid is assumed to be homogeneous (there are no inclusions or other fluids involved).

(h) Simulations are conducted individually to analyze the hydrodynamics behavior and efficiency of the tracer's distribution at every injection point.

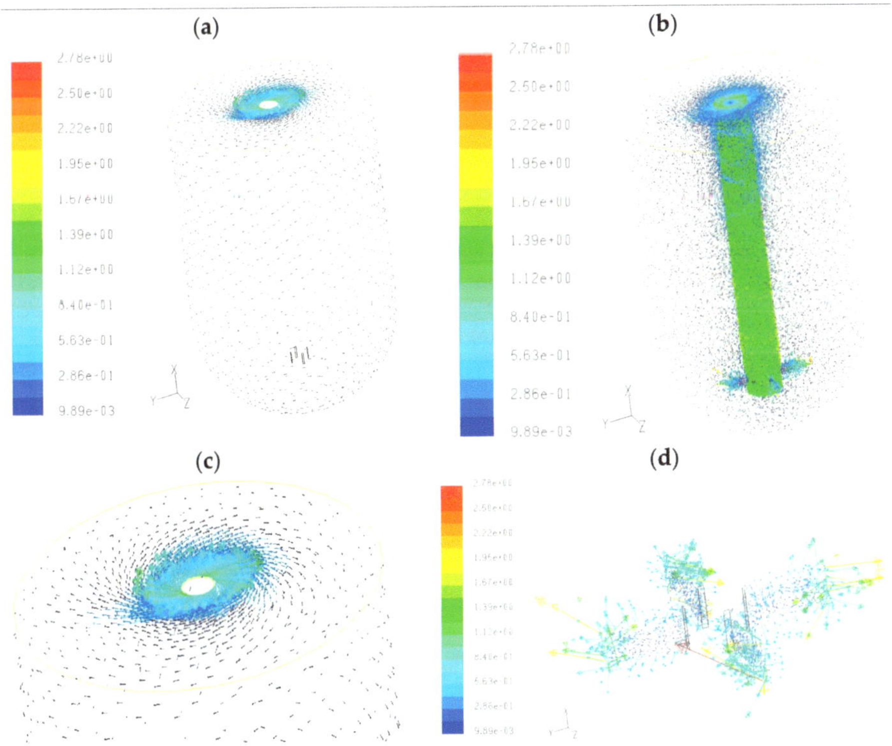

Figure 4. (**a**) Vectors profile inside the tank reactor considering the (**a**) stationary initial lead reactor; (**b**) final vector speed inside the tank; (**c**) velocity vectors at the top of the tank around the shaft of the impeller; (**d**) zoomed-in view of vectors near the impeller at the beginning of the stirring simulation.

The tracer, which is initially stationary, will be distributed all around the tank reactor as the simulation is executed; then, a certain percentage of the tracer will be in every zone of the tank, and as the simulation continues the tracer will tend to be distributed homogeneously and a final value of tracer will be obtained. The injected tracer is assumed as ideal with the same physical properties as the resident lead in the tank. This condition is often used by many authors who work on physical and computer simulations [12–17,19,27–30].

Industrial agents that promote the cleaning of lead were added by plunging solids or blowing dust particles. Thus, there were no injection points below the impeller, as shown in Figure 3a; this condition is geometrically complicated and is beyond normal practices.

5. Mathematical Model

The main parameter to evaluate the mixing efficiency is the tracer's diffusion in the entire reactor. This was evaluated by measuring the tracer's concentration at each of the monitoring points. Equations (1) and (2) were used to evaluate the tracer's concentration at every step time (Δt) during the simulation. Thus, when the tracer's concentration tended to be invariable, an ideal mix was obtained. Thus, Equations (1) and (2) were the criterion for determining the invariability of the tracer's concentration at every monitoring point; then, if the concentration at every monitoring point tended to a minimum (zero), the tracer was homogeneously distributed. Here, the superscripts (t) and (t − Δt) refer to the latest and previously calculated value of the tracer, respectively; and the subscripts (pm) and (tracer) were used to indicate that the tracer was measured at every monitoring point and as the contribution of all them, respectively. While Equation (2) indicates that the variation in the tracer's concentration tended to a minimum, which was the criterion for considering the tracer as homogeneously distributed.

$$\frac{dC}{dt} == \frac{C_{pm}^{t} - C_{pm}^{t-\Delta t}}{\Delta t} = \frac{C_{tracer}^{t} - C_{tracer}^{t-\Delta t}}{\Delta t} \tag{1}$$

$$\frac{\mathrm{dC}}{\mathrm{dt}} \to 0 \tag{2}$$

The Navier–Stokes equations are nonlinear partial differential equations that are used to describe fluid motion. Here, the nonlinearity was due to the acceleration of the fluid in motion as a consequence of the change in the velocity at each position and at every instant during the simulation. Then, Equation (3) can be rewritten using Cartesian coordinates, as shown in Equations (4)–(6). These equations were solved using the k-ε model to calculate the hydrodynamics behavior (velocity and path of fluid inside the tank).

$$\frac{\partial \rho}{\partial t} + \nabla(\rho v) = 0 \tag{3}$$

$$\rho\left(\frac{\partial u}{\partial t} + u\frac{\partial u}{\partial x} + v\frac{\partial u}{\partial y} + w\frac{\partial u}{\partial z}\right) = -\frac{\partial p}{\partial x} + \mu\left(\frac{\partial^2 u}{\partial x^2} + \frac{\partial^2 u}{\partial y^2} + \frac{\partial^2 u}{\partial z^2}\right) + \rho g_x \tag{4}$$

$$\rho\left(\frac{\partial v}{\partial t} + u\frac{\partial v}{\partial x} + v\frac{\partial v}{\partial y} + w\frac{\partial v}{\partial z}\right) = -\frac{\partial p}{\partial y} + \mu\left(\frac{\partial^2 v}{\partial x^2} + \frac{\partial^2 v}{\partial y^2} + \frac{\partial^2 v}{\partial z^2}\right) + \rho g_y \tag{5}$$

$$\rho\left(\frac{\partial w}{\partial t} + u\frac{\partial w}{\partial x} + v\frac{\partial w}{\partial y} + w\frac{\partial w}{\partial z}\right) = -\frac{\partial p}{\partial z} + \mu\left(\frac{\partial^2 w}{\partial x^2} + \frac{\partial^2 w}{\partial y^2} + \frac{\partial^2 w}{\partial z^2}\right) + \rho g_z \tag{6}$$

And, the continuity equation can be written as follows:

$$\frac{\partial \rho}{\partial t} + \frac{\partial(\rho u)}{\partial x} + \frac{\partial(\rho v)}{\partial y} + \frac{\partial(\rho w)}{\partial z} = 0 \tag{7}$$

When the flow is at a steady state, (ρ) does not change with respect to time. Then, the continuity equation is reduced to:

$$\frac{\partial(\rho u)}{\partial x}+\frac{\partial(\rho v)}{\partial y}+\frac{\partial(\rho w)}{\partial z}=0 \tag{8}$$

When the flow is incompressible, (ρ) is constant and does not change with respect to space. Then, the continuity equation is reduced to:

$$\frac{\partial u}{\partial x}+\frac{\partial v}{\partial y}+\frac{\partial w}{\partial z}=0 \tag{9}$$

The velocity components (the dependent variables to be solved) are typically named u, v, and w. This system is the most commonly and formally used, though comparatively more compact than other representations; nevertheless, this is still a nonlinear system of partial differential equations for which solutions are difficult to obtain. Thus, computational representations allow for the reproduction of geometrical elements and physical conditions in detail, and numerical methods can be programed to solve complex problems concerning the flow of fluid.

The general differential equation for the mass transfer of component "A", can be written in Cartesian coordinates as is shown in Equation (10); and, in cylindrical and spherical coordinates as Equations (11) and (12), respectively, which represent Fick's law equations; these equations were solved for the triangular cells used for the discretization with the finite element method and a k-ε model; turbulence and diffusion are obtained. A nested routine saved the tracer's concentration values at every monitoring point after every time step ($p_m^{t+\Delta t}$); then, the data were analyzed using Microsoft Excel. The previous equations can be used to solve regular structured or nonstructured meshes. The analyzed fluid was a mix of lead + tracer that was incorporated into the numerical solution as a multicomponent fluid problem, and Equation (13) was used to represent it [36], where D_{eff} was the diffusion and turbulent coefficient.

$$\frac{\partial c_A}{\partial t}=D_{AB}\left[\frac{\partial^2 c_A}{\partial x^2}+\frac{\partial^2 c_A}{\partial y^2}+\frac{\partial^2 c_A}{\partial z^2}\right] \tag{10}$$

$$\frac{\partial c_A}{\partial t}=D_{AB}\left[\frac{\partial^2 c_A}{\partial r^2}+\frac{1}{r}\frac{\partial c_A}{\partial r}+\frac{1}{r^2}\frac{\partial^2 c_A}{\partial \theta^2}+\frac{\partial^2 c_A}{\partial z^2}\right] \tag{11}$$

$$\frac{\partial c_A}{\partial t}=D_{AB}\left[\frac{1}{r^2}\frac{\partial}{\partial r}\left(\frac{\partial c_A}{\partial r}\right)+\frac{1}{r^2\sin\theta}\frac{\partial}{\partial \theta}\sin\theta\frac{\partial c_A}{\partial \theta}+\frac{1}{r^2\sin^2\theta}\frac{\partial^2 c_A}{\partial \varphi^2}\right] \tag{12}$$

$$\frac{\partial}{\partial t}(\rho_l C)+\nabla\bullet(\rho_l u_l C)=\nabla\bullet(\rho_l D_{eff}\nabla C) \tag{13}$$

6. Analysis and Efficiency of Mixing

A good mixing is considered to be obtained when the resident fluid has been homogeneously distributed throughout the entire tank. Here, the mechanical stirring system with the impeller was employed to promote mixing by adding movement to the resident fluid inside the tank.

The mixing efficiency is measured by monitoring the tracer's concentration (C_{tracer}) all around the tank. During the simulation, the tracer's concentration at each monitoring point was stored and saved (C^t_{pm}) at every time step (t + Δt) to understand the hydrodynamics inside the tank. Different tracer concentration curves resulted at the monitoring points as a function of the influence of the impeller, the walls, and the injection point. Some of the obtained curves oscillated and then were quickly damped, and others were quasi-parabolic or overdamped, which suggests exponential decay due to the different distributions of the tracer in every zone. Consequently, in order to understand the fluid flow behavior in every zone inside the tank, an analysis considering two different injection points is shown next.

One analyzed injection point (p_{i1}) was placed near the impeller shaft at a very high position near the tank's top surface, and one point (p_{i9}) was placed in the middle of the tank, as shown in Figure 3a. Figures 5–11 show the tracer's concentration in different zones inside the tank, where some curves are intentionally repeated in order to compare and then explain the hydrodynamics behavior of the liquid lead inside every zone of the tank compared with other points at vertical and lateral positions. All curves shown in these figures tended to an ideal value of the concentration $C_{tracer} = 0.00135$, which was the final tracer concentration value for considering total mixing, and the horizontal axes scales represent the required mixing time in s to obtain the final tracer concentration. Measurements of the tracer's concentration in the tank during the simulation with the condition $C_{tracer} > 0.00135$ indicate an excess of the tracer in the region, in contrast to the condition $C_{tracer} < 0.00135$, which indicates a deficit of the tracer. Figures 5a, 6a, 7a, 8a, 9a, 10a and 11a correspond to the injection at point P_{i1}, and Figures 5b, 6b, 7b, 8b, 9b, 10b and 11b are for point P_{i9}. Figures 5c, 6c, 7c, 8c, 9c, 10c and 11c illustrate the analyzed zone inside the tank. Moreover, the time scales were estimated to illustrate when the final tracer concentration was reached.

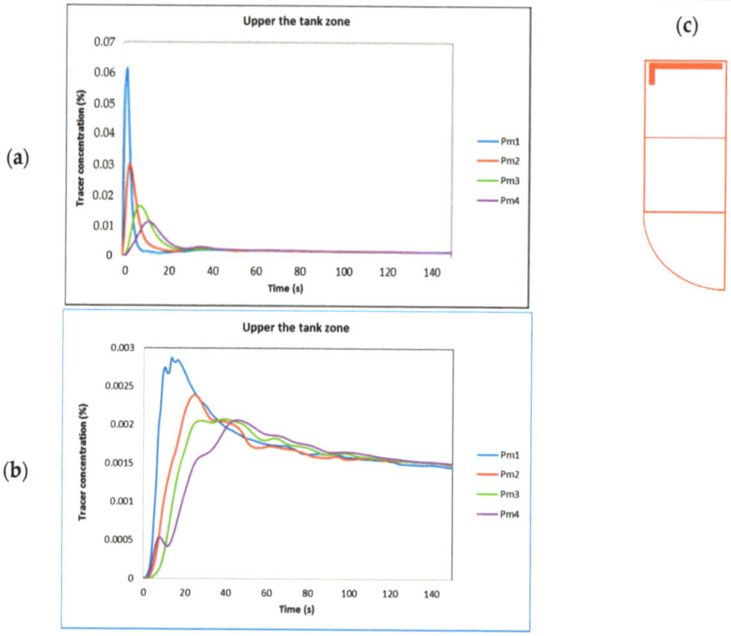

Figure 5. Curves of the analysis of the mixing efficiency near the top (upper zone) and cylinder wall: (**a**) injection point P_{i1}; (**b**) injection point P_{i9}; (**c**) analyzed region inside the tank.

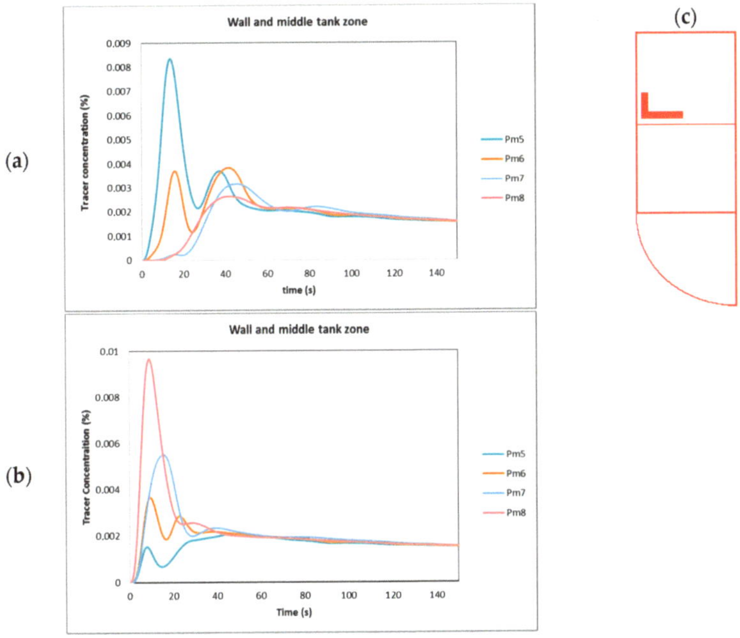

Figure 6. Curves for the analysis of the mixing efficiency of the wall and middle points: (**a**) injection point P_{i1}; (**b**) injection point P_{i9}; (**c**) analyzed region inside the tank.

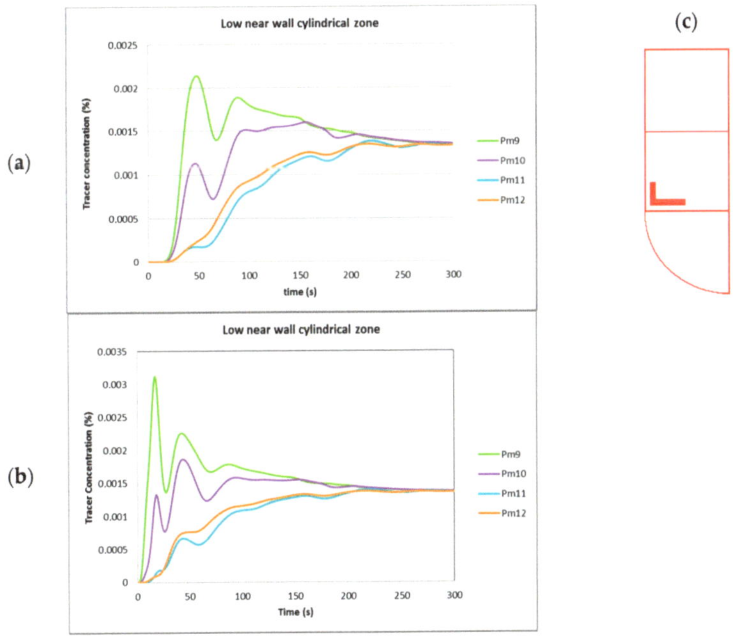

Figure 7. Curves for the analysis of the mixing efficiency at the middle and low cylindrical zone: (**a**) injection point P_{i1}; (**b**) injection point P_{i9}; (**c**) analyzed region inside the tank.

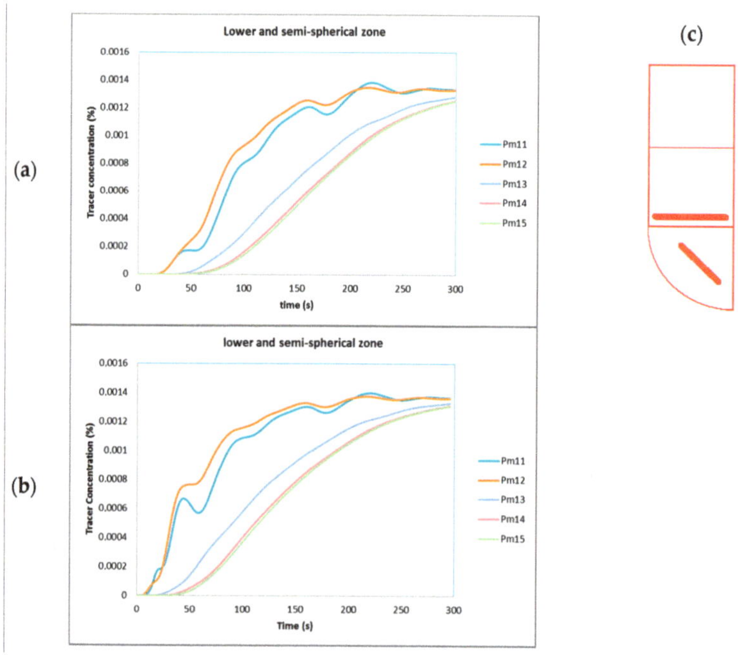

Figure 8. Curves for the analysis of the mixing efficiency at the border between the cylindrical and hemispherical zones: (**a**) injection point P_{i1}; (**b**) injection point P_{i9}; (**c**) analyzed region inside the tank.

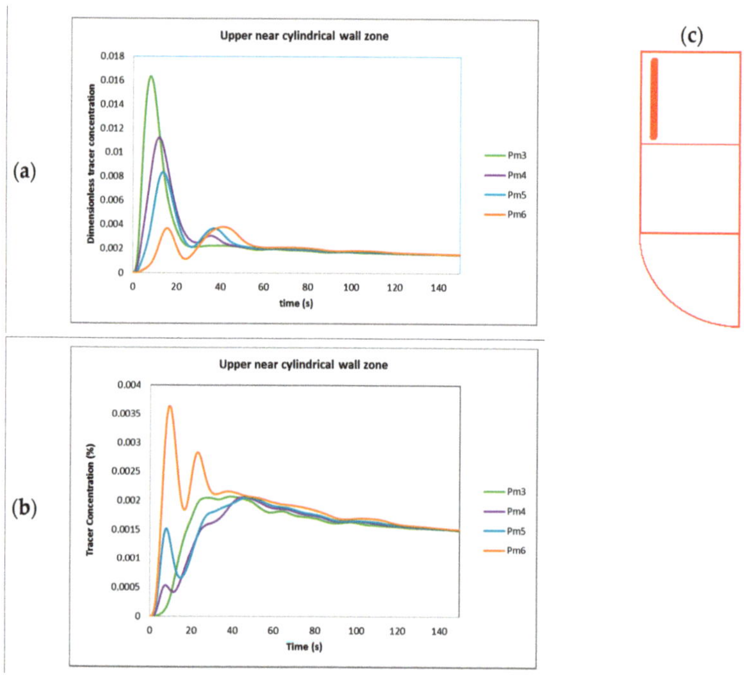

Figure 9. Curves for the analysis of the mixing efficiency on the upper near cylindrical wall zone: (**a**) injection point P_{i1}; (**b**) injection point P_{i9}; (**c**) analyzed region inside the tank.

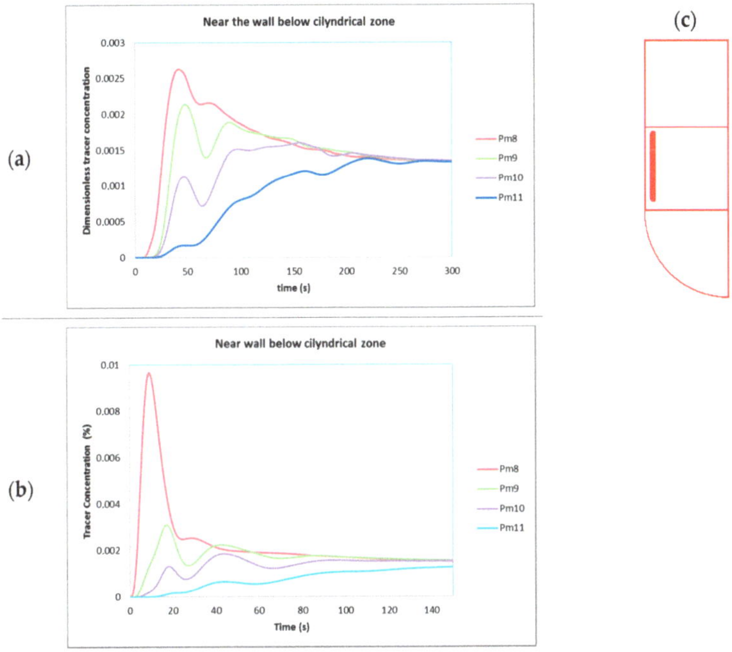

Figure 10. Curves for the analysis of the mixing efficiency on the low and cylindrical wall zone, just below the injection point: (**a**) injection point P_{i1}; (**b**) injection point P_{i9}; (**c**) analyzed region inside the tank.

The hydrodynamics behavior shown in Figure 5a,b corresponds to the top of the cylindrical section and near the upper corner of the tank. Here, a similar behavior can be identified; these curves exhibited an oscillating and irregular but quickly damped behavior; all of them had an excess of tracer concentration at the beginning, which tended to the final tracer concentration in Figure 5a. The highest tracer concentration was at the monitoring point (P_{m1}) because it was nearest to the injection point (P_{i1}); all these curves had an excess of tracer, but it was quickly distributed, and the final tracer concentration was quickly reached (no more than 50 s was required). In Figure 5b, some perturbations (suboscillating) can be observed, which may be attributed to the impact of the fluid against the flat top and cylindrical walls. Nevertheless, these small fluctuations were smoothed as the simulation continued to run. Additionally, in this figure, the curves had low tracer excess due to the injection point (P_{i9}) being placed in a position near the middle of the high point of the cylindrical body of the tank.

In Figure 6a, all points (p_{m5} to p_{m8}) exhibited an oscillating, smoothed behavior, but the curves tended to delay achieving the final tracer concentration. The points (P_{m5}, P_{m6} and P_{m8}) displayed a similar behavior because they were near the wall in the cylindrical zone, but p_{m7} was placed in the middle of the tank's body. Furthermore, the oscillations were reduced in of all these curves as the simulation time increased; all of these curves showed an excess of tracer concentration, but the values were minor, as these points were far from the injection point (p_{i1}), and the final tracer concentration was reached after 140 s.

Figure 6b shows the tracer concentration at the nearest monitoring points to injection point P_{i9}. Thus, P_{m8} showed the highest tracer concentration value (in excess), and the second and third places were for the points P_{m7} and P_{m6}, respectively. Here an oscillating behavior for all curves can also be observed, and the tracer excess was reduced as the monitoring point was far from the injection point. The instability due to the tracer excess was also quickly damped. This figure also shows the evolution of the tracer concentration horizontally in the middle position of the tank reactor. For the monitoring points P_{m7} and P_{m8}, both curves had very similar behavior, evidencing they were positioned horizontally. Finally, the final tracer concentration was quickly achieved for injection point P_{i9}.

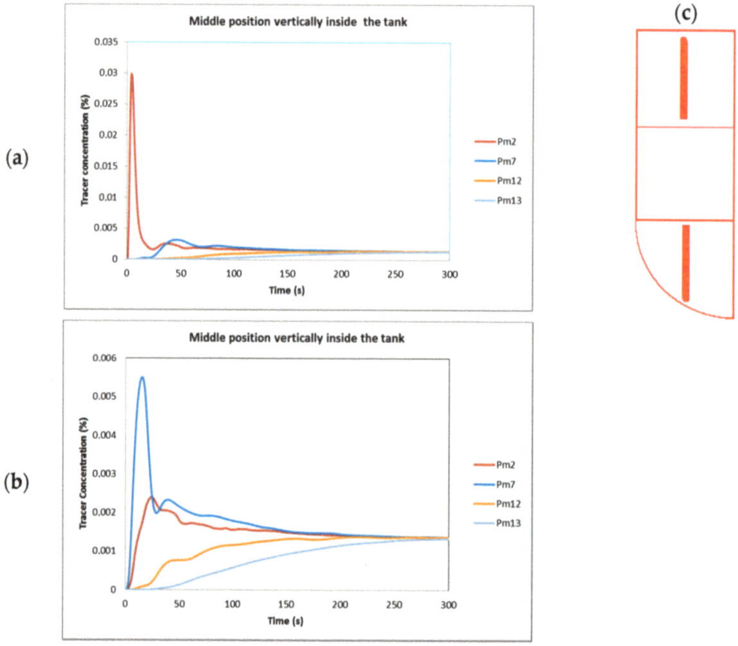

Figure 11. Curves for the analysis of the mixing efficiency in the middle vertical zone: (**a**) injection point P_{i1}; (**b**) injection point P_{i9}; (**c**) analyzed region inside the tank.

The next analyzed zone is in the middle but lower position on the tank. Here, the points P_{m9} to P_{m12} were placed near the wall of the cylindrical section of the tank near the border between the cylindrical and hemispherical zones; these points were far away from the injection point P_{i1} but just below injection point P_{i9}. Then, as shown in Figure 7a, all curves exhibited a similar oscillating but damped behavior. The points P_{m9} and P_{m10} showed a moderate excess of tracer concentration, but the curves for points P_{m11} and P_{m12} displayed a deficit of tracer concentration. All the curves were damped, and all of the curves tended to the final tracer concentration but had a delay of nearly 300 s, and it can be noticed that the main tracer deficit appeared between the cylindrical and the hemispherical zones.

In Figure 7b, a similar oscillating damped fluid behavior can be appreciated for all of the curves. The points with the highest tracer excess were those near the injection point (P_{i9}). The monitoring points P_{m9} and P_{m10} exhibited an excess of tracer and then a strong damping can be noticed. In contrast, the monitoring points P_{m11} and P_{m12} were placed on the border between the cylindrical and hemispherical zones of the tank and did not have any tracer excess; thus, the damping was not as strong; moreover, these points always remained with a deficit of tracer concentration until all of them slowly tended to the final tracer concentration. Moreover, the oscillations of the tracer concentrations were stronger for the monitoring points P_{m9} and P_{m10}, which were in the cylindrical region, than for the monitoring points P_{m11} and P_{m12}. Finally, the curves in Figure 7b showed a similar behavior as in Figure 7a, but the curves for points P_{m9} and P_{m10} displayed a major excess of tracer concentration, as these were near injection point P_{i9}.

Figure 8a,b illustrate the hydrodynamics behavior in the hemispherical region. All of the curves in these figures were quite similar and exhibited a deficit of tracer concentration.

However, an oscillating, damped behavior was displayed for the monitoring points P_{m11} and P_{m12} in both figures; in contrast, the curves for the points P_{m13}, P_{m14} and P_{m15} were overdamped, evidencing a very slow increment in the tracer's concentration. Moreover, the mixing times were longer. Thus, it is possible to affirm that this is a stagnant zone. In the same way in Figure 8b, the points were placed in the deepest positions of the tank reactor. Points P_{m11} and P_{m12} were on the border between the cylindrical and hemispherical zones. The oscillating behavior can be noticed, but the hydrodynamics in the hemispherical zone is different, as illustrated in the curves for the points P_{m13}, P_{m14} and P_{m15}. This zone is very difficult to access and the delay in distributing the tracer is long; furthermore, the tracer concentration is always deficient. Moreover, it is possible to notice that the monitoring points near the impeller were oscillating with a tracer deficit, and the points near the hemispherical wall were overdamped.

All the monitoring points tended to adopt the final tracer concentration. This is a reliable and feasible criterion for evaluating the distribution efficiency. The excess of tracer must be distributed, and the deficits must be filled all around the tank reactor for good mixing.

Excesses of tracer concentration were at the monitoring points near the corresponding injection points. But these excesses were frequently quickly distributed. In contrast, a deficit of the tracer occurred in cases where the monitoring points were placed very far from the tracer injection point or far from the impeller.

The evolution of the tracer's concentration in the cylindrical upper wall section is shown in Figure 9a,b. All curves for the points P_{m3} to P_{m6} exhibited similar oscillating behavior to a family of curves; for the injection point P_{i1}, the monitoring points with the highest and minimum excess of tracer were P_{m3} and P_{m6}, respectively. Although there was high tracer excess at the beginning, the final tracer concentration was quickly reached. For the injection point P_{i9}, the nearest monitoring point was P_{m6}; therefore, this point had the highest tracer excess. Here, again, an alternating oscillating behavior was also shown for all of the curves, and the damping decreased over the simulation time. Finally, all of the values tended to adopt the final tracer concentration but always with a tracer concentration in excess. Thus, the influence of the injection points can be noticed on the families of curves in Figure 9a,b; here, the curves are different for both injection points, evidencing different hydrodynamics behaviors.

The evolution of the tracer's concentration near the cylindrical lower wall section is shown in Figure 10a,b. In Figure 10a, all curves exhibit am oscillating damped behavior; here, the monitoring points P_{m8} to P_{m10} display an excess of tracer concentration, but the point P_{m11} show a deficit, as it is placed between the cylindrical and hemispherical zones. The excess of the tracer concentration at points P_{m8} and P_{m9}, as shown in Figure 10a, are minor in comparison with those for the injection point P_{i9} due to the injection point being far away. Finally, all curves tended to the final tracer concentration, but the delay was longer than the curves for the point P_{i9}. In comparison, in Figure 10b, the monitoring point P_{m8} was the nearest to the injection point; thus, it had the greatest tracer concentration. Moreover, the tracer concentration was reduced from P_{m8} to P_{m11} as a function of the distance from the injection point. The oscillating, alternating behavior can also be observed for all of the curves. Finally, only P_{m11} was, again, the only point with no tracer excess.

All of the curves with an irregular oscillating form were in the cylindrical zone, but these were damped; on the other hand, the curves with an overdamped behavior were in the hemispherical zone of the tank. The irregularity of the oscillating curves represents an excess of the tracer concentration and instability, but the tracer was quickly distributed at the monitoring point near the impeller's influence, while the oscillating curves with a deficit of tracer were in the zones where the distribution was complicated.

The overdamped curves were always with a deficit of tracer concentration, evidencing a very slow tracer distribution; thus, these represented stagnant zones. Thus, the hemispherical region of the tank requires longer times for mixing, maybe a geometrical modification of the original tank based on hydrodynamics behavior will accelerate the tracer distribution.

The evolution of the tracer's concentration vertically in the middle position of the tank reactor can be observed in Figure 11a,b. In Figure 11a, the monitoring points P_{m2} and P_{m7} exhibited an oscillating behavior. Although the tracer's excess was quickly distributed, there was a very large difference in the tracer's concentration for these points. On the other hand, the points P_{m12} and P_{m13} had deficit of tracer and showed an overdamped behavior; unfortunately, the tracer distributed slowly. In Figure 11b, the monitoring points P_{m2} and P_{m7} were higher than the injection point P_{i9} and showed tracer excess; in contrast, the points P_{m12} and P_{m13} were in the hemispherical zone, evidencing a very different behavior and without any tracer excess. The curves in the middle tank were transformed from an oscillating form to an overdamped one from the upper to the lower zone of the tank.

The contributions of all of the monitoring points can be added and averaged in order to obtain one single curve for representing the hydrodynamics behavior inside the tank and evaluating the mixing efficiency as a function of the injection points. Furthermore, the resulting curves can be derived and then the final tracer concentration can be subtracted to determine if the variation of the tracer concentration tended to zero (minimum variation); this means that the tracer was homogeneously distributed. In Figure 12a,b, the curves for the analyzed injection points are shown. Here, it is possible to confirm that the injection point P_{i9} was the best because the values of the tracer concentration were always minor compared to those for the point P_{i1}. Moreover, on the vertical axis, the derivate curve values were also minor. The average curves tended to the final tracer concentration and the derived curves tended to zero. The tendency of all curves was 0.00135, which was the volume fraction of the tracer distributed in every 3D cell where the monitoring points were placed. As a criterion for the homogeneous distribution of the tracer, it was considered that the rest of the cells used for discretization of the volume in the tank must contain the same tracer concentration. Moreover, the invariability was considered as a criterion to determine that the extra time of stirring was not necessary.

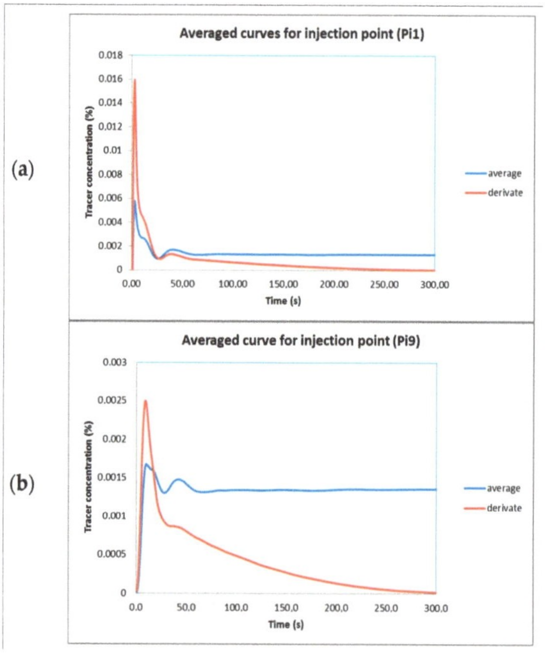

Figure 12. Average and derivative curves for the injection points: (**a**) P_{i1}; (**b**) P_{i9}.

A good mixing was obtained when the tracer concentration tended to be constant (i.e., invariable). According to this criterion, the best injection point was pi_{18}, and the worst was pi_{21}. As is shown in Figure 13, the final tracer concentration for pi_{18} was quickly stabilized. This curve also shows minimized fluctuations and short periods of oscillating behavior, which were shorter than those for the other injection points. In contrast, the curve for pi_{21} showed a very slow tracer dispersion. The slope of this curve was soft, resulting in a homogeneous and eventually very slow tracer distribution all around the tank. This point was very far from the impeller, and the injection here was not favorable due to the diffusion problems. In this figure, the scale was modified by dividing it with the final tracer concentration; then, all curves tended to 100; a percentage that means the final tracer concentration.

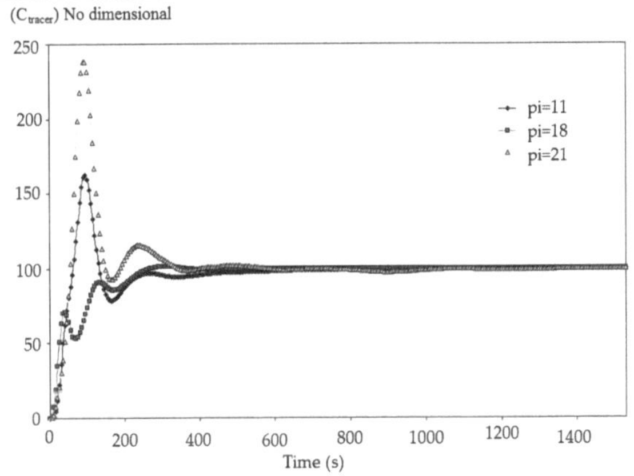

Figure 13. Tracer concentrations for the best, intermediate and worst injection points (Pi_{18}, Pi_{11} and Pi_{21}, respectively).

The hemispherical bottom of the tank is a critical zone for mixing because the tracer distribution here involves diffusion problems and accessibility to this zone is also complicated.

Moreover, a close up at the end of the figure can be taken to show fluctuations during mixing became very short but remained; consequently, these can be neglected due to the fluctuations being nonsignificant. This concept can be appreciated in the tracer concentration profiles shown in Figures 5–12.

7. Validation

For validation purposes, a physical model of the tank was built using transparent acrylic scaled 1:11.4 considering the lead density. The rotational speed remained constant, the fluid used was water and the tracer was NaCl. The model was sized as a function of the evaluation between the inertial and viscous forces under different velocities; then, the Reynolds number in Equation (14) was calculated at every monitoring point position and included in the computational solution, while the values of the tracer concentration were obtained from the physical modeling measured instantly using a particle indicator velocimeter (PIV).

$$\mathrm{Re} = \frac{\rho v D}{\mu} \tag{14}$$

Measurements at every monitoring point were taken using the instrument at every $\Delta t = 0.5$ s. In contrast, the time step used for the computer simulation at the smallest cells of the volume near the impeller blades was $\Delta t = 0.0052$ s resulting from the rotatory speed 1 min/32 RPM/360°. This time was considered good enough for following the evolution of the mixing process. Then, the results were saved every 10 steps to be applied on the wider mesh that discretized the tank volume.

The values of the tracer's concentration for the physical model and computational simulation were compared for validation of the Pi_{18}, Pi_{11} and Pi_{21}. These are shown in Figure 14a–c and compared to evaluate the fidelity between both models. In this figure, the difference is evident at the beginning of the simulation but reduces as the simulation continues over time; both curves tended to the final tracer concentration. Then, it is possible to mention that the global behavior was similar. Although there were some differences between the physical and computational modeling which could have affected the resulting curves.

(a) The tracer in the physical model was injected in one single amount by a blowing pipeline, as shown in Figure 15, but the impulse force was not considered in the computational model. The tracer used in the physical model was NaCl with a size between 5×10^{-4} and 7×10^{-4} m. Consequently, controlling the injection of the entire tracer in one single pulse in the physical model was difficult. The same occurs during real industrial practices;

(b) The time for acquiring data in the physical model was averaged every 25 s, and the tracer was measured at every monitoring point across a plane with the same width as a tetragonal cell used for discretization in the computational model;

(c) The curves of the tracer concentration (C_{tracer}) were the sum of all contributions of the monitoring points for the computer simulations measured independently, which were taken every 25 s for comparison in contrast to all previous figures, which were measured every 1 s.

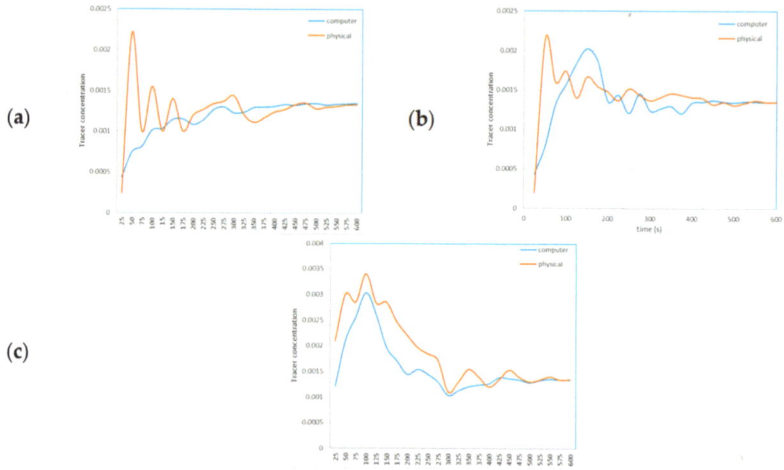

Figure 14. Comparisons between the physical modeling and computer simulation injection points: (**a**) pi_{18}; (**b**) pi_{11}; (**c**) pi_{21}.

Different mesh sizes were tested with computational modeling, including finer meshes of tetrahedral cells, hexahedral (cubic) and honey comb cells, but the final tracer concentration was so similar in all of the simulations; thus, it can be considered, again, as the moment when the tracer was homogeneously distributed.

Additionally, the errors and approach were obtained using Equations (15) and (16). Here, the subscripts pi correspond to the analyzed injection point, and the superscripts (t) and (t − Δt) refer to the latest and previous tracer concentration value during the simulation. The resulting graphics are shown in Figures 16a–c and 17a–c, respectively. These curves use the information on the tracer concentration in Figure 15a–c. It can be appreciated that approach improves as the tracer is homogenized inside the tank. Moreover, all of the curves tended to zero. Then, the dampening of the curves confirm that the tank was in a new stationary condition with the tracer homogeneously distributed all around.

$$error = 100\left[C_{pi}^{t} - C_{pi}^{t-\Delta t}\right] \quad (15)$$

$$error = 100\left[\frac{C_{pi-comp}^{\Delta t} - C_{pi-phys}^{\Delta t}}{C_{tracer-final}}\right] \quad (16)$$

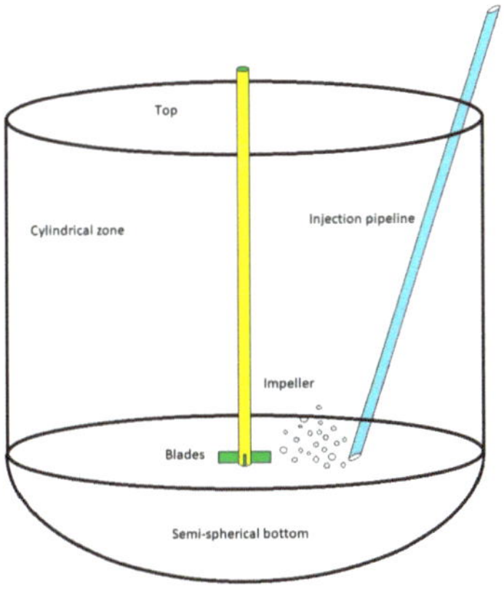

Figure 15. Sketch of the stirred tank built for the physical model and the injection of NaCl using a lance blowing.

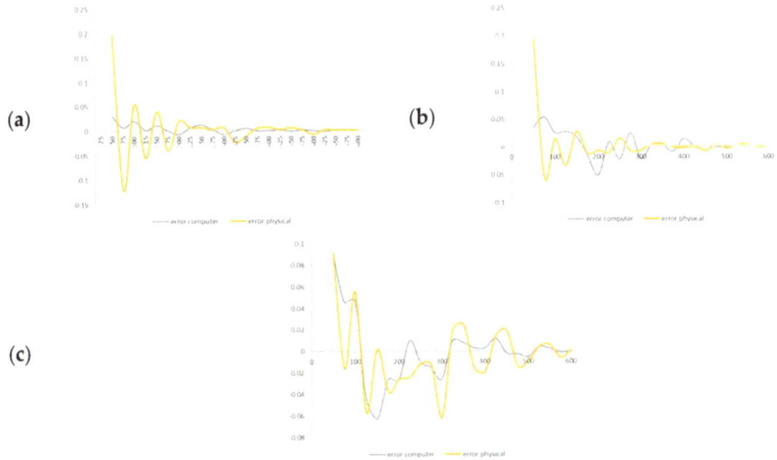

Figure 16. Calculated errors for comparisons between the physical modeling and computer simulation as a function of the latest tracer concentration for the injection points: (**a**) pi_{18}; (**b**) pi_{11}; (**c**) pi_{21}.

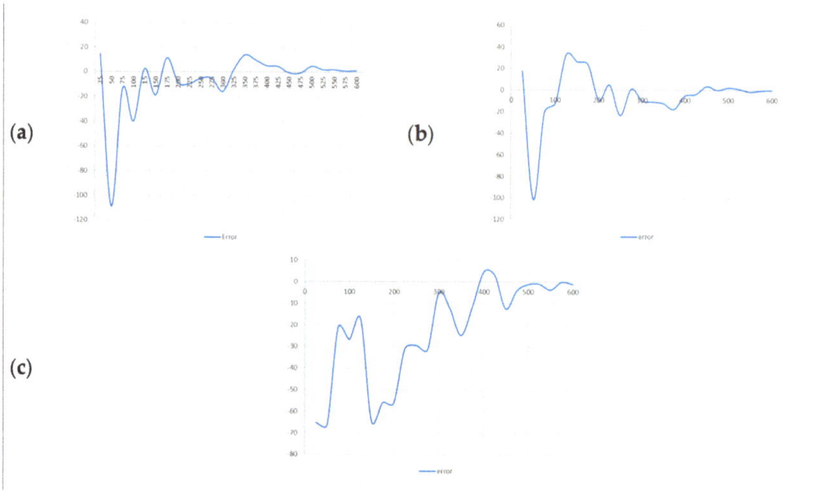

Figure 17. Calculated errors for comparisons between the physical modeling and computer simulation for the injection points: (**a**) pi_{18}; (**b**) pi_{11}; (**c**) pi_{21}.

8. Conclusions

In order to understand the hydrodynamics of fluid flow inside the tank it was observed that different tracer concentrations were measured as a function of the injection and monitoring positions. The tracer concentration measurement at monitoring points near the impeller evidenced its influence for promoting fluid displacement and tracer dispersion. Thus, after the analysis, it was possible to affirm the following facts:

The influence of the impeller was moderate at the monitoring points near the walls and at the top due to the fluid impacting against it, and the tracer was slowly distributed.

After analyzing all of the injection points, the best injection point was pi_{18}, which was placed near the impeller and promoted a quick tracer dispersion, profiting from the force provided by the impeller.

Diffusion problems were related to the geometrical accessibility of the fluid to zones in the bottom of the tank, and here the losses of kinetic energy caused the fluid to become stagnant.

During the validation, a comparison between the tracer concentration obtained from computer simulation and the physical model showed that the error between them was reduced when the stirring process dominated the hydrodynamics; consequently, the most difficult period to simulate was the beginning of the mixing, and there was instability affecting the calculation, but the phenomenon became predictable as the tracer was homogenized.

The oscillation of the tracer concentrations was major for the monitoring points in the cylindrical region of the tank, although these oscillations were reduced the nearer to the impeller.

The smoothing of the oscillating curves indicates that the variation in the tracer concentration (C_{tracer}) was reduced; thus, the distribution became more homogeneous and the mixing improved. Here, the points in the middle region of the tank had a major excess of tracer than the points near the walls.

Choosing an appropriate injection point can contribute to improving the mixing and reducing the working time dramatically, eliminating unnecessary work.

Simulations over very long times when the values of the tracer concentration were not significant are not required; consequently, long working times are not required either.

Finding the best injection point proved that the mixing time can be reduced and industrial practices can be improved only by modifying this condition; nevertheless, this is a first step. For future works, the author recommends to analyze fluid hydrodynamics behavior including a complementary study with different tracer properties in order to understand if there is any other force influencing mixing like buoyancy forces; this is a necessary study because many of the slags and chemical components added to melting metals are with different properties like weight, density and viscosity. Additionally, a simulation considering an incomplete integration of the tracer in one single injection or in programmed pulses will be in better concordance with real industrial practices, in addition to modifying the geometrical configurations of the tank and impeller, including the tank's geometrical form, number of blades, blades sizes and rotation speed of the impeller.

Funding: This research received no external funding.

Institutional Review Board Statement: Not applicable.

Data Availability Statement: Not applicable.

Acknowledgments: The authors wish to thank their institutions: Technological and Autonomous Institute of México (ITAM) and the Mexican Association of Culture and National Council of Science and Technology (CONACyT). This work is dedicated to the memory of my friend and colleague Pedro Vite-Martinez.

Conflicts of Interest: The author declares no conflict of interest.

References

1. Zhang, W.; Yang, J.; Wu, X.; Hu, Y.; Yu, W.; Wang, J.; Dong, J.; Li, M.; Liang, S.; Hu, J.; et al. A critical review on secondary lead recycling technology and its prospect. *Renew. Sustain. Energy Rev.* **2016**, *61*, 108–122. [CrossRef]
2. Li, Y.; Yang, S.; Taskinen, P.; He, J.; Liao, F.; Zhu, R.; Chen, Y.; Tang, C.; Wang, Y.; Jokilaakso, A. Novel recycling process for lead-acid battery paste without SO_2 generation—Reaction mechanism and industrial pilot campaign. *J. Clean. Prod.* **2019**, *217*, 162–171. [CrossRef]
3. Sun, Z.; Cao, H.; Zhang, X.; Lin, X.; Zheng, W.; Cao, G.; Sun, Y.; Zhang, Y. Spent lead-acid battery recycling in China—A review and sustainable analyses on mass flow of lead. *Waste Manag.* **2017**, *64*, 190–201. [CrossRef] [PubMed]
4. Prengaman, R.; Mirza, A. 20—Recycling concepts for lead–acid batteries. In *Lead-Acid Batteries for Future Automobiles*; Elsevier: Amsterdam, The Netherlands, 2017; pp. 575–598. [CrossRef]
5. Hildebrandt, T.; Osada, A.; Peng, S.; Moyer, T. 19—Standards and tests for lead–acid batteries in automotive applications. In *Lead-Acid Batteries for Future Automobiles*; Elsevier: Amsterdam, The Netherlands, 2017; pp. 551–573. [CrossRef]
6. Moseley, P.; Rand, D.; Garche, J. 21—Lead–acid batteries for future automobiles: Status and prospect. In *Lead-Acid Batteries for Future Automobiles*; Elsevier: Amsterdam, The Netherlands, 2017; pp. 601–618. [CrossRef]
7. Sloop, S.; Kotaich, K.; Ellis, T.; Clarke, R. RECYCLING | Lead–Acid Batteries: Electrochemical. In *Encyclopedia of Electrochemical Power Sources*; Elsevier: Amsterdam, The Netherlands, 2009; pp. 179–187. [CrossRef]
8. Tian, X.; Gong, Y.; Wu, Y.; Agyeiwaa, A.; Zuo, T. Management of used lead acid battery in China: Secondary lead industry progress, policies and problems. *Resour. Conserv. Recycl.* **2014**, *93*, 75–84. [CrossRef]
9. Blanpain, B.; Arnout, S.; Chintinne, M.; Swinbourne, D.R. Lead Recycling. In *Handbook of Recycling*; Elsevier: Amsterdam, The Netherlands, 2014; pp. 95–111. [CrossRef]
10. Bailey, C.; Kumar, S.; Patel, M.; Piper, A.W.; Fosdick, R.A.; Hance, S. Comparison between CFD and measured data for the mixing of lead bullion. In Proceedings of the 2nd International Conference on CFD in Minerals and Process Industries CSIRO, Melbourne, Australia, 6–8 December 1999; pp. 351–356.
11. Choi, B.S.; Wan, B.; Philyaw, S.; Dhanasekharan, K.; Ring, T.A. Residence Time Distributions in a Stirred Tank: Comparison of CFD Predictions with Experiment. *Ind. Eng. Chem. Res.* **2004**, *43*, 6548–6556. [CrossRef]
12. Guha, D.; Ramachandran, P.; Dudukovic, M. Flow field of suspended solids in a stirred tank reactor by Lagrangian tracking. *Chem. Eng. Sci.* **2007**, *62*, 6143–6154. [CrossRef]
13. Wang, F.; Mao, Z.-S.; Wang, Y.; Yang, C. Measurement of phase holdups in liquid–liquid–solid three-phase stirred tanks and CFD simulation. *Chem. Eng. Sci.* **2006**, *61*, 7535–7550. [CrossRef]
14. Hartmann, H.; Derksen, J.; Akker, H.v.D. Numerical simulation of a dissolution process in a stirred tank reactor. *Chem. Eng. Sci.* **2006**, *61*, 3025–3032. [CrossRef]
15. Huang, S.; Mohamad, A.; Nandakumar, K. Numerical Analysis of a Two-Phase Flow and Mixing Process in a Stirred Tank. *Int. J. Chem. React. Eng.* **2008**, *6*. [CrossRef]
16. Koh, P.T.L.; Xantidis, F. CFD Modelling in the scale-up of a stirred reactor for resin beads. In Proceedings of the 2nd International Conference on CFD in Minerals and Process Industries CSIRO, Melbourne, Australia, 6–8 December 1999; pp. 369–374.
17. Murthy, B.; Deshmukh, N.; Patwardhan, A.; Joshi, J. Hollow self-inducing impellers: Flow visualization and CFD simulation. *Chem. Eng. Sci.* **2007**, *62*, 3839–3848. [CrossRef]

18. Zhang, Z.; Chen, G. Liquid mixing enhancement by chaotic perturbations in stirred tanks. *Chaos Solitons Fractals* **2008**, *36*, 144–149. [CrossRef]
19. Colli, A.; Bisang, J. Evaluation of the hydrodynamic behaviour of turbulence promoters in parallel plate electrochemical reactors by means of the dispersion model. *Electrochim. Acta* **2011**, *56*, 7312–7318. [CrossRef]
20. Zhao, H.-L.; Zhang, Z.-M.; Zhang, T.-A.; Liu, Y.; Gu, S.-Q.; Zhang, C. Experimental and CFD studies of solid–liquid slurry tank stirred with an improved Intermig impeller. *Trans. Nonferrous Met. Soc. China* **2014**, *24*, 2650–2659. [CrossRef]
21. Gong, Y.; Tian, X.-M.; Wu, Y.-F.; Tan, Z.; Lv, L. Recent development of recycling lead from scrap CRTs: A technological review. *Waste Manag.* **2016**, *57*, 176–186. [CrossRef] [PubMed]
22. Tamburini, A.; Cipollina, A.; Micale, G.; Brucato, A.; Ciofalo, M. Influence of drag and turbulence modelling on CFD predictions of solid liquid suspensions in stirred vessels. *Chem. Eng. Res. Des.* **2013**, *92*, 1045–1063. [CrossRef]
23. Wadnerkar, D.; Tade, M.O.; Pareek, V.K.; Utikar, R.P. CFD simulation of solid–liquid stirred tanks for low to dense solid loading systems. *Particuology* **2016**, *29*, 16–33. [CrossRef]
24. Karcz, J.; Cudak, M.; Szoplik, J. Stirring of a liquid in a stirred tank with an eccentrically located impeller. *Chem. Eng. Sci.* **2005**, *60*, 2369–2380. [CrossRef]
25. Szalai, E.; Arratia, P.; Johnson, K.; Muzzio, F. Mixing analysis in a tank stirred with Ekato Intermig® impellers. *Chem. Eng. Sci.* **2004**, *59*, 3793–3805. [CrossRef]
26. Pinho, F.; Piqueiro, F.; Proença, M.; Santos, A. Turbulent flow in stirred vessels agitated by a single, low-clearance hyperboloid impeller. *Chem. Eng. Sci.* **2000**, *55*, 3287–3303. [CrossRef]
27. Vite-Martínez, P.; Durán-Valencia, C.; Cruz-Maya, J.; Ramírez-López, A.; López-Ramírez, S. Optimization of reagents injection in a stirred batch reactor by numerical simulation. *Comput. Chem. Eng.* **2014**, *60*, 307–314. [CrossRef]
28. Pan, H.; Geng, Y.; Dong, H.; Ali, M.; Xiao, S. Sustainability evaluation of secondary lead production from spent lead acid batteries recycling. *Resour. Conserv. Recycl.* **2018**, *140*, 13–22. [CrossRef]
29. Escudié, R.; Liné, A. Analysis of turbulence anisotropy in a mixing tank. *Chem. Eng. Sci.* **2006**, *61*, 2771–2779. [CrossRef]
30. Vite-Martinez, P.; Ramirez-Lopez, A.; Romero-Serrano, A.; Chavez-Alcala, F. Improving of Mixing Efficiency in a Stirred Reactor for Lead Recycling Using Computer Simulation. *Inz. Miner.* **2014**, *16*, 25–32.
31. Wadnerkar, D.; Utikar, R.P.; Tade, M.O.; Pareek, V.K. CFD simulation of solid–liquid stirred tanks. *Adv. Powder Technol.* **2012**, *23*, 445–453. [CrossRef]
32. Xie, L.; Luo, Z.-H. Modeling and simulation of the influences of particle-particle interactions on dense solid–liquid suspensions in stirred vessels. *Chem. Eng. Sci.* **2018**, *176*, 439–453. [CrossRef]
33. Gu, D.; Liu, Z.; Xie, Z.; Li, J.; Tao, C.; Wang, Y. Numerical simulation of solid-liquid suspension in a stirred tank with a dual punched rigid-flexible impeller. *Adv. Powder Technol.* **2017**, *28*, 2723–2734. [CrossRef]
34. Jauhiainen, A.; Jonsson, L.; Sheng, D.-Y. Modelling of alloy mixing into steel. The influence of porous plug placement in the ladle bottom on the mixing of alloys into steel in a gas-stirred ladle. A comparison made by numerical simulation. *Scand. J. Metall.* **2001**, *30*, 242–253. [CrossRef]
35. Chen, C.; Jonsson, L.T.I.; Tilliander, A.; Cheng, G.; Jönsson, P.G. A Mathematical Modeling Study of Tracer Mixing in a Continuous Casting Tundish. *Metall. Mater. Trans. A* **2014**, *46*, 169–190. [CrossRef]
36. Zhang, Y.; Chen, C.; Lin, W.; Yu, Y.; Dianyu, E.; Wang, S. Numerical Simulation of Tracers Transport Process in Water Model of a Vacuum Refining Unit: Single Snorkel Refining Furnace. *Steel Res. Int.* **2020**, *91*, 2000022. [CrossRef]
37. Ouyang, X.; Lin, W.; Luo, Y.; Zhang, Y.; Fan, J.; Chen, C.; Cheng, G. Effect of Salt Tracer Dosages on the Mixing Process in the Water Model of a Single Snorkel Refining Furnace. *Metals* **2022**, *12*, 1948. [CrossRef]
38. Morales, R.D.; Calderon-Hurtado, F.A.; Chattopadhyay, K. Demystifying Underlying Fluid Mechanics of Gas Stirred Ladle Systems with Top Slag Layer Using Physical Modeling and Mathematical Modeling. *ISIJ Int.* **2019**, *59*, 1224–1233. [CrossRef]
39. Rahimi, M.; Parvareh, A. CFD study on mixing by coupled jet-impeller mixers in a large crude oil storage tank. *Comput. Chem. Eng.* **2007**, *31*, 737–744. [CrossRef]
40. Sossa-Echeverria, J.; Taghipour, F. Computational simulation of mixing flow of shear thinning non-Newtonian fluids with various impellers in a stirred tank. *Chem. Eng. Process.—Process. Intensif.* **2015**, *93*, 66–78. [CrossRef]
41. Lassin, A.; Piantone, P.; Burnol, A.; Bodénan, F.; Chateau, L.; Lerouge, C.; Crouzet, C.; Guyonnet, D.; Bailly, L. Reactivity of waste generated during lead recycling: An integrated study. *J. Hazard. Mater.* **2007**, *139*, 430–437. [CrossRef]
42. Weidenhamer, J.D.; Clement, M.L. Evidence of recycling of lead battery waste into highly leaded jewelry. *Chemosphere* **2007**, *69*, 1670–1672. [CrossRef]
43. Milewska, A.; Molga, E. CFD simulation of accidents in industrial batch stirred tank reactors. *Chem. Eng. Sci.* **2007**, *62*, 4920–4925. [CrossRef]

fluids MDPI

Review

A Critical Review of CFD Modeling Approaches for Darrieus Turbines: Assessing Discrepancies in Power Coefficient Estimation and Wake Vortex Development

Saïf ed-Dîn Fertahi [1,2,*], Tarik Belhadad [3], Anass Kanna [3], Abderrahim Samaouali [1], Imad Kadiri [4] and Ernesto Benini [5]

1 "Thermodynamics and Energy" Research Team, Energy Research Center, Physics Department, Faculty of Science, Mohammed V University in Rabat, 4 Avenue Ibn Batouta, Rabat B.P. 1014, Morocco; abderrahimsamaouali@yahoo.fr
2 Laboratoire des Sciences pour l'Ingénieur Appliquées à la Mécanique et au Génie Electrique (SIAME)—Fédération IPRA, Université de Pau et des Pays de l'Adour (UPPA/E2S), EA4581, 64000 Pau, France
3 Engineering Sciences Laboratory, Polydisciplinary Faculty of Taza, Sidi Mohamed Ben Abdellah University of Fez (USMBA), Taza B.P. 1223, Morocco; tarik.belhadad@gmail.com (T.B.); anass.kanna@gmail.com (A.K.)
4 Laboratoire d'Etude des Matériaux Avancés et Applications (LEM2A), Ecole Supérieure de Technologie de Meknès, Université Moulay Ismail (UMI), Km 5, Route d'Agouray, N6, Meknes 50040, Morocco; i.kadiri@umi.ac.ma
5 Department of Industrial Engineering, University of Padova, Via Venezia, 1, 35131 Padova, Italy; ernesto.benini@unipd.it
* Correspondence: fertahi.sayfdin@gmail.com; Tel.: +212-668-365-373

Abstract: This critical review delves into the impact of Computational Fluid Dynamics (CFD) modeling techniques, specifically 2D, 2.5D, and 3D simulations, on the performance and vortex dynamics of Darrieus turbines. The central aim is to dissect the disparities apparent in numerical outcomes derived from these simulation methodologies when assessing the power coefficient (C_p) within a defined velocity ratio (λ) range. The examination delves into the prevalent turbulence models shaping C_p values, and offers insightful visual aids to expound upon their influence. Furthermore, the review underscores the predominant rationale behind the adoption of 2D CFD modeling, attributed to its computationally efficient nature vis-à-vis the more intricate 2.5D or 3D approaches, particularly when gauging the turbine's performance within the designated λ realm. Moreover, the study meticulously curates a compendium of findings from an expansive collection of over 250 published articles. These findings encapsulate the evolution of pivotal parameters, including C_p, moment coefficient (C_m), lift coefficient (C_l), and drag coefficient (C_d), as well as the intricate portrayal of velocity contours, pressure distributions, vorticity patterns, turbulent kinetic energy dynamics, streamlines, and Q-criterion analyses of vorticity. An additional focal point of the review revolves around the discernment of executing 2D parametric investigations to optimize Darrieus turbine efficacy. This practice persists despite the emergence of turbulent flow structures induced by geometric modifications. Notably, the limitations inherent to the 2D methodology are vividly exemplified through compelling CFD contour representations interspersed throughout the review. Vitally, the review underscores that gauging the accuracy and validation of CFD models based solely on the comparison of C_p values against experimental data falls short. Instead, the validation of CFD models rests on time-averaged C_p values, thereby underscoring the need to account for the intricate vortex patterns in the turbine's wake—a facet that diverges significantly between 2D and 3D simulations. In a bid to showcase the extant disparities in CFD modeling of Darrieus turbine behavior and facilitate the selection of the most judicious CFD modeling approach, the review diligently presents and appraises outcomes from diverse research endeavors published across esteemed scientific journals.

Keywords: Darrieus turbine; Computational Fluid Dynamics (CFD); power coefficient; vortex structures; turbulence models; parametric studies

Citation: Fertahi, S.e.-D.; Belhadad, T.; Kanna, A.; Samaouali, A.; Kadiri, I.; Benini, E. A Critical Review of CFD Modeling Approaches for Darrieus Turbines: Assessing Discrepancies in Power Coefficient Estimation and Wake Vortex Development. *Fluids* **2023**, *8*, 242. https://doi.org/10.3390/fluids8090242

Academic Editors: Jianping Zhang and D. Andrew S. Rees

Received: 15 July 2023
Revised: 22 August 2023
Accepted: 22 August 2023
Published: 25 August 2023

1. Motivations for a Review

This critical review aims to investigate the extent to which Computational Fluid Dynamics (CFD) modeling can effectively forecast and reproduce the intricate flow structures generated by Darrieus Vertical Axis Wind Turbine (VAWT). The study encompasses an extensive bibliometric analysis based on the Scopus database, exploring yearly published papers related to CFD modeling approaches for Darrieus turbines and shedding light on the growing interest within the scientific community. The review scrutinizes the influence of various CFD modeling approaches, such as 2D, 2.5D, and 3D simulations, on the performance and vortex structures of Darrieus within a specified velocity ratio λ range. Divergent numerical outcomes from 2D, 2.5D, and 3D CFD simulations are observed when assessing the power coefficient C_p relative to the λ range. The review comprehensively compiles results from over 250 published articles, elucidating the evolution of various coefficients, velocity contours, pressure fields, vorticity patterns, turbulent kinetic energy production and dissipation, streamlines, and Q-criterion of the vorticity, among other pertinent variables. Additionally, it accentuates the impact of commonly employed turbulence models on C_p values, substantiating its claims through illustrative figures.

The review also highlights the rationale behind the prevalence of 2D CFD modeling, considering its computational tractability compared to 2.5D or 3D alternatives when estimating turbine performance within the designated λ range. However, it stresses the need to balance computational resources with an accurate portrayal of the underlying flow phenomena. The limitations of 2D parametric studies in capturing turbulent flow structures arising from geometric modifications are elucidated through comprehensive CFD contour plots. Moreover, based on the literature review, the research offers valuable recommendations for mesh generation in CFD simulations of Darrieus turbines, including guidelines for mesh types, grid point selection as a function of Reynolds number, and considerations for simulation accuracy. A practical guide is suggested for constructing meshes tailored to VAWTs based on available knowledge and best practices. Definitions and equations of crucial quantities like C_p and C_l are explicitly included in the text to enhance clarity. Overall, this study provides a concise and comparative assessment of CFD modeling approaches for vertical axis wind turbines, specifically Darrieus, contributing to a comprehensive understanding and aiding researchers and engineers in selecting the most appropriate approach for accurate predictions.

2. Introduction

A bibliometric analysis is presented, focusing on a critical review of CFD modeling approaches applied to Darrieus turbines. The Scopus database was utilized through our institution's subscription, and relevant keywords and filters were applied to retrieve yearly-published papers from 2010 to 2023. The gathered metadata was used to construct Figure 1, illustrating the distribution of documents by year, type, author, and country/territory. Through this study, the aim is identification of discrepancies in power coefficient estimation and wake vortex development, while also shedding light on the current research trends and influential contributors in the field of CFD modeling for Darrieus turbines.

The CFD modeling of Darrieus turbines continues to attract considerable interest within the scientific community. In 2019, the scientific production on Darrieus turbines witnessed a five-fold increase compared to 2010 (Figure 1a). Over the past 13 years, it has been observed that the percentage of published reviews pertaining to Darrieus turbines remains notably lower (4.6%) when contrasted with the overall scientific contributions on the subject (95.4%) (Figure 1b).

Bianchini [1], with 30 scientific articles authored and co-authored in 2023, has earned recognition as a key reference in the domain of CFD modeling for Darrieus turbines (Figure 1b). Furthermore, Italian research laboratories have emerged as the leading contributors to numerical CFD investigations on Darrieus turbines. As of 2023, Italy holds the foremost position with 61 publications, followed closely by China with 59 publications, and the United States ranking third, with 43 publications (Figure 1c).

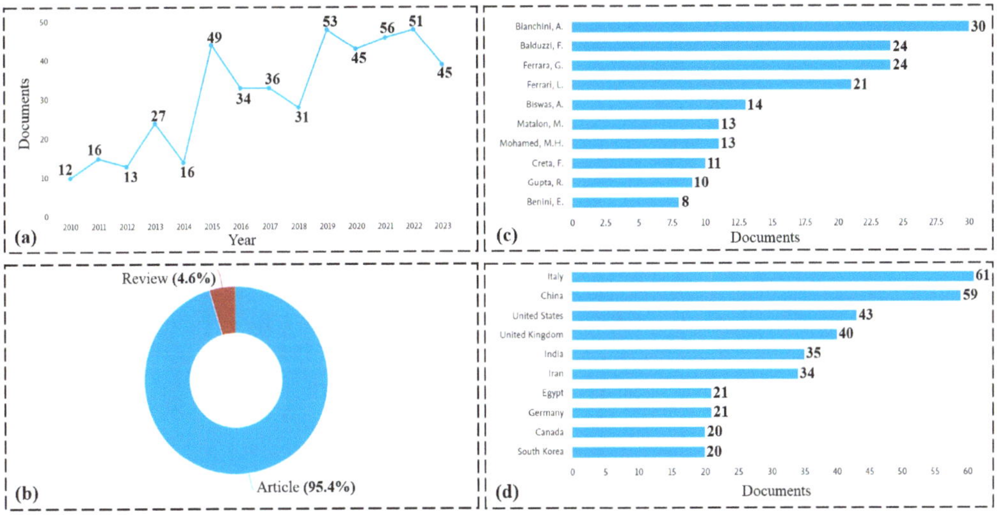

Figure 1. Bibliometric study based on Scopus database [2]. (**a**) Documents by year. (**b**) Documents by type. (**c**) Documents by author [2]. (**d**) Documents by country/territory.

The Darrieus wind turbine belongs to the family of VAWTs, and is characterized by its distinctive airfoiled blades and vertical axis of rotation. Figure 2a illustrates a scaled-down prototype of a typical Darrieus wind turbine, which has been specifically developed for the purpose of testing and validation. Such a prototype enables comprehensive analysis of the turbine's performance, aerodynamics, and structural behavior under diverse operational conditions. Through rigorous testing, a prototype provides invaluable insights into the feasibility and potential of the Darrieus design for efficient wind energy generation.

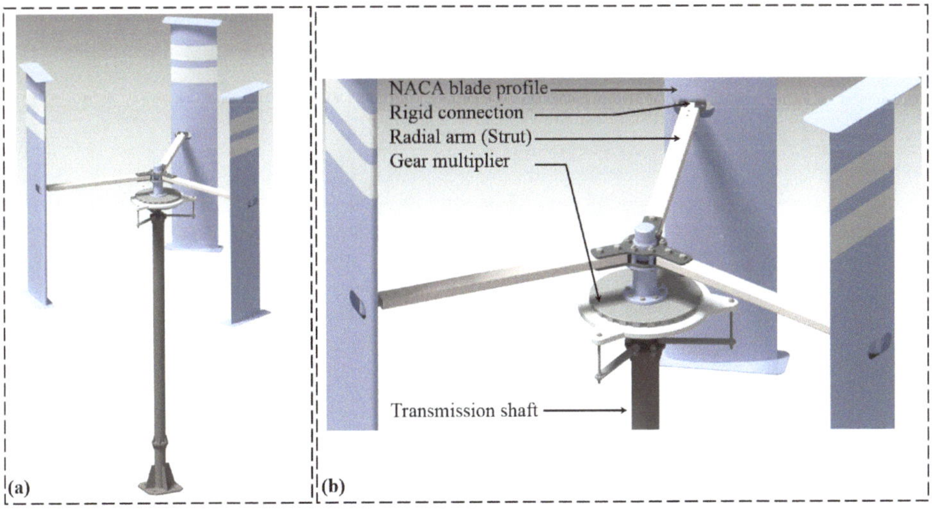

Figure 2. (**a**) Prototype of Darrieus wind turbine. (**b**) Mechanical transmission system of Darrieus.

Figure 2b focuses on the mechanical transmission system of Darrieus. The Darrieus assembly consists of multiple curved blades interconnected to a central hub. The primary function of the mechanical transmission system is to convert the rotational motion of Darrieus into usable mechanical or electrical energy. This system encompasses various components, including gears, shafts, and bearings, meticulously designed to optimize energy transfer and ensure the overall efficiency of the turbine. The mechanical transmission system plays a critical role in preserving the structural integrity and stability of the Darrieus wind turbine.

The Darrieus wind turbine, as depicted in these figures, represents a remarkable advancement in VAWT technology, and underscores its potential for harnessing wind energy. Its unique design confers numerous advantages, including compactness, scalability, and suitability for urban and decentralized applications. The prototype and mechanical transmission system portrayed in the figures serve as pivotal elements in the ongoing development and optimization of the Darrieus wind turbine, thereby contributing significantly to the progression of sustainable energy solutions.

The figures presented in Figure 3 depict various types of Darrieus wind turbines, each showcasing unique characteristics and advancements in VAWT technology. These different designs collectively contribute to the ongoing progress in VAWT technology, and highlight the potential for efficient harnessing of wind energy.

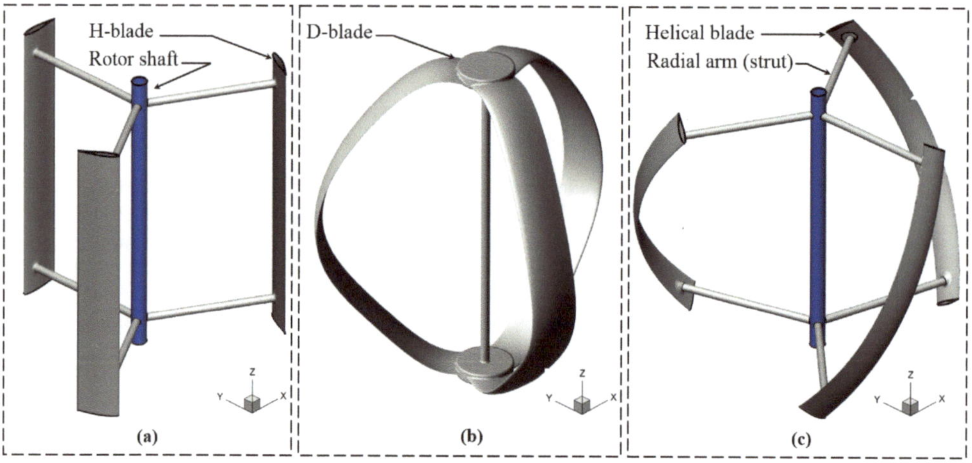

Figure 3. (**a**) H-Darrieus wind turbine, (**b**) D-Darrieus wind turbine, and (**c**) Helical-Darrieus wind turbine.

Figure 3a illustrates the H-Darrieus wind turbine. In Figure 3b, the D-Darrieus wind turbine is showcased. This variant features a distinctive double-curved blade design that significantly enhances the aerodynamic performance and overall efficiency of the turbine. The advanced blade configuration enables increased power generation, making the D-Darrieus turbine an appealing choice for wind energy applications. Figure 3c displays the Helical-Darrieus wind turbine, which introduces helical-shaped blades to the traditional Darrieus design. This innovative modification offers several advantages, including improved torque generation and reduced noise and vibration levels compared to conventional straight-bladed Darrieus turbines. The helical configuration optimizes the turbine's performance and further contributes to the development and optimization of VAWT systems.

The development and application of VAWTs are highlighted in a series of studies, as depicted in Figure 4. A 30 kW working prototype of a VAWT, presented in Figure 4a, has been developed and studied by Martin and Rosander [3], showcasing the potential of VAWTs as a viable solution for wind energy generation. In urban areas, wind energy applications have garnered significant attention, and the Eiffel Tower in France serves as a landmark example where VAWTs have been implemented (Figure 4b). Li et al. [4] conducted experimental studies on a VAWT integrated into the Eiffel Tower, demonstrating the feasibility and effectiveness of utilizing wind energy in urban settings.

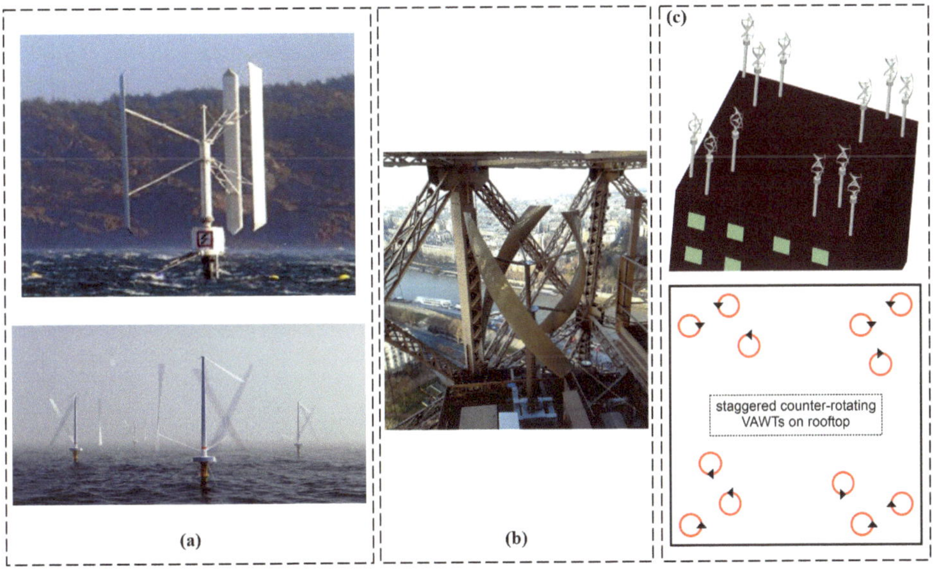

Figure 4. (**a**) S1 30 kW working prototype (reproduced with permission from Martin Rosander, *Case Study: SeaTwirl—The Future of Offshore Floating Wind Turbines*, published by ANSYS, Inc, 2018 [3]), (**b**) application of wind energy in urban areas (VAWT applied in the Eiffel Tower, France) (reproduced with permission from Ye Li, *Energy Conversion and Management*, published by ELSEVIER, 2021 [4]), and (**c**) schematic of 12 Darrieus helical VAWTs installed in a staggered arrangement on a rooftop. Each pair of adjacent turbines is counter-rotating (reproduced with permission from Rezaeiha, *Energy Conversion and Management*, published by ELSEVIER, 2020 [5]).

Furthermore, Figure 4c presents a schematic representation of twelve Darrieus helical VAWTs installed in a staggered arrangement on a rooftop. Each pair of adjacent turbines is designed to counter-rotate, aiming to maximize energy extraction and optimize system efficiency, as proposed by Rezaeiha et al. [5]. Collectively, these studies, depicted in the corresponding figures, provide valuable insights into the development and application of VAWTs in various contexts. The findings contribute to the advancement of wind energy technology, particularly in urban areas and landmark structures. Through these investigations, the potential and effectiveness of VAWTs as sustainable energy solutions are further emphasized, promoting the integration of wind energy into the global energy mix.

A schematic diagram in Figure 5a [6] illustrates the scour topography surrounding a Darrieus-type tidal current turbine. This diagram provides a visual representation of the erosion patterns and changes in the seabed caused by the operation of the turbine. Understanding the scour topography is crucial in the design and installation of tidal current turbines to ensure their long-term stability and performance. In Figure 5b [7], an overview of the forces acting on an airfoil is presented. This diagram describes the various aerodynamic forces, such as lift and drag, exerted on the airfoil as it interacts with the flow of air. Accurate understanding of these forces is essential for optimizing the performance and efficiency of turbine systems based on airfoils. The schematic representation of the conventions used for Darrieus in the current study is depicted in Figure 5c [8]. This diagram outlines the key parameters and reference frames employed in the analysis and evaluation of the Darrieus. The established conventions provide a standardized framework for comparing and assessing the performance of different Darrieus designs.

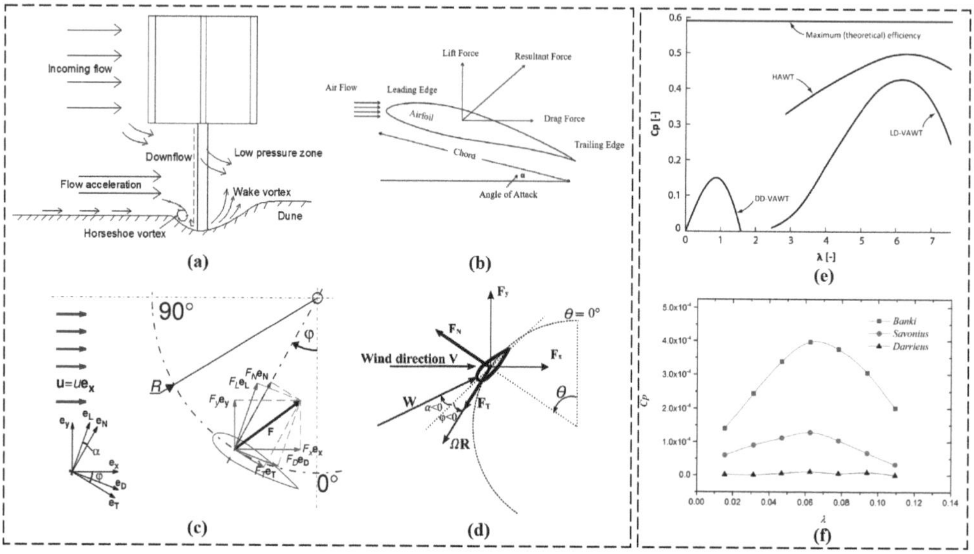

Figure 5. (**a**) Schematic diagram of scour topography around Darrieus-type tidal current turbine (reproduced with permission from Wei Haur Lam, *Energy Conversion and Management*, published by ELSEVIER, 2018 [6]), (**b**) forces acting on an airfoil (reproduced with permission from Ozkan, *Renewable Energy*, published by ELSEVIER, 2020 [7]), (**c**) schematic representation of the conventions used for Darrieus in the current study (reproduced with permission from Daroczy, *Energy*, published by ELSEVIER, 2015 [8]), (**d**) rotating directions and forces as a function of azimuthal angle (reproduced with permission from Nguyen, *Journal of Wind Engineering and Industrial Aerodynamics*; published by ELSEVIER, 2020 [9]), (**e**) typical power coefficients for HAWT, DD-VAWT and LD-VAWT and the Betz limit C_p = 0.59 (reproduced with permission from Trivellato, *Renewable and Sustainable Energy Reviews*, published by ELSEVIER, 2015 [10]), and (**f**) power coefficients for different Darrieus wind turbines (reproduced with permission from Mao, *Renewable Energy*, published by ELSEVIER, 2020 [11]).

Figure 5d [9] depicts the rotating directions and forces exerted on Darrieus as a function of the azimuthal angle. The diagram highlights the variations in forces experienced by the Darrieus as it rotates through different positions. Understanding these force dynamics is crucial for optimizing the efficiency and stability of turbine operation. In Figure 5e [10], the typical power coefficients (C_p) for horizontal axis wind turbines (HAWT), double-Darrieus vertical axis wind turbines (DD-VAWT), and large-diameter vertical axis wind turbines (LD-VAWT) are presented. The diagram includes the Betz limit (C_p = 0.59), representing the maximum theoretical efficiency of wind energy conversion. Comparing the power coefficients of different turbine types aids in assessing their performance and potential for energy generation.

Figure 5f [11] provides a comparative overview of the coefficients of power for different Darrieus designs. The diagram allows for a comprehensive analysis of the power coefficients of various Darrieus configurations, facilitating an understanding of their efficiency and performance characteristics.

Several CFD studies have been carried out on VAWTs since the 1990s, to assess and improve their aerodynamic performance within a specific tip speed ratio (TSR) range [12]. The CFD studies can be 2D [13], 2.5D [14], or 3D [15], and generally depend on the available computational resources [16]. The state of the art on VAWTs shows that the results from 2D modeling, more particularly the evolution of the power coefficient with respect to the TSR range, can be in a good agreement with the experimental studies of wind tunnels, admitting an acceptable margin of relative error compared to the experimental results [17,18]. Therefore, researchers are encouraged to perform 2D numerical studies, which is the case for several research laboratories, to suggest new optimized geometric designs [19], which basically have a 3D nature, nevertheless modeled by 2D studies, disregarding the structures of complex 3D flow that can develop during the continuous operating regime of the VAWTs [20]. In fact, one of the aims of the present review is to assess the predictive capability of multi-fidelity CFD tools from 2D to 3D approaches, and to highlight advantages and drawbacks of such approaches with respect to both global turbine performance prediction and scrutinity of local flow properties.

2.1. 2D CFD Studies

A novel generation of Darrieus turbines was developed by Wang et al. [21]. This innovative design features automatically deformable blades capable of achieving desired geometries during the operating cycle, thereby enhancing aerodynamic performance. The authors conducted a numerical simulation campaign utilizing the United CFD code. For instance, they showed that at low tip speed ratios below the optimum value, turbines with two and four deformable blades showed a maximum power coefficient increase of 7.51% and 8.07%, respectively, compared to conventional turbines with high solidity. Balduzzi et al. [22] performed 2D CFD simulations of Darrieus wind turbines to investigate critical issues and determine the essential parameters necessary for obtaining reliable unsteady results of a real Darrieus VAWT that had been previously tested in a wind tunnel. Furthermore, Hand et al. [23] conducted a comparative analysis of URANS and DES modeling methods to simulate the aerodynamics of an isolated NACA0018 airfoil undergoing Darrieus pitching motion, specifically dynamic stall at high Reynolds numbers. Their research demonstrated that the Spalart–Allmaras model offers a favorable balance between accuracy and computational requirements.

Arab et al. [24] performed 2D simulations to investigate the self-starting characteristics of the Darrieus turbine, taking into account its moment of inertia. It is worth noting that the most commonly used method by research teams to describe the aerodynamics of the Darrieus turbine involves applying various rotational speeds corresponding to tip speed ratios (TSRs). This approach, known as the forced method, does not consider the inherent inertia of the turbine. In a similar context, Celik et al. [25] focused on studying the start-up capability of the Darrieus, and found that increasing the turbine's inertia did not significantly affect the final rotational speed. Bakhumbsh et al. [26] investigated the impact of a micro-cylinder as a passive controller on Darrieus wind turbine performance. Their findings revealed that the turbine's performance decreases when the micro cylinder has a larger diameter or is positioned on the suction side or downstream of the blade leading edge. Tigabu et al. [27] examined the effects of turbine inertia on the starting performance of a vertical-axis hydrokinetic turbine, indicating that increasing the inertia reduces the peak but slows down the acceleration of the Darrieus turbine. Finally, Nguyen et al. [28] evaluated the effect of pitch angle on the power and hydrodynamics of a Darrieus turbine, reporting a power enhancement of up to 2.3% for positive pitch angles compared to a reference zero pitch.

2.2. 2.5D CFD Studies

Li et al. [14] conducted a study to evaluate the accuracy and feasibility of three different CFD approaches for analyzing the aerodynamics of straight-bladed VAWTs under high angle of attack flow conditions. The investigated approaches included 2D unsteady Reynolds-averaged Navier–Stokes (URANS), 2.5D URANS, and 2.5D large eddy simulation (LES). The results showed that the 2.5D LES approach provided the closest agreement with experimental data, as it captured the realistic three-dimensional vortex diffusion after flow separation. This led to accurate predictions of aerodynamic coefficients during static or dynamic stall situations.

Rezaeiha et al. [29] performed a 2D CFD analysis of dynamic stall in Darrieus turbines using the Scale-Adaptive Simulation (SAS) model. The SAS approach was compared against the URANS and hybrid RANS/LES methods. Notably, SAS has emerged as an improved URANS formulation that bridges the gap between the less accurate URANS and the computationally expensive hybrid RANS/LES approaches for highly separated unsteady flows. Furthermore, He et al. [30] investigated the suitability and accuracy of various CFD modeling techniques for predicting the aerodynamics of Darrieus turbines. The studied approaches included 2D URANS, 2.5D URANS, 2.5D LES, 3D URANS, and 3D LES. A comparison with wind tunnel results revealed that 2.5D LES offered an efficient approach for investigating the aerodynamic behaviors of Darrieus turbines. This choice considered the trade-off between accuracy and computational cost when compared to 2.5D LES, 3D LES, and 3D URANS methods.

2.3. 3D CFD Studies

Rossetti et al. [31] conducted a comprehensive study to assess the start-up capability of Darrieus turbines by comparing various numerical approaches. These approaches included the Blade Element Momentum (BEM) method, as well as two- and three-dimensional computational fluid dynamics (CFD) simulations. In a similar vein, Castelli et al. [32] focused on investigating the deformation patterns of Darrieus turbine blades. They achieved this by employing a combination of CFD simulations to evaluate the aerodynamic and mechanical inertial contributions. Notably, the successful execution of these investigations was made possible through the development and validation of a fluid–structure interaction (FSI) coupling code by the research team.

Furthermore, Le et al. [33] highlighted the limitations of using the sliding mesh technique in 2D and 3D CFD simulations of VAWTs. They specifically pointed out that such simulations fail to account for three crucial parameters: self-starting capability, torque fluctuations, and rotation per minute (RPM). These parameters, which are typically observed and measured during laboratory experiments, cannot be replicated solely through CFD simulations employing prescribed tip speed ratios (TSRs).

In a different study, Joo et al. [34] conducted three-dimensional unsteady Reynolds-averaged Navier–Stokes (URANS) simulations to examine the aerodynamic characteristics of a two-bladed Darrieus VAWT. Their main focus was to investigate the effects of varying solidities and rotational speeds on the turbine's power coefficient. The accuracy of their findings was confirmed by comparing the numerical results with experimental data obtained from wind tunnel tests.

Moving on to more advanced simulations, Patil et al. [20] performed three-dimensional large-eddy simulations (LES) of Darrieus. They utilized the WALE subgrid model and a dedicated block-structured mesh to analyze the flow field around the turbine. The accuracy of their LES approach was verified by comparing the simulation results with experimental data, specifically normal and tangential forces, provided by TU Delft Aerospace Engineering University.

Addressing the practical implications, Dessoky et al. [35] emphasized the suitability of Darrieus for both low wind velocity conditions and urban areas. However, they acknowledged the need for comprehensive investigations into the complex aerodynamic and aeroacoustic characteristics of these turbines, requiring a combination of experimental and numerical studies. In line with this, Dessoky et al. [36] performed three-dimensional CFD simulations using a high-order Delayed Detached-Eddy Simulation (DDES) model to gain insights into the turbulent aerodynamics and noise mechanisms of Darrieus turbines.

Furthermore, Clary et al. [37] developed an optimally controlled RANS Actuator Force Model (AFM) for efficient computations of arrays of Darrieus tidal turbines. Their approach incorporated detailed three-dimensional force distributions derived from blade-resolved URANS simulations. Expanding on the practical applications, Su et al. [38] highlighted the significance of VAWTs for offshore wind energy conversion as a complementary power source. They conducted three-dimensional CFD simulations to investigate a novel V-shaped blade profile for Darrieus turbines. The objective was to enhance aerodynamic performance at moderate TSRs, mitigating the undesirable blade tip effects commonly observed in conventional straight blades.

In a related context, Tian et al. [11] explored the potential of harnessing turbulent wind energy from the strong wakes generated by high-speed moving vehicles. To broaden the scope of research in this area, they conducted three-dimensional CFD simulations to evaluate the suitability of VAWTs for highway wind energy recovery. Moreover, Ahmad et al. [39] focused on the design optimization of a double-Darrieus hybrid vertical-axis wind turbine. They proposed a methodology that enables the achievement of an optimum design configuration of the hybrid VAWT, along with a high-fidelity analysis in the early design phase. Notably, their findings indicated a self-starting speed as low as 2.81 m/s and a rated power of 1.522 kW at a wind speed of 7.5 m/s for their proposed hybrid wind turbine.

Table 1 presents a summary of computational resources used for Computational Fluid Dynamics (CFD) simulations in modeling the aerodynamic flow around Darrieus wind turbines. Various numerical approaches were employed by different researchers, such as 2D Unsteady Reynolds-Averaged Navier–Stokes (URANS), Large Eddy Simulation (LES), and Fluid–Structure Interaction (FSI). The table provides insights into the processing parameters used in each study.

For instance, Hand et al. [23] utilized a parallel computing setup with 12 Intel Xeon E5-2695 v2 CPUs, running at 2.40 GHz, and 64 GB of RAM for 2D URANS simulations. On the other hand, Hashem et al. [40] employed eight CPUs with a clock speed of 3.40 GHz for their URANS simulations.

The computational effort varied among studies, with cycle times ranging from 25 min [40] to 65.34 h [23]. Patil et al. [20] used 1024 CPUs connected via InfiniBand, with each node featuring two Xeon E5-2630 v3 CPUs and 256 GB of RAM for their 3D LES simulations, taking 48 h per cycle.

Table 1. CFD calculations demand substantial computational resources to simulate the aerodynamic flow field around Darrieus wind turbines.

Authors	Numerical Modeling	Numerical Approach	Processing	CPU	Nbr of CPUs	Clock Speed (GHz)	RAM (GB)	Wall Clock Time	Research Institute
Hand et al. [23]	2D	URANS	Parallel calculations	12 Intel Xeon E5-2695 v2 Ivy Bridge	12	2.40	64	65.34 h/cycle	Irish Centre for High-End Computing
Hashem et al. [40]				(-)	8	3.40	(-)	25 min/cycle	(-)
Trivellato et al. [10]				(-)	(-)	2.33	(-)	(-)	(-)
Asr et al. [41]				Core i7 processor	8	1.73	(-)	(-)	(-)
Mohammed et al. [42]				(-)	8	2.80	(-)	70 min/cycle	(-)
Bianchini et al. [1]				Intel gold	16	(-)	(-)	38 h/cycle	(-)
Bedon et al. [43]	3D	URANS		Quad-core Intel i7-860 processor	(-)	(-)	64	(-)	(-)
Patil et al. [20]		LES		172 nodes connected with an InfiniBand. 2 Xeon E5-2630 v3 per node	1024	(-)	256	48 h/cycle	(-)
Castelli et al. [32]		FSI		(-)	8	2.33	(-)	15 days for the simulations	(-)
Rezaeiha et al. [29]	2D; 2.5D	URANS, SAS, SBES		Intel(R) Xeon(R) E5-2690 v3	72	(-)	320	741.9 h/cycle (2.5D SBES model)	(-)

The table also highlights that some studies did not provide detailed CPU specifications, making it challenging to assess their computational resources accurately. Nevertheless, it is evident that CFD simulations for Darrieus wind turbines demand substantial computing power, and researchers must carefully consider their numerical approach and processing parameters to achieve reliable results.

Table 2 provides an overview of scientific investigations conducted between 2010 and 2020 to enhance the aerodynamic performance of Darrieus wind turbines using Computational Fluid Dynamics (CFD) modeling. These studies have employed both two-dimensional (2D) and three-dimensional (3D) simulations to analyze the complex flow behavior around the turbine blades. Different NACA four-digit series airfoils were explored in various investigations to assess their impact on aerodynamic efficiency, while the turbine's diameter was also varied to study its influence on overall performance.

To accurately model turbulence effects, researchers utilized various turbulence models, such as k-ω SST, k-ε, Realizable k-ε, LES (SGS), Spallart Allmaras, TSST, and DDES, depending on the specific focus of each study. Meshing played a crucial role in achieving accurate and computationally efficient results, and the number of meshing elements varied across studies. Here are some examples of the number of meshing elements used in certain studies:

1. Rossetti et al. [31] used 8,870,576 meshing elements for their 3D simulations with an SAS model and a y-plus value below 1;
2. Malipeddi et al. [44] performed 2D simulations with a k-ω SST model and a total of 443,155 nodes to study the effect of duct integration;
3. Marco et al. [45] used 111,000 meshing elements for their 2D simulations with a Realizable k-ε model to study low-head hydropower production capability;
4. Joo et al. [34] used a meshing with 1,200,000 elements for their 3D simulations while studying the effect of solidities and rotation speeds;
5. Dessoky et al. [36] conducted 3D simulations with y-plus values below 1 and a meshing consisting of 41,312,000 elements to study the aerodynamics and aeroacoustics of Darrieus wind turbines equipped with "wind-lens" technology.

The dimensionless wall distance (y^+) was carefully controlled to ensure appropriate near-wall flow resolution for reliable simulations.

The scientific investigations explored diverse objectives, including the analysis of self-starting capability, the effect of airfoil selection on efficiency, low-head hydropower production capability, the influence of blade geometries (helical, straight, and J-shaped blades), solidities, and rotation speeds. Some studies also focused on examining the interaction of multiple turbines in a wind farm layout, dynamic stall phenomena, noise assessment, and the impact of various technologies like cavity airfoils, winglets, and Gurney flaps on turbine performance.

These investigations collectively contribute to a comprehensive understanding of Darrieus wind turbine aerodynamics and their implications for designing and operating more efficient and sustainable vertical axis wind turbines, ultimately leading to improved wind energy generation. The knowledge gained from these studies is essential for advancing the field and harnessing wind resources more effectively.

Table 2. Advancements in aerodynamic performance of Darrieus wind turbines: a decade of scientific research (2010–2020).

Authors	Year	CFD Modeling (2D, 3D)	NACA 4-Digit Series	Diameter of Darrieus (m)	Turbulence Model	y-Plus (y^+)	Meshing Elements	Scientific Investigations to Enhance the Aerodynamic Performance of the Darrieus Wind Turbine
Rossetti et al. [31]	2012	3D	0018	0.75	SAS	$y^+ < 1$	8,870,576	Self-start capability
Malipeddi et al. [44]	2012	2D	63_3-018	1.60	k-ω SST	$y^+ < 5$	443,155 nodes	Effect of the duct integration
Mohamed et al. [46]	2012	2D	20 profiles	1.03	k-ε	$30 < y^+ < 50$	95,000	Effect of 20 airfoils on the overall efficiency
Marco et al. [45]	2013	2D	0018	0.39	Realizable k-ε	first cell height 0.23 mm	111,000	Low-head hydropower production capability

Table 2. *Cont.*

Authors	Year	*CFD* Modeling (2D, 3D)	NACA 4-Digit Series	Diameter of Darrieus (m)	Turbulence Model	y-Plus (y^+)	Meshing Elements	Scientific Investigations to Enhance the Aerodynamic Performance of the Darrieus Wind Turbine
Mohamed et al. [47]	2014	2D	25 airfoils	2	k-ω SST	$y^+ = 3.70$	202,426 nodes	Low speed energy conversion capability
Le et al. [33]	2014	2D	0020	3	k-ω SST	$y^+ < 5$	12,945	Effect of helical and straight blades
Joo et al. [34]	2015	3D	0022, 0012	2	Realizable k-ε	-	1,200,000	Effect of solidities and rotating speeds
Balduzzi et al. [48]	2015	2D	0018	1.70	k-ω SST	-	817,422	CFD modeling methodology
Wang et al. [49]	2015	2D	0021, 0018	2	Spallart Allmaras	-	115,000	Influence of the blade vortex interactions
Chen et al. [50]	2015	2D	0015	1.25	k-ω SST	$y^+ = 1$	400,000	Combined effect of the opening ratio and its location
Sengupta et al. [51]	2015	2D	0018, S815, EN0005	Aspect ratio	RNG k-ε	-	-	Effect of high solidity symmetrical and unsymmetrical blades
Arab et al. [24]	2016	2D	4415	0.60	k-ω SST	$y^+ = 0.53$	-	Aerodynamic performance and self-starting characteristics (effect of inertia)
Balduzzi et al. [16]	2016	3D	0021	1.03	k-ω SST	$y^+ < 1$	64,000,000	3D reliable analysis for the calibration of lower fidelity CFD models
Asr et al. [41]	2016	2D	2418, 0018, 4418, 2418	0.75	k-ω SST	-	74,000	Self-start capability (combined effect of the pitch angle and NACA profiles)
Bedon et al. [43]	2016	2D	0018	35	k-ω SST	$y^+ < 1$	669,000	Performance-optimized NACA airfoil for a 500 kW Darrieus wind turbine
Mabrouk et al. [52]	2016	2D	0021	1.03	k-ε	-	119,128	Dynamic mechanical vibrations
Zamani et al. [53]	2016	2D	0015	3.70	k-ω SST	$y^+ < 1$	1,397,827	Starting torque enhancement of a 3 kW Darrieus
Zamani et al. [54]	2016	3D	0015	2.50	k-ω SST	$y^+ = 1$	12,519,254	3D modeling of a J-shaped Darrieus
Qamar et al. [55]	2017	2D	0012, 4312	2.50	k-ω standard	-	400,000	Effect of six solidity values
Bianchini et al. [56]	2017	2D	0020	1	k-ω SST	$y^+ < 1$	593,880	Reliability of 2D CFD simulations
Shaheen et al. [57]	2017	2D	0021	1.03	k-ω SST	$y^+ < 1$	173,768	Design of a wind turbine farm constituted from 3 Darrieus
Sobhani et al. [58]	2017	2D	0021	1.03	k-ω SST	-	505,762	Effect of dimpled NACA0021 blades
Shaaban et al. [59]	2017	2D	0021	1.03	k-ε EWT	$y^+ = 3.5$	320,032	Six turbines cluster analysis
Dessoky et al. [35]	2018	3D	0021, 0018	Aspect ratio	k-ω SST, DDES	$y^+<1$	16,905,216	Aerodynamic and aeroacoustic CFD modeling of Darrieus equipped with wind-lens technology
Ali et al. [60]	2018	3D	0015	2	LES (SGS)	$y^+ < 1$	1,600,000	Effect of dynamic tangential velocity
Patel et al. [61]	2019	2D	0021, 0011	1.03	k-ω SST	$y^+ = 1$	880,000	Effect of three Gurney Flaps technologies located at the trailing edge of Darrieus
Barnes et al. [62]	2019	2D	0015	3.40	Transition k-ω SST	$y^+ < 1$	120,000	Mutual effect of 15 Darrieus turbines constituting a wind farm
Dessoky et al. [36]	2019	3D	0021, 0018	0.20	DDES	$y^+ < 1$	41,312,000	Turbulence and noise assessment
Xu et al. [63]	2020	3D	0015	1.70	DDES	$y^+ < 1$	6,500,000	Effect of winglet on the aerodynamic performance
Jain et al. [64]	2020	2D	0009, 0012, 0015, 0018, 0021	1.20	TSST	$y^+ < 1$	354,169	Effect of the ratio of blade thickness to the chord length on the dynamic stall phenomenon

3. Methodology

A meticulous state-of-the-art study on Darrieus turbines was conducted to achieve the objective of our research, which is a critical review of CFD modeling approaches for Darrieus turbines. Over 250 scientific publications extracted from the ScienceDirect platform were evaluated to compile this critical review. To begin with, an introduction was drafted to underscore the significance of wind energy in comparison to other renewable green energy sources. It also aimed to highlight the substantial interest generated within the scientific community towards CFD modeling studies. In the same section, the three primary categories of CFD modeling applied to Darrieus turbines were presented: 2D, 2.5D, and 3D modeling. Parametric studies conducted using CFD within these categories were discussed, specifically targeting an improvement in the power coefficient (Cp).

Following the introduction, the subsequent section involved the presentation and discussion of various methodologies available in the existing literature for modeling turbulent flow around Darrieus turbines. This allowed for a comprehensive overview of the available approaches. To support the critical review, a set of meticulously prepared figures, totaling 16, was introduced. These figures were utilized to facilitate discussions on the observed discrepancies in power coefficient estimation and wake vortex development for Darrieus turbines. As a result, the figures formed the cornerstone of this critical review.

Moreover, the necessary schemes employed for conducting and ensuring reproducibility in 2D, 2.5D, and 3D CFD modeling were extensively addressed. The discussion encompassed the importance of methodological consistency and accuracy in computational simulations. In addition, considerable attention was dedicated to the exploration of diverse meshing techniques employed to discretize the flow domain surrounding Darrieus turbines in the different modeling dimensions (2D, 2.5D, and 3D). These techniques were thoroughly examined to assess their impact on the accuracy of the results. The investigation further delved into the comprehensive exploration of the various boundary layer models implemented by research teams to accurately capture the near-wall viscosity effect on the Darrieus blade surfaces. This discussion shed light on the importance of capturing boundary layer phenomena in the modeling process. Moving forward, the shape, dimensions and boundary conditions applied to the fluid domain surrounding Darrieus turbines in each modeling dimension (2D, 2.5D, and 3D) were carefully examined. This analysis provided insights into the parameterization choices and their implications for the accuracy of the results.

Furthermore, the critical review encompassed the presentation of several parametric studies proposed by various authors to enhance the performance of Darrieus turbines. These studies contributed to a comprehensive understanding of the research objective and underscored the significance of conducting a critical review to assess discrepancies in power coefficient estimation and wake vortex development for Darrieus turbines. In order to highlight the impact of turbulence models on key aerodynamic performance indicators of Darrieus turbines, multiple graphs were presented. These graphs featured numerical results for Cp, Cm, Cl, and Cd, which were compared with reference experimental studies, including the influential work by Castelli et al. [65]. The aggregation and comparison of these results accentuated the uniqueness and novelty of this critical review.

To evaluate the effects of geometric modifications on conventional Darrieus configurations for improved aerodynamic performance, various CFD contours were showcased. These contours formed the foundation for assessing the disparities between different types of CFD modeling, particularly in the 2D, 2.5D, and 3D dimensions. The presentation of these findings further emphasized the significance of this critical review in assessing discrepancies in power coefficient estimation and wake vortex development for Darrieus turbines. Lastly, the critical review included examples of the experimental approach employed to characterize Darrieus turbines using Particle Image Velocimetry (PIV). This demonstration highlighted the importance of validating commercially available computational codes that are widely utilized by research teams and laboratories. Such validation is crucial when proposing geometric modifications to enhance the aerodynamic efficiency of Darrieus turbines.

4. Discussion

4.1. General Form of the Solved Governing Equations

When numerically modeling a Darrieus wind turbine using the governing equations presented in Equations (1)–(4), several key assumptions are incorporated to streamline the simulation process while providing a foundational framework for accurate results. The simulation considers the unsteady-state operation, capturing transient effects due to wind speed variations and turbine rotation. The fluid is treated as incompressible, with negligible density fluctuations from pressure changes. A Newtonian fluid assumption is adopted, implying constant dynamic viscosity irrespective of shear rate. Thermal effects, chemical reactions, and external interactions are omitted. Turbulence effects are accommodated through appropriate turbulence models, assuming predominantly turbulent flow. Inlet conditions are uniform, external forces are neglected, and idealized symmetrical geometry is employed. The blades are rigid, disregarding fluid-induced deformations. Interactions between blades, temperature variations, and phase change phenomena are excluded. FSI simulations assume linear elastic behavior of solid structures and employ a monolithic coupling approach for strong fluid–structure interaction. Collectively, these assumptions facilitate a comprehensive CFD analysis of both the aerodynamic and mechanical aspects of the Darrieus wind turbine.

The general form of the solved governing equations commonly used in published papers related to Darrieus VAWTs encompasses the fundamental principles of fluid dynamics and conservation laws. These equations provide a basis for understanding and simulating the aerodynamic behavior of VAWTs. In the context of CFD studies, the equations can be expressed as follows:

1. Continuity Equation: The continuity equation represents the conservation of mass and is expressed as Equation (1):

$$\frac{\partial \rho}{\partial t} + \nabla \cdot (\rho \mathbf{V}) = 0 \tag{1}$$

   where ρ is the fluid density, $\mathbf{V}$ is the velocity vector, and t represents time;

2. Navier–Stokes Equations: These equations describe the conservation of momentum and are divided into three components along the x (Equation (2)), y (Equation (3)), and z (Equation (4)) axes:

   a. X-Component:

$$\frac{\partial(\rho u)}{\partial t} + \nabla \cdot (\rho u \mathbf{V}) = -\frac{\partial P}{\partial x} + \nabla \cdot (\mu \nabla u) + \rho g_x \tag{2}$$

   b. Y-Component:

$$\frac{\partial(\rho v)}{\partial t} + \nabla \cdot (\rho v s.\mathbf{V}) = -\frac{\partial P}{\partial y} + \nabla \cdot (\mu \nabla v) + \rho g_y \tag{3}$$

   c. Z-Component:

$$\frac{\partial(\rho w)}{\partial t} + \nabla \cdot (\rho w \mathbf{V}) = -\frac{\partial P}{\partial z} + \nabla \cdot (\mu \nabla w) + \rho g_z \tag{4}$$

   where u, v, and w are the velocity components in the respective directions, P is pressure, μ is dynamic viscosity, and g_x, g_y, and g_z represent acceleration due to gravity along the x, y, and z axes;

3. Turbulence Modeling Equations: The choice of turbulence modeling approach depends on factors such as computational resources, desired level of accuracy, and specific flow characteristics. These turbulence models enhance the predictive capabilities of simulations, enabling a more comprehensive understanding of the aerodynamic behavior and performance of Darrieus VAWTs;

(a) RANS with Turbulence Models (k-ε, k-ω, etc.): Different turbulence models, such as k-ε, and k-ω, are used within the RANS framework to capture turbulence effects. These models introduce transport equations for turbulence kinetic energy and its dissipation rate;
(b) Unsteady Reynolds-Averaged Navier–Stokes (URANS): URANS simulations provide a time-averaged representation of turbulent flows. The URANS equations introduce additional turbulence terms to the governing equations to account for Reynolds stresses. This approach is computationally efficient for industrial applications;
(c) Hybrid URANS-LES (Large Eddy Simulation): The hybrid approach combines URANS and LES techniques. URANS is used in regions where turbulence is well-developed and computationally expensive LES is employed in areas with complex flow structures. This balances accuracy and computational cost;
(d) Detached Eddy Simulation (DES): DES is a compromise between URANS and LES. It employs URANS in the near-wall region and LES in the outer flow. This provides accurate predictions for separated flows and regions with turbulent eddies;
(e) Large Eddy Simulation (LES): LES resolves large-scale turbulent structures and models only the small-scale turbulence. It is computationally demanding, but provides highly detailed information about the turbulent flow behavior.

These governing equations are typically solved numerically using techniques like finite volume, finite element, or finite difference methods, depending on the numerical solver employed. The solutions provide insights into the flow patterns, pressure distributions, and forces acting on the Darrieus VAWT blades, aiding in the analysis and optimization of its aerodynamic performance.

In the pursuit of modeling acoustic phenomena arising from the operation of a Darrieus wind turbine, several specific considerations come into play. The governing equations traditionally employed, such as the Navier–Stokes equations for fluid dynamics, can be extended to account for acoustic interactions. The acoustic wave equations, coupled with the equations of mass conservation, momentum, and energy, allow for the description of sound wave propagation resulting from the motion of the blades and interactions with the surrounding atmospheric flow. Models encompassing absorption, diffusion, and reflection of acoustic waves may also be incorporated to accommodate specific characteristics of the surrounding terrain and acoustic environment.

Conversely, when the focus shifts towards mechanical aspects through Fluid–Structure Interaction (FSI) modeling, the governing equations assume an even more intricate form. In addition to the Navier–Stokes equations and structural dynamics equations, FSI coupling equations are introduced to address bidirectional interactions between fluid motions and structural deformations. The equations of mass conservation, momentum, and energy are coupled with the solid deformation equations, thereby enabling the simulation of interactions between the Darrieus wind turbine blades and the surrounding fluid. Advanced numerical techniques, such as time and space partitioning methods, are often required to effectively and accurately solve these coupled equations, providing essential insights into the mechanical and fluid responses of the wind turbine under varying operational conditions.

Performance Indicators

First of all, the tip speed ratio (TSR) is given by Equation (5). It presents the ratio between the tangential velocity and the upstream far-field flow velocity which is designated by U_∞.

$$TSR = \lambda = \frac{R\omega}{U_\infty} \tag{5}$$

As the turbine rotates with the angular velocity ω, the attack velocity at a blade during a revolution is given by Equation (7) [66]. The relative flow velocity can be obtained by Equation (7) [40].

$$\vec{W} = \vec{U}_\infty - \vec{\omega} \wedge \overrightarrow{OM} \tag{6}$$

$$\begin{cases} W = U_\infty \sqrt{(cos(\theta+\lambda)^2 + sin^2(\theta))} \\ = U_\infty \sqrt{1 + 2\lambda cos(\theta) + \lambda^2} \\ = \sqrt{V_c^2 + V_n^2} \end{cases} \tag{7}$$

where V_c represents the chordal velocity component and V_n represents the normal velocity component.

- $V_n = V_a sin(\theta)$ is the normal component of the relative velocity, which is perpendicular to the cord line;
- $V_c = R\omega + V_a cos(\theta)$ is the tangential component of the relative velocity, which is parallel to the cord line.

The relation between the angle of attack α, the azimuth angle θ, and the tip speed ratio λ can be defined from the velocity triangle analysis. It is expressed as shown in Equation (8) [40].

$$\alpha = tan^{-1}\frac{sin(\theta)}{(\lambda + cos(\theta))} \tag{8}$$

The Reynolds number at which the blade operates can be calculated using the following Equation [61]:

$$Rc = \frac{\rho W c}{\mu} \tag{9}$$

The value of torque coefficient (C_m) can be obtained from the instantaneous values of torque coefficients (C_{ti}) by integration over an entire revolution (Equation (10)), after reaching a periodic steady state [67]. Generally, 2D and 3D CFD simulations of Darrieus are carried out until the numerical solution reaches a periodic steady state, wherein no cycle-to-cycle variation in the solution is observed. θ $(= \omega t)$ is the azimuthal position or phase of the turbine.

$$C_m = \frac{1}{2\pi}\int_0^{2\pi} C_{ti}(\theta) d\theta \tag{10}$$

The performance coefficient (C_p) is calculated from the torque coefficient (C_m) using Equation (12). The torque coefficient is expressed using Equation (11) [24].

$$C_m = \frac{T_m}{\frac{1}{2}\rho U_\infty^2 RA} \tag{11}$$

Finally, the power coefficient could be expressed using Equation (12) [24]. In addition, the swept area of Darrieus is generally a priori expressed as shown in Equation (13).

$$C_p = \lambda C_m = \frac{T_m \omega}{\rho U_\infty^3 HR} \tag{12}$$

$$A = H.D \tag{13}$$

The values of C_p obtained for each tip speed ratio are compared with the experimental values to validate the CFD used codes. For this, standard deviation of computational data from the experimental data is determined using Equations (14) and (15) [17,18].

$$\sigma^2 = \frac{1}{n}[\sum_{i=1}^{n}(C_{pi} - \bar{C_P})] \tag{14}$$

$$\bar{C}_p = \frac{1}{n}\sum_{i=1}^{n} C_{pi} \tag{15}$$

4.2. CFD Modeling Approaches

Numerical modeling through CFD is a powerful approach for studying complex aerodynamic flows around Darrieus wind turbines. Each of the three modeling approaches—2D, 2.5D, and 3D—offers specific advantages and disadvantages to consider based on the study objectives.

The 2D modeling stands as a versatile approach for investigating the aerodynamics of Darrieus wind turbines. Its distinct advantages include computational efficiency and low resource costs. The simplicity of a two-dimensional representation accelerates simulations and allows for the exploration of various configurations and conditions within constrained time frames [40]. Furthermore, 2D modeling, by reducing geometry to a single dimension, demands fewer computational resources, making it well-suited for preliminary studies and swift analyses [23]. Nevertheless, its notable disadvantages stem from geometric simplification and the lack of three-dimensional consideration. This omission of the third dimension can lead to reduced accuracy, particularly in regions where three-dimensional effects are prominent [1]. Additionally, the deficiency in capturing complex features and three-dimensional vortices limits the model's ability to fully comprehend intricate flow dynamics around Darrieus turbines.

The 2.5D CFD modeling approach emerges as a compelling compromise between computational efficiency and enhanced spatial information. Its advantages encompass a partial consideration of three-dimensional effects while maintaining a balance of simplification. This intermediate approach yields more accurate results compared to pure 2D modeling, especially concerning vortex and swirl phenomena [29]. While more resource-intensive than 2D modeling, 2.5D modeling remains less demanding than full-fledged 3D simulations [14]. However, this approach's drawbacks include increased complexity and geometric limitations. The incorporation of partial three-dimensional effects elevates resource requirements and computational time, albeit to a lesser extent than full 3D modeling [30]. Geometric constraints may still result in an underestimation of intricate 3D interactions within specific flow regions.

For a comprehensive understanding of Darrieus wind turbine aerodynamics, 3D CFD modeling provides a level of realism that surpasses the aforementioned approaches. Its merits encompass a maximum representation of three-dimensional flow geometry and intricate physics. This capacity to capture complex phenomena like flow separation, three-dimensional vortices, and blade interactions offers an unparalleled level of fidelity [20]. Enhanced precision derives from the thorough consideration of three-dimensional effects, rendering the 3D approach the most accurate among the three methods, particularly in regions with substantial three-dimensional flow [63]. However, these benefits come at the cost of elevated resource requirements. Three-dimensional modeling necessitates significantly higher computational resources, encompassing extended calculation times and increased computational power [36]. Moreover, the computational complexity associated with the 3D approach demands expertise in mesh preparation, boundary condition handling, and result interpretation [43].

In the realm of Darrieus wind turbine research, the selection of a CFD modeling approach hinges on a balance between accuracy, computational resources, and research goals. The trade-offs presented by each approach must be meticulously considered to align with the objectives of the investigation.

A close-up view of the mesh surrounding the airfoil, enabling a detailed analysis of the fluid flow characteristics and aerodynamic behavior within a 2D modeling approach, is presented in Figure 6a [68]. The performance of the airfoil and its interactions with the flow field are illuminated by the detailed mesh representation. It should be noted, however, that the 2D modeling approach is limited in capturing three-dimensional effects and flow phenomena, which can potentially impact the accuracy of predictions and their applicability to real-world scenarios.

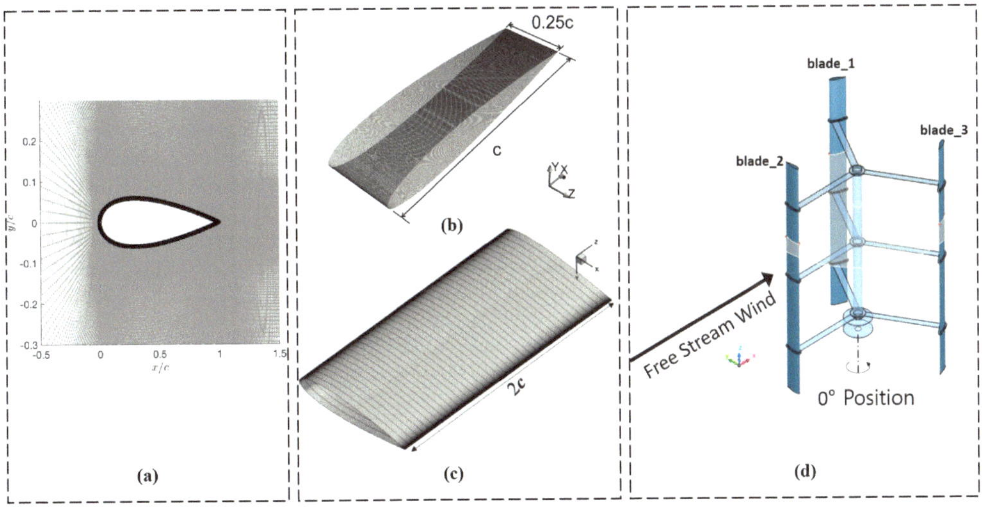

Figure 6. (**a**) 2D modeling: close-up view of the mesh around the airfoil (reproduced with permission from Atlaschian, *Renewable Energy*, published by ELSEVIER, 2021 [68]), (**b**) 2.5D modeling: computational grid of the turbine for the SAS simulation with a total of 19,753,402 quadrilateral cells (reproduced with permission from Reazeiha, *Energy Conversion and Management*, published by ELSEVIER, 2019 [29]), (**c**) 2.5D modeling: mesh distribution for single static NACA0018 airfoil (reproduced with permission from Zhu, *Renewable Energy*, published by ELSEVIER, 2013 [14]), (**d**) 3D modeling: definition of blade numbers and 0 degree azimuthal position (reproduced with permission from Jang, *Energy Conversion and Management*, published by ELSEVIER, 2018 [60]).

In Figure 6b, a 2.5D modeling approach is employed to demonstrate the computational grid of the turbine utilized for the SAS simulation [29]. The high-resolution grid design, consisting of a large number of cells, facilitates precise modeling of the turbine's geometry and flow domain, leading to accurate predictions of its performance and aerodynamic characteristics. Nonetheless, the 2.5D modeling approach assumes axisymmetric flow and disregards spanwise variations, which may restrict the representation of flow phenomena occurring in the spanwise direction.

The mesh distribution for a single static NACA0018 airfoil within a 2.5D modeling approach is illustrated in Figure 6c [14]. The optimized mesh distribution ensures an accurate representation of flow features and captures the intricate aerodynamic behavior of the airfoil. However, it is important to critically evaluate the sensitivity of the results to the mesh resolution and assess the convergence of the simulations to ensure reliable predictions of performance parameters. Furthermore, Figure 6d depicts the definition of blade numbers and the 0-degree azimuthal position within a 3D modeling approach [60]. These parameters are crucial in accurately representing the geometry and orientation of the turbine blades in three-dimensional space. However, it is important to consider the potential impact of simplifications made in the modeling approach and the assumptions associated with blade geometry and azimuthal position on the accuracy of the results.

4.3. Mesh Generation around Darrieus for CFD Modeling

Creating a mesh for a Darrieus wind turbine involves a systematic approach to ensure precise and dependable simulations. The methodology outlined below facilitates the generation of meshes for 2D, 2.5D, and 3D modeling. The initial step encompasses importing the Darrieus wind turbine geometry into a compatible CAD software, ensuring clean and accurate definitions for distinct components such as blades and hub. Subsequently, the selection of an appropriate meshing strategy hinges on the modeling approach and desired accuracy level. In the context of 2D simulations, the creation of a structured or unstructured mesh along blade profiles and around Darrieus takes center stage. Emphasis is placed on establishing boundary layers near surfaces to accurately capture near-wall flow behavior. Meanwhile, for 2.5D simulations, an extension of the 2D mesh into the third dimension, preserving a simplified geometry representation, is pursued. This can involve extruding the 2D mesh or applying a prism layer to effectively capture boundary layer effects. In the realm of 3D simulations, a comprehensive three-dimensional mesh encompassing the entire wind turbine geometry, encompassing blades, hub, and the surrounding flow domain, is meticulously generated. Special attention is devoted to refining the mesh around blade surfaces, near-wall regions, and potential flow separation zones. To ensure mesh quality, diagnostic tools within the chosen CFD software are employed to identify suboptimal elements. Mesh parameter adjustments are made to optimize quality. Boundary conditions are rigorously defined for inlet, outlet, and walls, aligning with the modeling approach. As part of refinement, mesh studies evaluate result sensitivity to size, with incremental refinements applied to key areas to optimize accuracy. This methodology yields well-structured meshes, forming the basis for insightful aerodynamic analyses of Darrieus wind turbines using CFD software.

The decision on the modeling nature of CFD to characterize Darrieus wind turbine aerodynamics involves choosing between 2D, 2.5D, or 3D modeling, each offering distinct advantages and complexities. Once the nature is determined, selecting an appropriate modeling technique becomes crucial. Techniques like the Multiple Reference Frame (MRF) method, or dynamic methods such as Sliding Mesh or Six Degrees of Freedom (6DOF), provide viable options [69,70]. Each technique requires an appropriate mesh and configuration for accurate results. Special attention must be paid to the small dimensions of the blade trailing edge, affecting mesh cells significantly. Ensuring mesh quality is vital to preserve flow behavior accuracy. Mesh refinement strategies include condensing cells near walls, within boundary layers, and in wake zones. Starting simulations with moderately coarse meshes facilitates efficient wake region identification for subsequent refinement. Partitioning the fluid domain aids structured mesh generation in the wake zone, enhancing computational efficiency. For 2D modeling, triangular cells near the wall are recommended due to blade curvature. A mixed mesh with triangular and quadrilateral elements within the boundary layer separation region is advised. Transitioning to a quadrilateral mesh in the wake zone optimizes resource utilization. In 2.5D and 3D calculations, the mosaic meshing technique proves advantageous, ensuring seamless transitions between boundary layer and flow domain meshes, yielding high-quality hexahedral mesh elements [69]. Careful consideration of these modeling and meshing aspects is pivotal for accurate and insightful Darrieus wind turbine aerodynamic characterizations, advancing the understanding of its behavior and performance comprehensively.

In all cases, the validation of mesh independence and the assurance of convergence to a stable solution are essential. It should be borne in mind that meshing constitutes a critical phase in CFD simulations, and the impact of accuracy and result reliability is significantly influenced by the selection of the modeling approach and mesh quality.

The variations in the torque coefficient for one blade as a function of Darrieus angular position at a fixed tip speed ratio (TSR) of 1.0 are presented in Figure 7a [24]. Fluctuations in the torque coefficient throughout the blade's rotation are observed through the analysis. However, it is of crucial importance to critically assess the methodology employed, including the choice of grid resolutions and their impact on torque predictions. The influence of grid resolution on the accuracy and reliability of the torque coefficient results should be thoroughly evaluated to ensure robust conclusions.

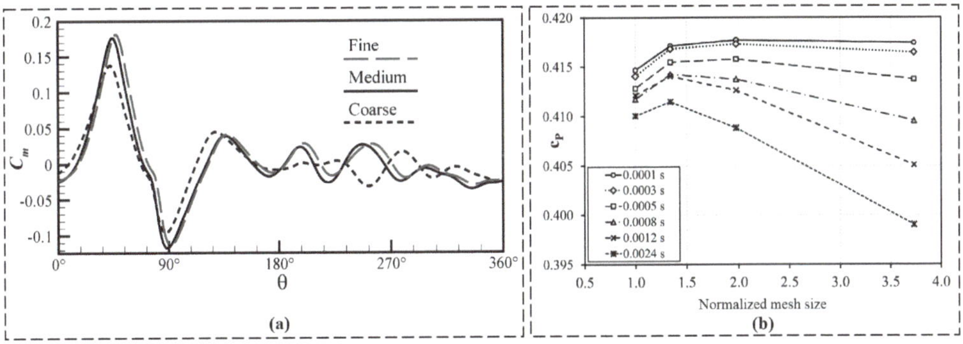

Figure 7. (**a**) The variations in the torque coefficient for one blade, as a function of Darrieus angular position, for the three grid resolutions at TSR=1.0 (reproduced with permission from Arab, *Renewable Energy*, published by ELSEVIER, 2017 [24]). (**b**) Sensitivity analysis at $\lambda = 3.3$: power coefficient (C_p) as a function of the normalized mesh size (reproduced with permission from Ferrari, *Energy*, published by ELSEVIER, 2016 [48]).

Figure 7b shows a sensitivity analysis conducted at TSR = 3.3, investigating the power coefficient (C_p) as a function of the normalized mesh size [48]. The power coefficient, a key metric for assessing wind turbine efficiency, is examined while systematically varying the mesh size. While this sensitivity analysis provides valuable insights into the impact of mesh size on power predictions, the methodology employed must be critically analyzed. Factors such as the range of mesh sizes considered, the convergence criteria used, and the selection of other modeling parameters should be carefully evaluated to ensure reliable and meaningful results.

The insights gained from these figures, as depicted in Figure 7, contribute to the understanding of wind turbine aerodynamics and efficiency. However, it is important to critically evaluate the limitations and potential sources of uncertainty associated with the analysis. Factors such as the modeling assumptions, turbulence models, and boundary conditions employed in the simulations should be critically assessed to ascertain the reliability of the findings.

Moreover, it is crucial to consider the broader context of these findings within the existing body of literature. Conducting a critical comparison of the results with experimental data and other numerical studies can provide a more comprehensive understanding of the observed trends and their significance. Additionally, incorporating a sensitivity analysis that considers multiple influencing factors can strengthen the validity of the conclusions drawn from the figures.

4.3.1. 2D Grid

Figure 8a presents the rotating domain of the mesh, emphasizing the computational region where the simulation takes place [8]. The wind-lens concept is depicted in Figure 8b [35], demonstrating its potential benefits in enhancing wind turbine performance. The importance of the rotating zone and the mesh design in capturing flow features are demonstrated in Figure 8c,d [24,45]. The significance of turbine spacing and layout in array configurations is highlighted in Figure 8e [71]. The mesh quality and resolution are demonstrated in Figure 8f [71].

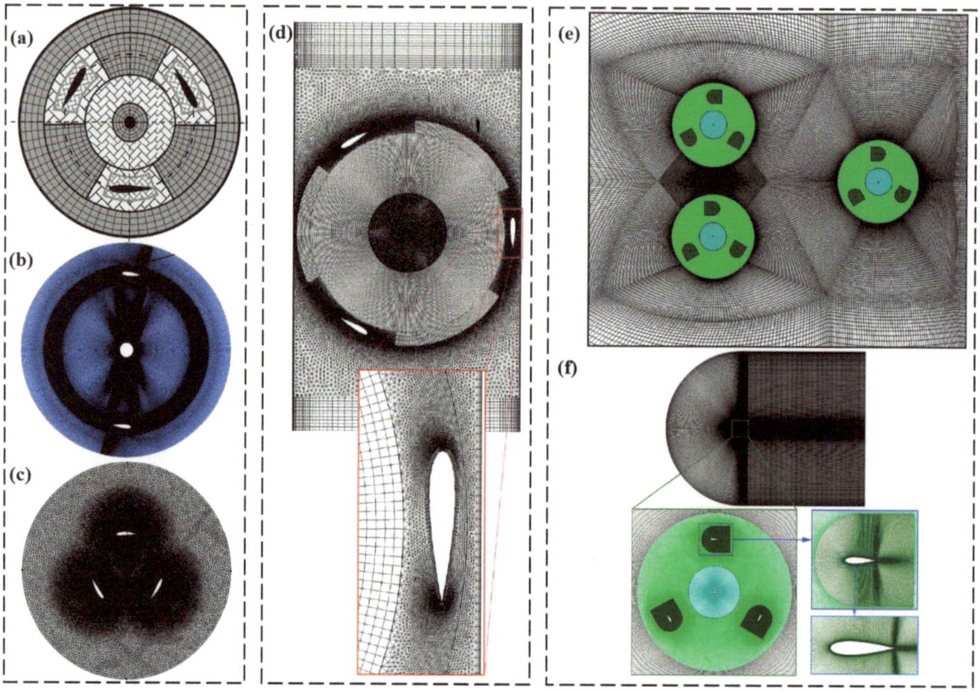

Figure 8. (**a**) Rotating domain of the mesh (reproduced with permission from Daroczy, *Energy*, published by ELSEVIER, 2015 [8]), (**b**) wind-lens (reproduced with permission from Dessoky, *Energy*, published by ELSEVIER, 2019 [35]), (**c**) the rotating zone (reproduced with permission from Arab, *Renewable Energy*, published by ELSEVIER, 2017 [24]), (**d**) details of the mesh in the zone swept by the Darrieus (reproduced with permission from Torresi, *Procedia Computer Science*, published by ELSEVIER, 2013 [45]), (**e**) mesh distribution of VAWTs in array configurations (reproduced with permission from Ni, *Energy*, published by ELSEVIER, 2021 [71]), and (**f**) computational mesh distributions (reproduced with permission from Ni, *Energy*, published by ELSEVIER, 2021 [71]).

4.3.2. Boundary Layer around the Darrieus Turbine Blades

The mesh distribution and grid refinement are presented in Figure 9a,b [43,71]. The near-wall grid resolution and boundary layer development are emphasized in Figure 9c [23]. The boundary layer generation on the blade wall is depicted in Figure 9d [58]. The optimization of the worst elements at the blade tips is shown in Figure 9e [72]. The mesh sensitivity analysis is presented in Figure 9f [22]. The structured grid configuration in close proximity to the turbine blade is showcased in Figure 9g [18].

4.3.3. 3D Grid

Figure 10a depicts the sectional 3D computational mesh utilized in the simulations [35]. The computational mesh used in the simulations is further illustrated in Figure 10b, emphasizing its background [35]. Figure 10c displays the mesh structure of the 3D computational domain [54]. The computational mesh around the combined Darrieus and the front view of the mesh near the blades are depicted in Figure 10d [17]. Figure 10e provides additional details of the computational mesh [16]. Lastly, Figure 10f shows the generated mesh for three-dimensional (3D) models of the turbine using the Sliding Mesh Interface (SMI) [73] and includes a zoomed-in view of Darrieus and the 3D mesh on the blade tip [31].

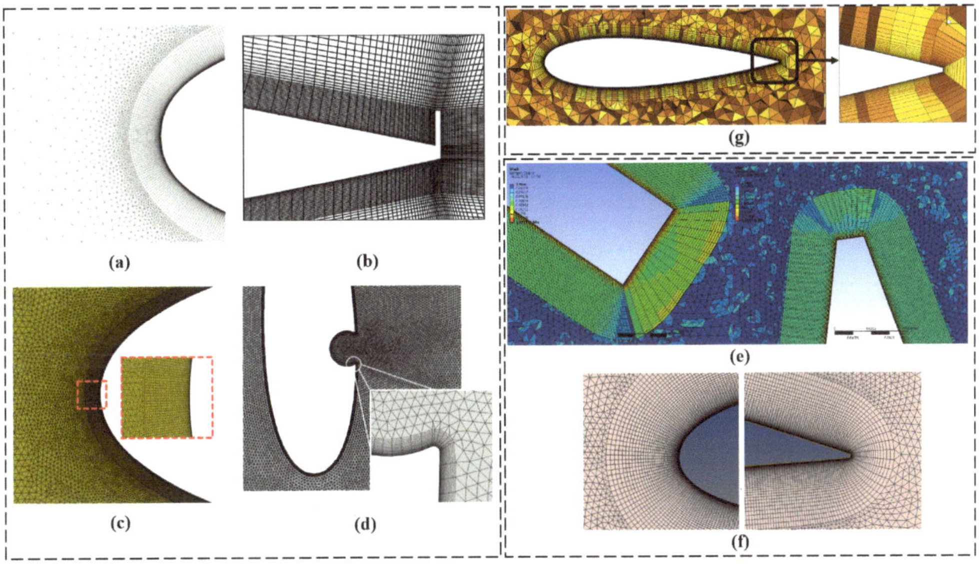

Figure 9. (**a**) Representation of the mesh near the blade walls and for the rotating domain (hybrid mesh near the airfoil leading edge) (reproduced with permission from Bedon, *Renewable Energy*, published by ELSEVIER, 2016 [43]), (**b**) mesh distribution of the airfoil with GF (reproduced with permission from Ni, *Energy*, published by ELSEVIER, 2021 [71]), (**c**) computational grid (near wall grid resolution) (reproduced with permission from Hand, *Computers & Fluids*, published by ELSEVIER, 2017 [23]), (**d**) generation of the boundary layer on the blade wall (reproduced with permission from Sobhani, *Energy*, published by ELSEVIER, 2017 [58]), (**e**) the worst elements of the hybrid grid structure lied on the tips of the blades (reproduced with permission from Hosseini, *Energy conversion and management*, published by ELSEVIER, 2019 [72]), (**f**) mesh details of the fourth grid (G4)—Mesh sensitivity analysis (reproduced with permission from Balduzzi, *Renewable Energy*, published by ELSEVIER, 2016 [22]), and (**g**) structured grid close to the turbine blade (reproduced with permission from Karimian, *Energy*, published by ELSEVIER, 2020 [18]).

4.4. Computational Domain and Boundary Condition Assignment

Figure 11a presents the computational domain and boundary conditions employed in the simulations [24]. It is crucial to critically evaluate the chosen domain size and boundaries to ensure accurate representation of the flow physics and reliable predictions of turbine performance. In the case of a ducted turbine, Figure 11b illustrates the schematic of the computational domain [44]. It is important to critically evaluate the chosen duct design and its influence on flow behavior, considering the impact on turbine performance and potential trade-offs. Figure 11c depicts the computational domain of a cluster consisting of six turbines [59]. It is necessary to critically evaluate the chosen cluster layout and its implications on flow interactions, including the sensitivity to turbine spacing and the impact of turbine wakes. Principal dimensions and boundary conditions of the computational domain are provided in Figure 11d [54]. It is crucial to critically evaluate their adequacy for capturing the flow physics and turbine behavior, considering the representativeness of the dimensions and the influence of domain truncation. Figure 11e focuses on mesh refinement in the wake region [56], highlighting its importance. It is important to critically evaluate the chosen mesh refinement strategy, including the criteria for refinement, the impact on wake capture, and computational cost considerations.

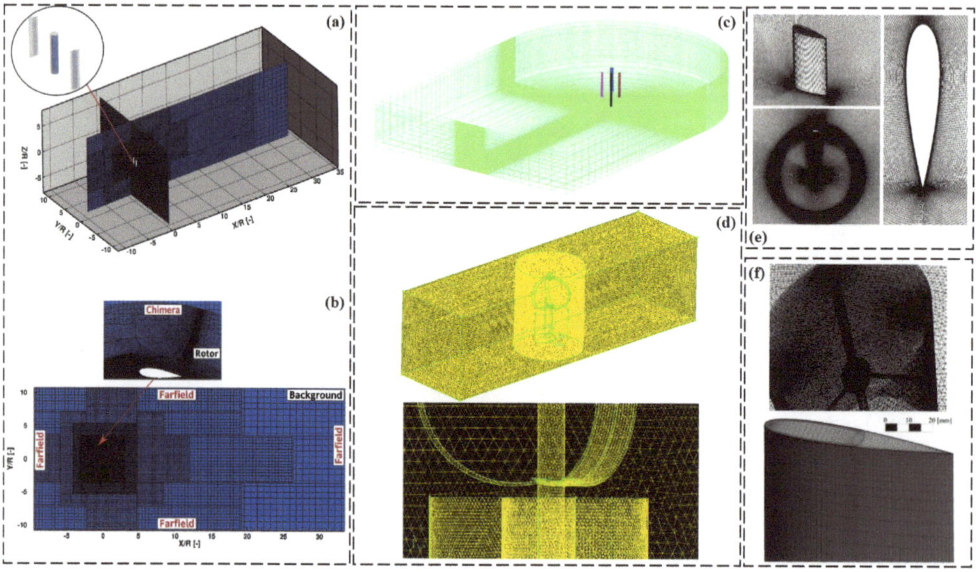

Figure 10. (**a**) Sectional 3D computational mesh used in the simulations (reproduced with permission from Dessoky, *Energy*, published by ELSEVIER, 2019 [35]), (**b**) computational mesh used in the simulations: background (reproduced with permission from Dessoky, *Energy*, published by ELSEVIER, 2019 [35]), (**c**) mesh structure of the 3D computational domain (reproduced with permission from Zamani, *Energy*, published by ELSEVIER, 2016 [54]), (**d**) computational mesh around combined Darrieus and the front view of the mesh very near to the blades (reproduced with permission from Ghosh, *Journal of the Energy Institute*, published by ELSEVIER, 2015 [17]), (**e**) some details of the computational mesh (reproduced with permission from Balduzzi, *Energy*, published by ELSEVIER, 2017 [16]), and (**f**) SB-VAWT: generated mesh for the three-dimensional (3D) models of turbine using Sliding Mesh Interface (SMI) (zoomed Darrieus view) and 3D mesh on blade tip (reproduced with permission from Siddiqui, *Renewable Energy*, published by ELSEVIER, 2021 and Rossetti, *Renewable Energy*, published by ELSEVIER, 2013 [31,73]).

Figure 12a presents the computational domain, including boundary conditions and the grid [34]. It is crucial to critically evaluate their appropriateness and representativeness, considering factors such as domain size, shape, and sensitivity of results to boundary conditions. Figure 12b provides the dimensions of the 3D domain [31]. It is important to critically evaluate their impact on the simulation results, including their representativeness, sensitivity to variations, and scaling effects. Figure 12c depicts the computational domain of the experimental apparatus and its boundary conditions [35]. It is necessary to critically evaluate their alignment with the physical setup and test conditions, considering the accuracy of the representation and fidelity of the description. Figure 12d presents the geometrical parameters of the computational domain and an isometric view of the rotating zone [18]. It is important to critically evaluate the selected geometrical parameters, their accuracy, and their impact on the flow behavior, considering the limitations associated with the isometric view.

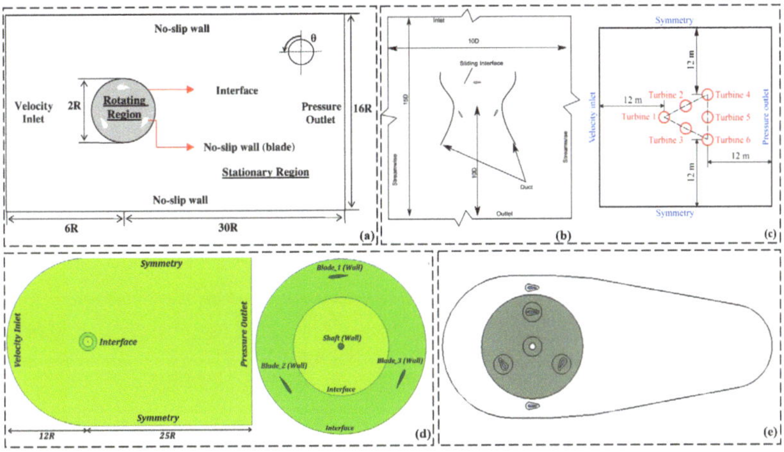

Figure 11. (**a**) Computational domain and boundary conditions (reproduced with permission from Arab, *Renewable Energy*, published by ELSEVIER, 2017 [24]), (**b**) schematic of the computational domain in case of ducted turbine (reproduced with permission from Malipeddi, *Renewable Energy*, published by ELSEVIER, 2012 [44]), (**c**) computational domain of a 6 turbine cluster (reproduced with permission from Shaaban, *Renewable Energy*, published by ELSEVIER, 2018 [59]), (**d**) principal dimensions and boundary conditions of computational domain (reproduced with permission from Zamani, *Energy*, published by ELSEVIER, 2016 [54]), and (**e**) mesh refinement in the wake (reproduced with permission from Bianchini, *Energy Conversion and Management*, published by ELSEVIER, 2017 [56]).

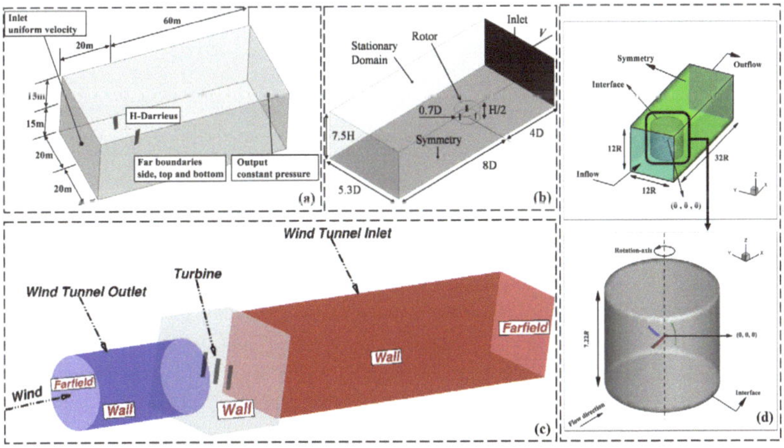

Figure 12. (**a**) Computational domain, including boundary conditions (non-scaled) and the grid computational domain (reproduced with permission from Joo, *Energy*, published by ELSEVIER, 2015 [34]), (**b**) 3D domain dimension (reproduced with permission from Rossetti, *Renewable Energy*, published by ELSEVIER, 2013 [31]), (**c**) experimental apparatus computational domain (reproduced with permission from Dessoky, *Energy*, published by ELSEVIER, 2019 [35]), and (**d**) geometrical parameters of the computational domain (isometric view of the rotating zone) (reproduced with permission from Karimian, *Energy*, published by ELSEVIER, 2020 [18]).

Table 3 presents an overview of Darrieus wind turbine localization on either side of the flow field, along with specific boundary conditions used in CFD modeling. The studies listed in the table focus on 2D and 3D simulations, examining the positioning of the turbines in relation to the upstream and downstream distances, width, and height. Additionally, some investigations include a sliding zone to analyze the movement of the turbines during the simulation.

Table 3. Positioning of Darrieus wind turbines on both sides of the flow field.

Authors	CFD Modeling	Upstream	Downstream	Width	Height	Sliding Zone	Inlet	Outlet	Left and Right Sides	Top and Bottom Sides
Balduzzi et al. [48]	2D	40 D	100 D	60 D	-	2 D	velocity inlet	pressure outlet	-	-
Asr et al. [41]		2 D	5.20 D	3.60 D	-	3 D			-	-
Mohammed et al. [42]		12.50 D	12.50 D	25 D	-	1.25 D			-	-
Hosseini et al. [72]		10 D	14 D	19.20 D	-	2 D			-	-
Mohammed et al. [74]		7.50 D	17.50 D	15 D	-	-			-	-
Mabrouk et al. [75]		15 D	24 D	20 D	-	-			-	-
Sengupta et al. [76]		5 D	10 D	10 D	-	-			-	-
Joo et al. [34]	3D	20 m	60 m	40 m	30 m	-			walls	walls
Rossetti et al. [31]		8 D	8 D	5.30 D	7.50 H	1.40 D			symmetry	symmetry
Jacob et al. [67]		1.10 m	10 D	1.50 D	1.70 m	-			walls	wall
Ali et al. [60]		7.50 D	20 D	15 D	8.50 D	-			Slip walls	Slip walls

In terms of boundary conditions, all studies utilized a velocity inlet to simulate the incoming wind flow. For the outlet, a pressure boundary condition was applied to represent the pressure at the exit of the computational domain. The table also specifies the boundary conditions on the left and right sides of the flow field, which were either symmetry or slip walls, depending on the study.

Here are some examples of the Darrieus localization and boundary conditions from the table:

1. Balduzzi et al. [48] conducted 2D simulations with upstream and downstream distances of 40 D and 100 D, respectively. The width of the domain was set to 60 D, and the height was not specified. A sliding zone was included in their investigation. The inlet boundary condition was velocity, and the outlet was pressure. The left and right sides of the flow field had unspecified boundary conditions;
2. Joo et al. [34] performed 3D simulations with an upstream distance of 20 m and a downstream distance of 60 m. The width of the domain was set to 40 m, and the height was 30 m. The simulation included walls as boundary conditions on the top and bottom sides of the flow field;
3. Rossetti et al. [31] conducted 3D simulations with an upstream distance of 8 D and a downstream distance of 8 D. The width of the domain was set to 5.30 D, and the height was 7.50 H. The sliding zone was included in their investigation. The left and right sides of the flow field had symmetry boundary conditions;
4. Ali et al. [60] performed 3D simulations with an upstream distance of 7.50 D and a downstream distance of 20 D. The width of the domain was set to 15 D, and the height was 8.50 D. The left and right sides of the flow field had slip walls as boundary conditions.

These examples demonstrate the various localization and boundary condition setups used in the CFD modeling of Darrieus wind turbines, providing insights into how researchers approach and analyze the turbine's performance in different flow field configurations.

4.5. Parametric Studies on Darrieus Turbines Aiming to Enhance Their Power Coefficient

A Darrieus wind turbine is depicted in Figure 13a [40], showcasing its configuration. A critical evaluation of its advantages and limitations compared to other designs is necessary, considering factors such as compactness, scalability, and suitability for various applications. The straight-bladed Darrieus design is shown in Figure 13b [41], highlighting its enhanced aerodynamic performance. However, a critical evaluation of its real-world performance, structural integrity, and power generation capabilities is crucial. Figure 13c presents the final assembly of the Darrieus turbine with specified dimensions [72], which requires a critical evaluation of the accuracy and relevance of these dimensions, considering manufacturing tolerances and material properties. The schematic representation of the experimental setup is depicted in Figure 13d [61], necessitating a critical evaluation of its ability to replicate real-world conditions and measure turbine performance accurately. Figure 13e presents a computer-aided design (CAD) model of the assessed Darrieus turbine [56], requiring a critical evaluation of the CAD model's fidelity in capturing the turbine's behavior and its impact on performance predictions. The schematic view of the architecture of the analyzed Darrieus turbines is shown in Figure 13f [19], requiring a critical evaluation of design choices, including structural integrity, material selection, and system efficiency.

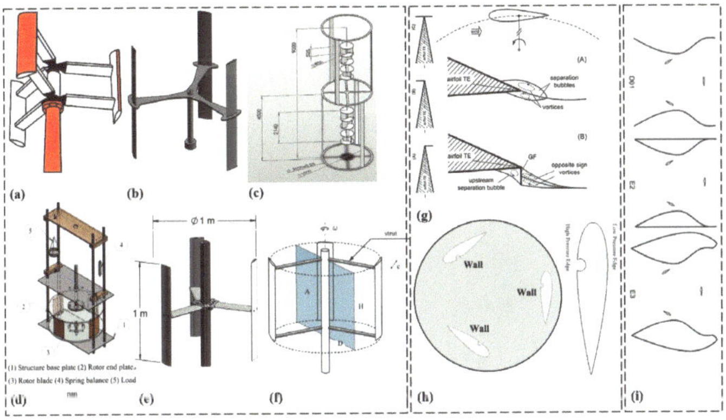

Figure 13. (**a**) Darrieus wind turbine (reproduced with permission from Hashem, *Energy*, published by ELSEVIER, 2018 [40]), (**b**) straight-bladed Darrieus schematic (reproduced with permission from Asr, *Energy*, published by ELSEVIER, 2016 [41]), (**c**) the final assembly (dimensions in mm) (reproduced with permission from Hosseini, *Energy conversion and management*, published by ELSEVIER, 2019 [72]), (**d**) schematic representation of the experimental set up (reproduced with permission from Patel, *International journal of marine energy*, published by ELSEVIER, 2017 [61]), (**e**) CAD of the assessed Darrieus turbine (reproduced with permission from Bianchini, *Energy Conversion and Management*, published by ELSEVIER, 2017 [56]), (**f**) schematic view of the architecture of the analyzed Darrieus turbines (reproduced with permission from Bianchini, *Energy conversion and management*, published by ELSEVIER, 2015 [19]), (**g**) different compared GF configurations: (left (A) GF in configuration; (B) GF out configuration; (C) GF both configuration). (right (A) flow field around smooth airfoil; (B) hypothesized flow field around an airfoil with a Gurney flap (reproduced with permission from Bianchini, *Energy conversion and management*, published by ELSEVIER, 2019 [1])), (**h**) computational domain and boundary conditions (reproduced with permission from Sobhani, *Energy*, published by ELSEVIER, 2017 [58]), and (**i**) the different external shapes for which the simulations were performed (reproduced with permission from Malipeddi, *Renewable Energy*, published by ELSEVIER, 2012 [44]).

A comparison of different configurations of Gurney flaps (GFs) and the conventional blade mounting in Darrieus is depicted in Figure 13g [1], necessitating a critical evaluation of their effectiveness in improving turbine performance, considering flow conditions, GF design parameters, and aerodynamic behavior. Figure 13h illustrates the computational domain and boundary conditions [58], requiring a critical evaluation of their representativeness, accuracy, and impact on simulation results. The influence of different external shapes on turbine performance is presented in Figure 13i [44], requiring a critical assessment of the range of shapes considered, their impact on performance parameters, and their practical implementation.

The schematic representation of a Darrieus with a J-shaped section is depicted in Figure 14a [54], emphasizing the optimized curved profile of the turbine blades. A critical evaluation of the J-shaped section's performance gains, starting torque, power generation efficiency, and structural integrity is essential. Figure 14b shows a combined three-bladed Darrieus and three-bladed bucket Savonius configuration [17], which requires a critical evaluation of its real-world performance, aerodynamic losses, mechanical complexity, and system efficiency compared to other Darrieus designs. The geometric details of a wind turbine model used for CFD simulation are presented in Figure 14c [60], requiring a critical evaluation of the model's fidelity in capturing the turbine's behavior and the impact of geometric simplifications.

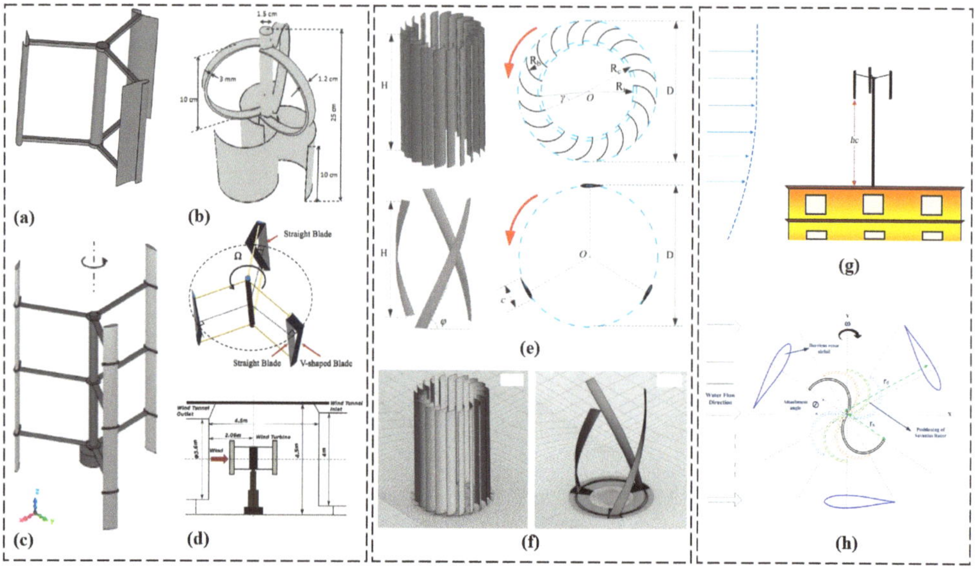

Figure 14. (**a**) The schematic representation of a Darrieus VAWT with the J-shaped section (reproduced with permission from Zamani, *Energy*, published by ELSEVIER, 2016 [54]), (**b**) combined three-bladed Darrieus and three-bladed bucket Savonius (reproduced with permission from Ghosh, *Journal of the Energy Institute*, published by ELSEVIER, 2015 [17]), (**c**) geometric details of wind turbine model for CFD simulation (perspective view) (reproduced with permission from Ali, *Energy Conversion and Management*, published by ELSEVIER, 2018 [60]), (**d**) the sketch map of the VAWT with V-shaped blade (transformation of the three-dimensional model)/schematic sketch of experimental apparatus (reproduced with permission from Dessoky, *Energy*, published by ELSEVIER, 2019 [35]), (**e**) three VAWTs in the simulation: Banki turbine (Top Configuration); Darrieus (Bottom Configuration) (reproduced with permission from Tian, *Renewable Energy*, published by ELSEVIER, 2020 [11]), (**f**) details of the computation mesh: grids around the Banki turbine; grids around the Darrieus (reproduced with permission from Tian, *Renewable Energy*, published by ELSEVIER, 2020 [11]), (**g**) SB-VAWT: the schematic of turbine placed on a rooftop illustrating variations in height; five levels of height are considered, i.e., 1.0c (Case i), 2.5c (Case ii), 4.0c (Case iii), 7.5c (Case iv), and 10.0c (Case v) (reproduced with permission from Siddiqui, *Renewable Energy*, published by ELSEVIER, 2021 [73]), and (**h**) schematic of a hybrid Darrieus (reproduced with permission from Saini, *Ocean Engineering*, published by ELSEVIER, 2020 [15]).

The sketch map of the VAWT with V-shaped blades is shown in Figure 14d [35], along with a schematic sketch of the experimental apparatus, necessitating a critical evaluation of their representativeness, scaling effects, instrumentation limitations, and potential sources of uncertainty. Figure 14e compares three VAWTs (Banki turbine and Darrieus) used in simulations [11], requiring a critical assessment of the simulation setup, accuracy of simulation models, boundary conditions, and validation against experimental data. Figure 14f provides details of the computation mesh [11], requiring a critical evaluation of the chosen mesh configuration, resolution, mesh-induced flow disturbances, and computational efficiency. The schematic of a rooftop-mounted SB-VAWT with variations in height is depicted in Figure 14g [73], requiring a critical evaluation of the practical implications, wind flow characteristics, structural considerations, aesthetics, and safety. The schematic of a hybrid Darrieus configuration is presented in Figure 14h [15], requiring a critical assessment of its performance gains, and trade-offs. Factors such as the aerodynamic interaction between different Darrieus components, mechanical complexity, and overall system efficiency should be carefully assessed to determine the practical implications and potential limitations of this configuration. By conducting a thorough analysis, the reliability and significance of the hybrid Darrieus configuration can be properly evaluated.

4.6. Assessment of Discrepancies in Darrieus Power Coefficient

The experimental validation based on Castelli et al. is presented in Figure 15a [8], serving as a reference to validate the numerical simulations. However, a critical evaluation of the experimental setup, methodology, and potential limitations is crucial.

Similarly, a comparison between the numerical results and experimental data of Takao et al. is shown in Figure 15b [24], allowing an assessment of the accuracy and effectiveness of different turbulence models. It is important to critically evaluate the robustness, limitations, and applicability of these turbulence models. Figure 15c illustrates the effect of different turbulence models on the simulation results [50], highlighting the need to critically evaluate their implications and limitations. A comparison between experimental and simulated power coefficients (C_p) at different Tip Speed Ratios (TSRs) is depicted in Figure 15d [30], necessitating a critical evaluation of the potential sources of discrepancy.

The lift and drag coefficients of the NACA0018 airfoil at a Reynolds number of 300,000 are displayed in Figure 15e [14], requiring a critical assessment of the limitations and applicability of the chosen airfoil model. In Figure 15f, the tangential force coefficients of a single blade against the azimuthal angle for various CFD schemes at a Tip Speed Ratio (TSR) of 2 are compared [30], necessitating a critical assessment of the assumptions and limitations of each CFD scheme.

The power coefficient (C_p) curves between 2D and 3D unsteady Reynolds-averaged Navier–Stokes (URANS) CFD simulations are compared in Figure 15g [35], requiring a critical evaluation of the assumptions and limitations associated with each approach. In Figure 15h, a comparison of the power coefficient (C_p) curve of a Darrieus VAWT model is presented [35], necessitating a critical evaluation of the experimental setup and measurement uncertainties. Lastly, the experimental and unsteady Reynolds-averaged Navier–Stokes (URANS) blade-resolved simulation power coefficient (C_p) curves are compared in Figure 15i [37], requiring a critical evaluation of the limitations and assumptions associated with the URANS modeling approach.

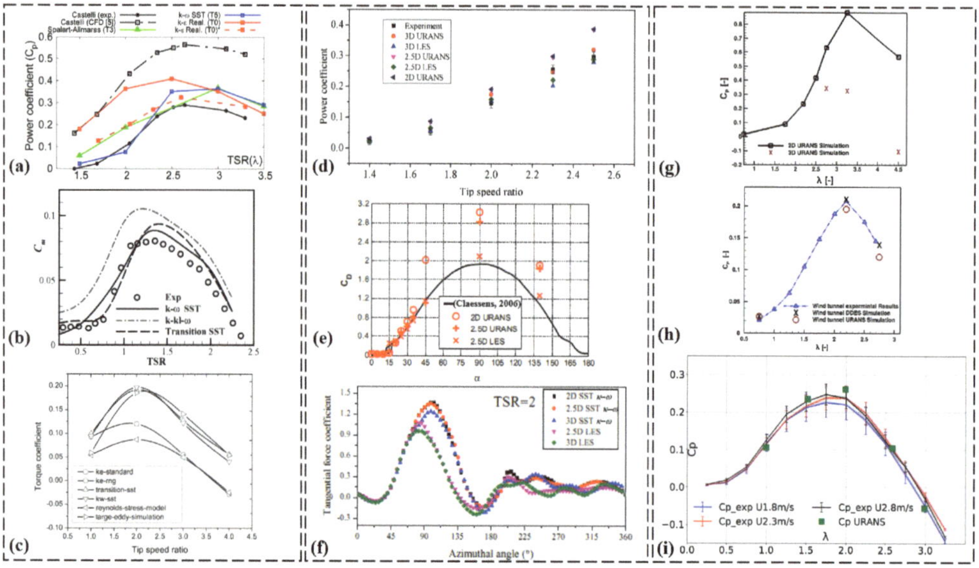

Figure 15. (**a**) Experimental validation based on Castelli et al. [65](exp) (reproduced with permission from Daroczy, *Energy*, published by ELSEVIER, 2015 [8]), (**b**) the numerical results comparison with the experimental data of Takao et al., using various turbulence models (reproduced with permission from Arab, *Renewable Energy*, published by ELSEVIER, 2017 [24]), (**c**) effect of different turbulence models on the simulation results (reproduced with permission from Chen, *Energy*, published by ELSEVIER, 2015 [50]), (**d**) comparison of experimental and simulated C_p at different TSRs (reproduced with permission from He, *Renewable Energy*, published by ELSEVIER, 2020 [30]), (**e**) lift and drag coefficients of NACA0018 airfoil for Re = 300,000 [77] (reproduced with permission from Li, *Renewable energy*, published by ELSEVIER, 2013 [14]), (**f**) tangential force coefficients of a single blade against the azimuthal angle for various CFD schemes (TSR = 2) (reproduced with permission from He, *Renewable Energy*, published by ELSEVIER, 2020 [30]), (**g**) comparison of the power coefficient (C_p) curves between 2D and 3D URANS CFD simulations (reproduced with permission from Dessoky, *Energy*, published by ELSEVIER, 2019 [35]), (**h**) comparison of power coefficient (C_p) curve of Darrieus VAWT model (reproduced with permission from Dessoky, *Energy*, published by ELSEVIER, 2019 [35]), and (**i**) comparison of the experimental and URANS blade-resolved simulation C_p curves (reproduced with permission from Clary, *Ocean Engineering*, published by ELSEVIER, 2020 [37]).

In Figure 16a, a comparison is made between numerical results and experimental data by Bravo et al. [53], necessitating a critical assessment of the experimental setup and methodology. Similarly, a comparison between numerical and experimental results is presented in Figure 16b [58], requiring a critical evaluation of the experimental uncertainties and limitations. The present computational model is validated by comparing it to experimental data and other computational results, as shown in Figure 16c [40], necessitating a critical evaluation of the assumptions and limitations of the model itself.

A comparison of the power coefficient between computational results and experimental values is presented in Figure 16d [71], requiring a critical evaluation of the potential sources of discrepancy between the computational and experimental data. By conducting thorough critical evaluations of these factors, the reliability, accuracy, and predictive capabilities of the numerical simulations and models can be ensured, enhancing the overall confidence in the results and their practical implications.

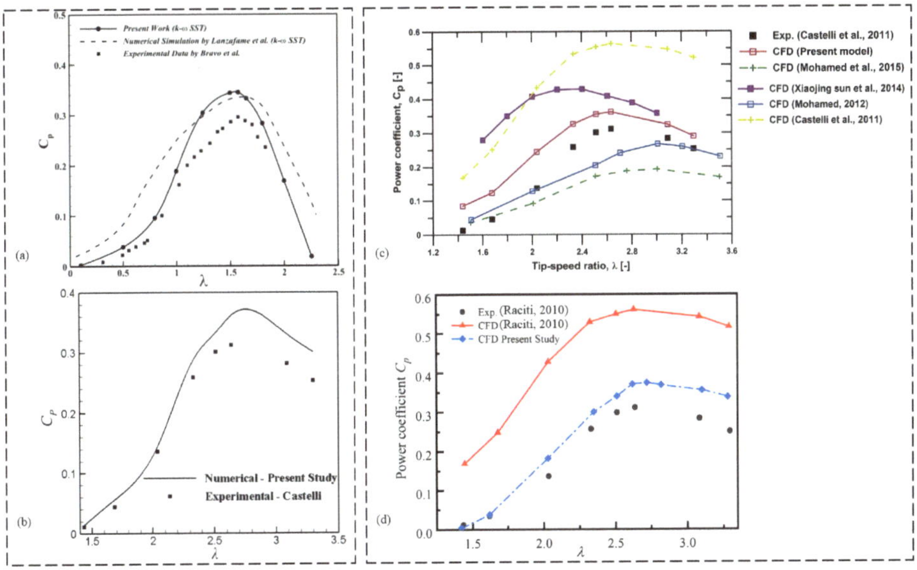

Figure 16. (**a**) Comparison of numerical results and experimental data by Bravo et al. (reproduced with permission from Zamani, *Renewable Energy*, published by ELSEVIER, 2016 [53]), (**b**) comparison of numerical and experimental results (reproduced with permission from Sobhani, *Energy*, published by ELSEVIER, 2017 [58]), (**c**) validation of present computational model, compared to experimental data for open Darrieus turbine and other computational results for Darrieus [40,46,47,65,78] (reproduced with permission from Hashem, *Energy*, published by ELSEVIER, 2018 [40]), and (**d**) power coefficient comparison between computation results and experiment values [79] (reproduced with permission from Ni, *Energy*, published by ELSEVIER, 2021 [71]).

Figure 17a presents a comparison of the torque coefficient in the first five revolutions with the test result at $\lambda = 2.29$ [63]. While this comparison allows for the evaluation of the accuracy of the numerical model in predicting the torque coefficient, it is important to critically assess the experimental conditions and measurement uncertainties. Factors such as instrumentation limitations, experimental setup variations, and potential sources of error, should be critically evaluated to ensure the reliability and significance of the experimental data as a validation reference.

The lift coefficient of the NACA0018 airfoil for Re = 300,000 is depicted in Figure 17b [14]. While this comparison between the numerical results and experimental data allows for the evaluation of the accuracy of the computational model, it is important to critically assess the experimental uncertainties and limitations. Factors such as measurement techniques, calibration procedures, and potential sources of systematic error should be critically evaluated to ensure the reliability and validity of the experimental data for validating the numerical model.

In Figure 17c, a comparison is made between the averages of the total power coefficient from experimental data by Elkhoury et al. and the present 3D results [18]. While this comparison aims to assess the accuracy of the computational model in predicting the power coefficient, it is important to critically evaluate the experimental uncertainties and potential sources of discrepancy. Factors such as measurement techniques, data acquisition procedures, and uncertainties in the experimental setup should be critically assessed to ensure the reliability and validity of the experimental data for validating the computational model.

The results of the SB-VAWT are benchmarked against experimental and numerical results by Howel et al. in Figure 17d. The numerical results demonstrate better agreement with reality compared to Howel et al.'s results at higher TSRs. However, it is crucial to critically evaluate the assumptions and limitations associated with both the numerical and experimental approaches. Factors such as modeling simplifications, numerical approximations, variations in the experimental setup, and uncertainties in the measurements should be thoroughly assessed. By critically evaluating these factors, the accuracy and reliability of the numerical simulations and experimental measurements can be ensured, leading to a more comprehensive understanding of the turbine's performance and enhancing confidence in the obtained results.

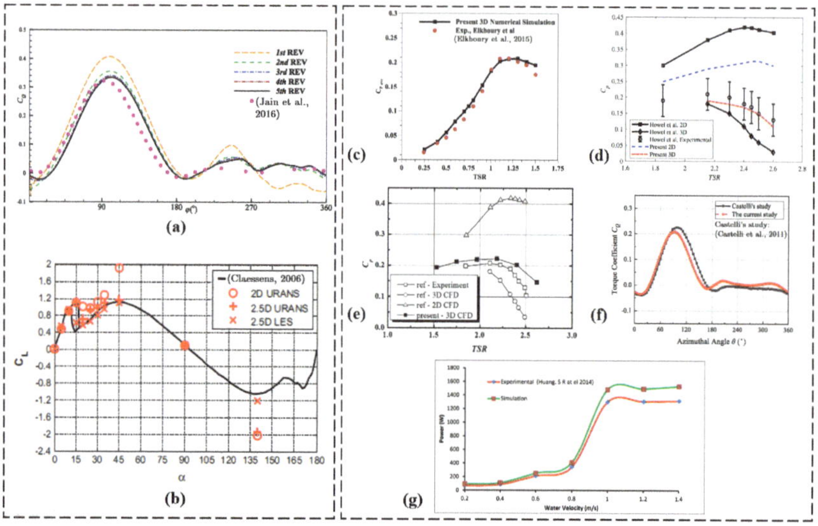

Figure 17. (**a**) Comparison of the torque coefficient in the first five revolutions with the test result, $\lambda = 2.29$ [80] (reproduced with permission from Xu, *Energy Conversion and Management*, published by ELSEVIER, 2020 [63]), (**b**) lift coefficient of NACA0018 airfoil for Re = 300,000 [77] (reproduced with permission from Li, *Renewable energy*, published by ELSEVIER, 2013 [14]), (**c**) comparison between averages of the total power coefficient of experimental data by Elkhoury et al. [81] and 3D results presented in [18] (reproduced with permission from Karimian, *Energy*, published by ELSEVIER, 2020 [18], (**d**) SB-VAWT: results are benchmarked with Howel et al. [73] experimental and numerical results (reproduced with permission from Siddiqui, *Renewable Energy*, published by ELSEVIER, 2021 [73]). Experimental results have been reported at a wind speed of 5.07 m/s. The numerical results are within the ±20% error range and more realistic than Howel et al. results at higher TSRs (reproduced with permission from Siddiqui, *Renewable Energy*, published by ELSEVIER, 2021 [73]), (**e**) power coefficient comparison between the computational and experimental results by Robert et al. (reproduced with permission from Joo, *Energy*, published by ELSEVIER, 2015 [34]), (**f**) comparison of the dynamic torque coefficient at tip speed ratio $\lambda = 3.1$ (Castelli et al. [65]) (reproduced with permission from Su, *Applied Energy*, published by ELSEVIER, 2020 [38]), and (**g**) computational model validation with published experimental data at different values of TSR (reproduced with permission from Tunio, *Renewable Energy*, published by ELSEVIER, 2020 [82]).

The power coefficient is compared between the computational and experimental results by Robert et al. in Figure 17e. This comparison enables the assessment of the accuracy and predictive capability of the computational model. However, it is crucial to critically assess the experimental setup and potential sources of discrepancy between the computational and experimental data. Factors such as measurement techniques, experimental uncertainties, and limitations in the experimental conditions should be thoroughly evaluated. By critically analyzing these factors, the reliability and validity of the experimental data for validating the numerical model can be ensured, leading to a more comprehensive understanding of the turbine's performance and enhancing confidence in the computational results.

The dynamic torque coefficient at a tip speed ratio (λ) of 3.1 is compared in Figure 17f, based on Castelli et al.'s findings [38]. This comparison aims to validate the accuracy of the numerical model in predicting the dynamic torque coefficient. However, it is crucial to critically assess the experimental conditions and potential sources of discrepancy between the numerical and experimental data. Factors such as instrumentation limitations, measurement uncertainties, and variations in the experimental setup should be thoroughly evaluated. By critically analyzing these factors, the reliability and significance of the experimental data as a validation reference can be ensured, enabling a more comprehensive understanding of the turbine's performance and improving confidence in the accuracy of the numerical model's predictions.

The computational model is validated with published experimental data at different values of tip speed ratio (TSR) in Figure 17g, as described by Tunio et al. [82]. This validation process aims to assess the accuracy and reliability of the computational model. However, it is crucial to critically evaluate the experimental uncertainties and limitations associated with the published data. Factors such as measurement techniques, calibration procedures, and potential sources of error should be thoroughly assessed. By critically analyzing these factors, the reliability and validity of the experimental data for validating the computational model can be ensured, leading to improved confidence in the accuracy of the model's predictions and a more comprehensive understanding of the turbine's performance.

4.7. Turbulence Quantities Obtained through CFD Modeling

Figure 18a compares the shed vortex centers' path for a three-bladed Darrieus using experimental visualization by Strickland JH et al. and FEVDTM prediction [13], allowing for the evaluation of the accuracy and predictive capability of the FEVDTM model. However, it is important to critically assess the experimental visualization techniques and potential sources of uncertainty. Similarly, Figure 18b presents the dimensionless velocity magnitude around the blades for Zone 1 [25], providing insights into the behavior and distribution of the dimensionless velocity magnitude.

However, a critical assessment of the numerical model's assumptions and limitations is necessary. Contours of vorticity magnitude are illustrated in Figure 18c [45], offering insights into the distribution and strength of vorticity. However, the numerical model's capability to accurately capture the vorticity phenomenon should be critically assessed. Figure 18d displays the normalized velocity in the wake at TSR = 1.4 based on the computed velocity field [56], providing information about the velocity distribution in the wake region. Still, the numerical model's ability to capture the wake characteristics accurately should be critically evaluated. Dimensionless vorticity contours are depicted in Figure 18e for TSR = 6.0, vs. = 312 degrees, and h_{GF} = 2%c [1], offering insights into the vorticity distribution and strength. However, the numerical model's ability to accurately capture the vorticity phenomenon should be critically assessed.

In Figure 19a, velocity contours around a nine Savonius wind turbine farm are illustrated [57], providing a visual representation of the flow field and the distribution of wind velocities within the farm. However, the numerical modeling approach and the accuracy of the velocity predictions should be critically assessed. Similarly, a velocity contour map for a 3 × 5 farm with spacing x = 5, y = 10 is shown in Figure 19b [62], providing detailed information about the velocity distribution within the farm. However, a critical assessment of the numerical modeling assumptions and limitations is necessary. By conducting thorough critical evaluations of these factors, the reliability, accuracy, and predictive capabilities of the numerical simulations and models can be ensured, enhancing the overall confidence in the results and their practical implications.

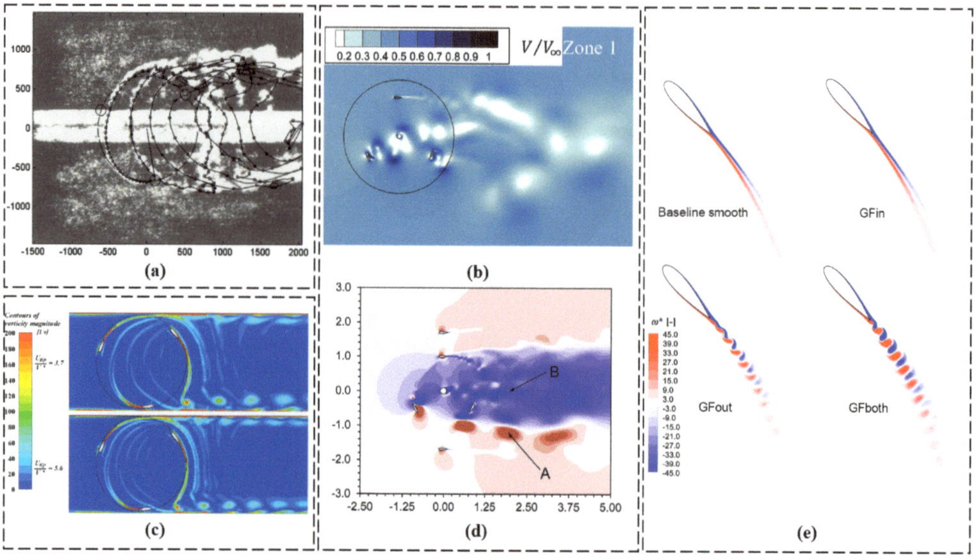

Figure 18. (**a**) Shed vortex centers path for a three-bladed Darrieus. Comparison of experimental visualization (Strickland JH et al.) and FEVDTM prediction (reproduced with permission from Ponta, *Renewable energy*, published by ELSEVIER, 2001 [13]), (**b**) dimensionless velocity magnitude around the blades for Zone 1 (reproduced with permission from Celik, *Journal of Wind Engineering and Industrial Aerodynamics*, published by ELSEVIER, 2020 [25]), (**c**) contours of vorticity magnitude (reproduced with permission from Marco, *Procedia Computer Science*, published by ELSEVIER, 2013 [45]), (**d**) normalized velocity in the wake at TSR = 1.4 (computed velocity field) (A: corresponds to powerful vortices that detach upwind from the blades and are subsequently carried downstream by the flow. The strength of these vortices might have been overestimated in CFD simulations compared to experimental observations. B: corresponds to an excessive reduction in velocity within the wake of the tower, a phenomenon that is not observed in experimental results) (reproduced with permission from Bianchini, *Energy Conversion and Management*, published by ELSEVIER, 2017 [56]) and (**e**) dimensionless vorticity contours (TSR = 6.0, vs. = 312 degree, h_{GF} = 2%c) (reproduced with permission from Bianchini, *Energy conversion and management*, published by ELSEVIER, 2019 [1]).

Figure 19c displays the turbulence intensity of a six-turbine cluster at X/D = 1.5 and λ = 3.3 [59]. While the turbulence intensity distribution provides insights into the level of turbulence generated within the cluster, it is important to critically assess the turbulence modeling approach and its ability to accurately capture the turbulence characteristics. Factors such as turbulence model selection, grid resolution, and numerical stability should be critically evaluated to ensure the accuracy and reliability of the turbulence intensity results.

The wake velocity contours at the tip speed ratio of 1.42 are depicted in Figure 19d [71]. While the wake velocity contours offer insights into the velocity distribution behind the turbines and the wake structure, it is important to critically assess the numerical modeling approach's ability to capture the wake dynamics accurately. Factors such as turbulence modeling, grid resolution, and boundary conditions should be critically evaluated to ensure the accuracy and reliability of the wake velocity contours in representing the wake characteristics.

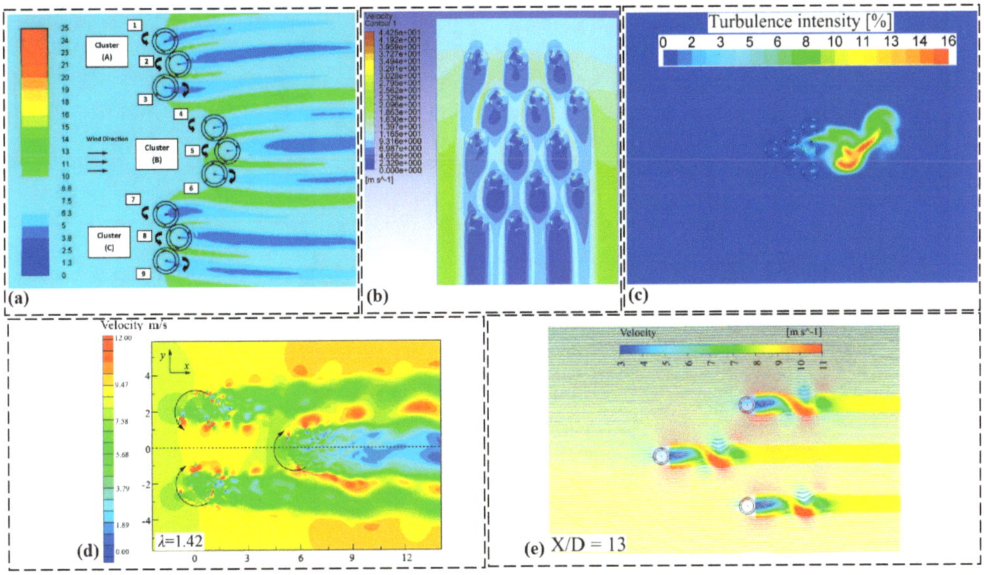

Figure 19. (**a**) Velocity contours around a nine savonius wind turbine farm (reproduced with permission from Shaheen, *Sustainable Energy Technologies and Assessments*, published by ELSEVIER, 2017 [57]), (**b**) velocity contour map for a 3 × 5 farm with spacing x = 5, y = 10 (reproduced with permission from Barnes, *Renewable energy*, published by ELSEVIER, 2019 [62]), (**c**) turbulence intensity of the six turbine cluster at X/D = 1.5 and λ = 3.3 (reproduced with permission from Shaaban, *Renewable Energy*, published by ELSEVIER, 2018 [59]), (**d**) the wake velocity contours at the tip speed ratio of 1.42 (reproduced with permission from Ni, *Energy*, published by ELSEVIER, 2021 [71]), and (**e**) pathlines for the three equilateral turbine cluster at X/D = 13 (reproduced with permission from Shaaban, *Renewable Energy*, published by ELSEVIER, 2018 [59]).

Figure 19e shows the pathlines for a three-equilateral turbine cluster at X/D = 13 [59]. While the pathlines provide visualizations of the flow patterns and particle trajectories, it is important to critically assess the numerical modeling approach's capability to accurately predict the flow behavior within the cluster. Factors such as numerical stability, grid resolution, and turbulence modeling should be critically evaluated to ensure the accuracy and reliability of the pathlines in representing the flow behavior and turbine interactions.

Figure 20a compares the wake of the Darrieus VAWT for an open turbine and a cycloidal surface diffuser wind lens [35]. This figure presents the wake characteristics and vortices generated behind the turbine. Figure 20b presents a comparison of the tip vortex between J-shaped blades and conventional blades [54]. This comparison offers insights into the vortex characteristics and differences between the blade configurations. Figure 20c shows the Q-criterion iso surfaces for three configurations of the rotating Darrieus [20]. These iso surfaces highlight the vortex structures and shedding dynamics. Figure 20d provides visualizations of vorticity magnitude and vortex shedding near the turbine at specific operating conditions [18,82]. These visualizations offer insights into the vorticity distribution and shedding patterns. Figure 20e presents visualizations of vorticity contours and distribution on 2D planes and horizontal cut planes [17,73]. These visualizations provide information about the vorticity behavior within the turbine and its surroundings.

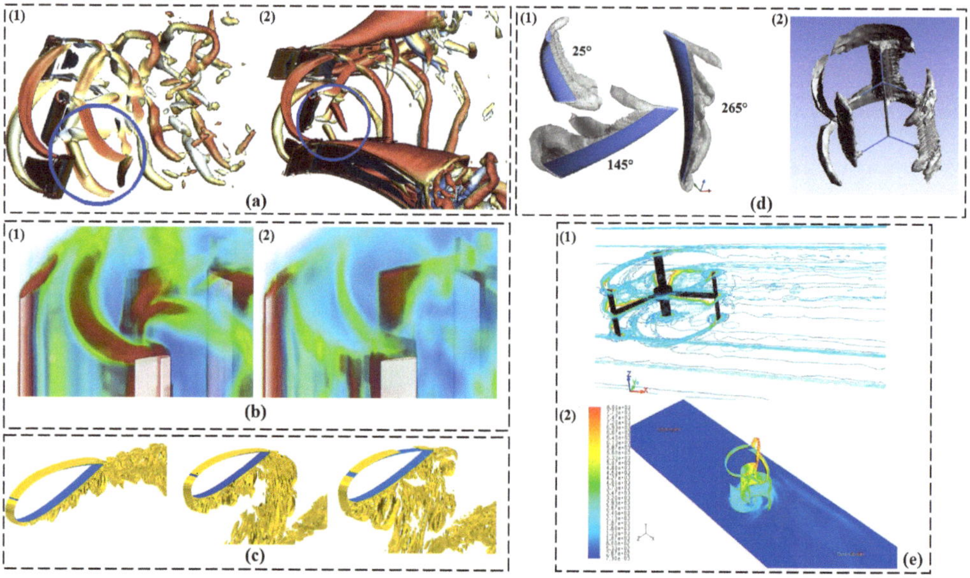

Figure 20. (**a**) Wake of the Darrieus VAWT for an open turbine (**1**) and a cycloidal surface diffuser wind lens (**2**). (reproduced with permission from Dessoky, *Energy*, published by ELSEVIER, 2019 [35]), (**b**) comparison of tip vortex generated by the J-shaped blades (**1**) and conventional blades (**2**) at a same azimuth position (reproduced with permission from Zamani, *Energy*, published by ELSEVIER, 2016 [54]), (**c**) Q-criterion (isosurface 300 s^{-2}) for three configurations of the rotating Darrieus modeled by LES (reproduced with permission from Patil, *Energy*, published by ELSEVIER, 2018 [20]), (**d**) (**1**) contour of the vorticity magnitude (3-PB configuration): iso surface of 65 s^{-1} at a wind speed of 7 m/s, rotational speed of 120 rpm, and azimuth angle of 25 deg. (**2**) Vortex shedding visualization for near turbine at vorticity of 16 ms^{-1} at inflow 1 m/s (reproduced with permission from Karimian, *Energy*, published by ELSEVIER, 2020, and from Tunio, *Renewable Energy*, published by ELSEVIER, 2020 [18,82]), and (**e**) (**1**) SB-VAWT: Contours of vorticity on 2D planes while turbine is mounted at 7.5c and operating at Tip Speed Ratio (TSR) of 3.0. (**2**) Vorticity distribution (in per s) on horizontal cut planes at y = 0.05 m (reproduced with permission from Siddiqui, *Renewable Energy*, published by ELSEVIER, 2021, and from Ghosh, *Journal of the Energy Institute*, published by ELSEVIER, 2015 [17,73]).

In Figure 21a, iso-surface visualizations of the wake of the Darrieus VAWT are compared for different turbulence resolving approaches and speed ratios [36]. These visualizations aim to showcase the wake characteristics and flow structures. Figure 21b presents a 3D view of the dimensional vorticity for a specific azimuth angle [36]. This visualization offers insights into the vorticity distribution and flow structures around the turbine. These visualizations and comparisons provide valuable information about wake characteristics, vortices, and flow structures in relation to turbine performance. Critical evaluations of the numerical modeling approaches, experimental techniques, and associated uncertainties enhance the reliability and accuracy of the presented insights.

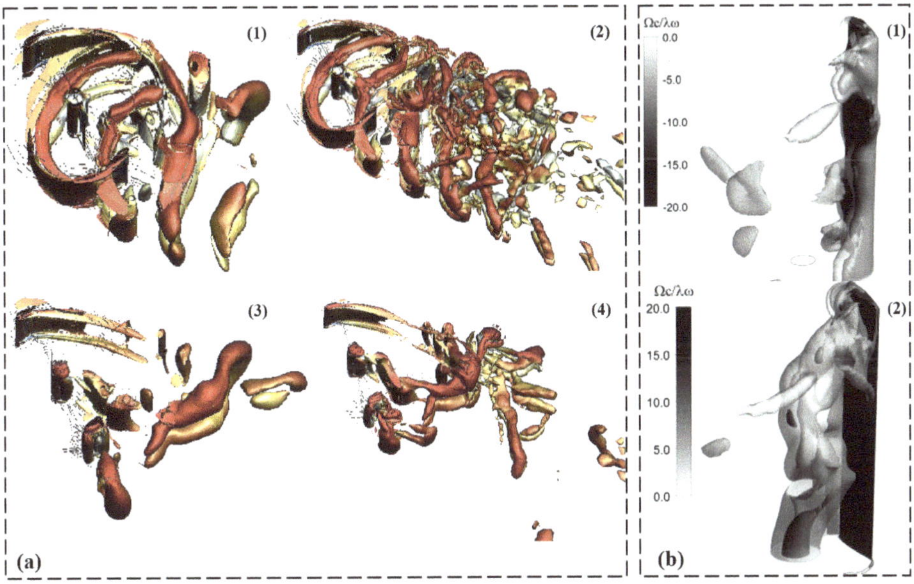

Figure 21. (**a**) (**1**) Wake of the Darrieus VAWT with different turbulence resolving approaches and speed ratio: URANS-JST/λ = 2.75, (**2**) DDES-WENO/λ = 2.75, (**3**) URANS-JST/λ = 0.75 and (**4**) DDES-WENO/λ = 0.75 (reproduced with permission from Dessoky, *Renewable Energy*, published by ELSEVIER, 2019 [36]), and (**b**) 3D view of the dimensional vorticity ($\frac{\Omega_c}{\lambda V}$) for θ = 120 deg; (**1**) positive vorticity; (**2**) negative vorticity (reproduced with permission from Rossetti, *Renewable Energy*, published by ELSEVIER, 2013 [31]).

4.8. Significance of Experimental Studies in Validating CFD Codes Numerically

A photograph of the two hybrid turbines used in the experiments is presented in Figure 22a [67]. This picture offers a visual representation of the physical turbines utilized in the study. However, it is important to critically evaluate the extent to which the experimental setup accurately represents real-world conditions. Factors such as scale effects, boundary conditions, and the influence of external factors should be critically considered when assessing the turbines' performance and characteristics based on the experimental data. By conducting a thorough analysis of these factors, the reliability and applicability of the experimental results can be ensured, providing valuable insights into the performance and behavior of the hybrid turbines.

Figure 22b illustrates the experimental setup of the model and instrumentation [6]. While the diagram shows the arrangement and positioning of the model turbine within the experimental setup, it is important to critically assess the experimental setup's capability to accurately reproduce real-world operating conditions. Factors such as flow uniformity, turbulence intensity, and measurement uncertainties should be critically evaluated to ensure the reliability and accuracy of the experimental data and subsequent analysis.

The connection between Darrieus, the generator, and the support system is depicted in Figure 22c [83]. The diagram highlights the mechanical interface and structural components involved in this connection. However, it is crucial to critically evaluate the design and reliability of the connection system. Factors such as structural integrity, power transmission efficiency, and maintenance requirements should be critically assessed to ensure the efficient and reliable operation of the turbine. By conducting a thorough evaluation of these factors, the turbine's overall performance, durability, and safety can be enhanced, leading to optimized power generation and long-term operational success.

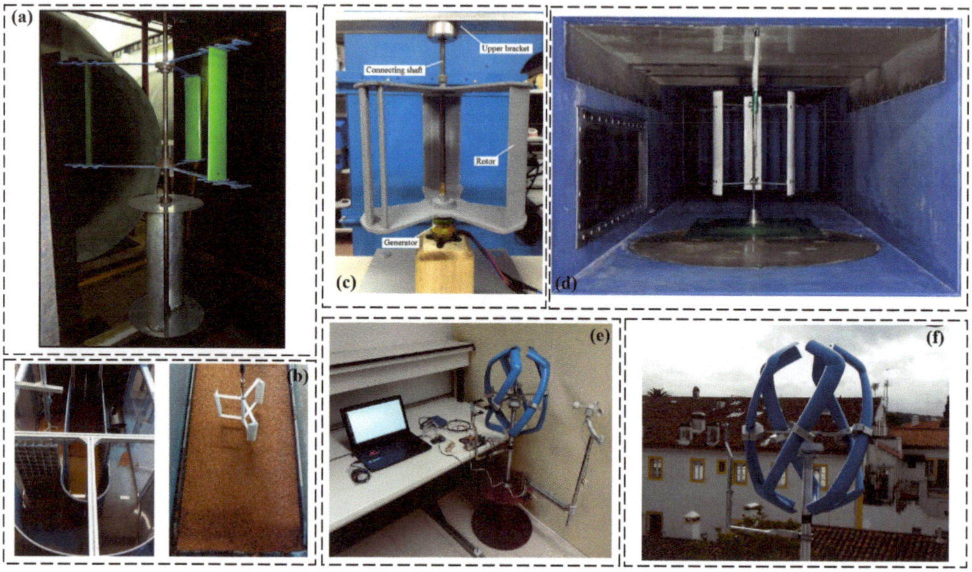

Figure 22. (**a**) Photograph showing the two hybrid turbines used in the experiments (reproduced with permission from Jacob, *Renewable Energy*, published by ELSEVIER, 2019 [67]), (**b**) experimental setup of model and instrumentation (reproduced with permission from Sun, *Energy Conversion and Management*, published by ELSEVIER, 2018 [6]), (**c**) connection of the Darrieus-generator and support system (reproduced with permission from Scungio, *Energy Conversion and Management*, published by ELSEVIER, 2016 [83]), (**d**) solid plate and Darrieus turbine installed in three different wind tunnels (wind tunnel-C, blockage ratio: 24.70%) (reproduced with permission from Jeong, *Journal of Wind Engineering and Industrial Aerodynamics*, published by ELSEVIER, 2018 [84]), (**e**) prototype and sensors (reproduced with permission from Pereira, *Energy*, published by ELSEVIER, 2018 [85]), and (**f**) new Darrieus VAWT design prototype (reproduced with permission from Batista, *Renewable and Sustainable Energy Reviews*, published by ELSEVIER, 2015 [86]).

The experimental setup for studying the aerodynamic performance of a solid plate and a Darrieus turbine in three different wind tunnels, including wind tunnel-C with a blockage ratio of 24.70%, is showcased in Figure 22d [84]. The image provides a visual representation of the wind tunnel configurations used to investigate the turbine's performance. However, it is crucial to critically assess the influence of blockage on the turbine's aerodynamic characteristics. Factors such as flow distortion, boundary layer effects, and the representativeness of the wind tunnel conditions should be critically evaluated to ensure the validity and applicability of the results obtained. By conducting a thorough analysis of these factors, the accuracy and reliability of the experimental findings can be ensured, leading to a better understanding of the turbine's performance under various flow conditions and enabling the development of more efficient and reliable wind energy systems.

The image in Figure 22e presents an overview of a prototype turbine and the associated sensors used for data collection in a study [85]. The image visually represents the physical turbine prototype and the sensors deployed for monitoring various parameters during operation. However, it is crucial to critically assess the accuracy and reliability of the sensors in capturing the intended measurements. Factors such as calibration procedures, measurement uncertainties, and sensor placement should be critically evaluated to ensure the accuracy and validity of the collected data. By thoroughly assessing these factors, the quality and reliability of the data can be ensured, enabling more accurate evaluations of the turbine's performance and facilitating improvements in its design and operation.

Figure 22f shows a new prototype of a Darrieus VAWT design [86]. While the image presents the physical prototype of the turbine and its unique design features, it is important to critically assess the performance and efficiency claims associated with this new design. Factors such as computational analysis, experimental validation, and comparative studies should be critically evaluated to assess the potential improvements offered by the new design.

Stereoscopic particle image velocimetry (PIV) is indeed a sophisticated experimental technique commonly used in the field of fluid mechanics to measure and analyze flow characteristics. However, it is important to critically assess the limitations and challenges associated with the technique to ensure the reliability and validity of the experimental findings.

In Figure 23a, a schematic representation of the TUDelft Open Jet Facility is depicted [87]. The diagram provides an overview of the experimental setup used in the study. However, it is crucial to critically evaluate the representativeness of the facility in relation to real-world conditions. Factors such as scale effects, boundary conditions, and the influence of the open jet setup on the flow characteristics should be carefully considered when interpreting the results obtained from experiments conducted in this facility. By critically assessing these factors, the validity and applicability of the experimental findings can be ensured, enabling a more accurate understanding of the aerodynamic behavior and performance of the tested turbine or airfoil.

Figure 23b describes the wind turbine model used in the experiments, with dimensions provided in millimeters, while the image represents a scaled-down version of a wind turbine, it is important to critically assess the accuracy of the scaling and the representativeness of the model. Factors such as Reynolds number effects, model fidelity, and the extrapolation of results to full-scale turbines should be critically evaluated to ensure the applicability of the findings.

In Figure 23c, the configuration of the stereoscopic PIV setup is depicted, illustrating the arrangement of cameras and light sources [87]. However, it is crucial to critically assess the accuracy and limitations of the measurement technique employed. Factors such as calibration errors, measurement uncertainties, and the potential influence of seeding particles on the flow should be carefully evaluated to ensure the accuracy and reliability of the velocity measurements obtained through stereoscopic PIV. By critically assessing these factors, the validity and significance of the experimental data can be ensured, enabling a more accurate analysis of the flow characteristics and behavior of the tested turbine or airfoil.

Figure 23d represents the planar PIV configuration as an alternative to stereoscopic PIV, while planar PIV provides valuable information about the flow characteristics in a specific plane, it is important to critically evaluate the limitations of this technique. Factors such as the assumption of planar flow, potential flow distortion, and limitations in capturing three-dimensional flow structures should be critically considered when interpreting the results obtained using planar PIV.

The contours of normalized out-of-plane vorticity for the combined field of views of the horizontal plane are displayed in Figure 23e. While the contours provide visual insights into the vortical structures and their distribution within the horizontal plane, it is important to critically evaluate the accuracy and resolution of the vorticity measurements. Factors such as image processing techniques, noise levels, and limitations in capturing small-scale vortices should be critically assessed to ensure the reliability of the vorticity visualization.

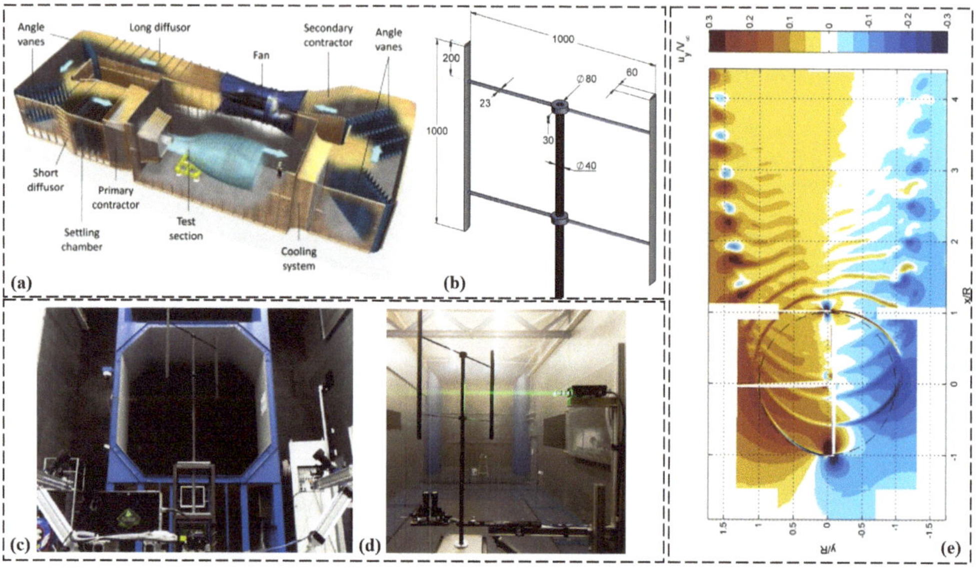

Figure 23. Stereoscopic particle image velocimetry. (**a**) Schematic of the TUDelft Open Jet Facility (reproduced with permission from Tescione, *Renewable Energy*, published by ELSEVIER, 2014 [87]). (**b**) Wind turbine model (dimensions in mm) (reproduced with permission from Tescione, *Renewable Energy*, published by ELSEVIER, 2014 [87]). (**c**) Stereoscopic PIV configuration (reproduced with permission from Tescione, *Renewable Energy*, published by ELSEVIER, 2014 [87]). (**d**) Planar PIV configuration (reproduced with permission from Tescione, *Renewable Energy*, published by ELSEVIER, 2014 [87]). (**e**) Contours of normalized out of plane vorticity for the combined field of views of the horizontal plane (reproduced with permission from Tescione, *Renewable Energy*, published by ELSEVIER, 2014 [87]).

5. The Viewpoint Regarding CFD Modeling of Darrieus Turbines

This research article advocates for the adoption of 3D modeling and advanced meshing techniques in the study of Darrieus wind turbines, highlighting the limitations of 2D modeling and the importance of comprehensive investigations. Based on expertise and findings from various research teams, it is evident that, while 2D CFD studies offer reduced computational time, they oversimplify the complex three-dimensional flow field, limiting the representation of crucial physical characteristics. This critical consideration calls for the allocation of additional computational resources to overcome this modeling challenge.

The utilization of powerful workstations with substantial RAM storage, high CPU frequency, and efficient graphics cards is essential in conducting CFD modeling for wind turbine analysis. These resources facilitate fluid domain mesh generation, expedite CFD calculations, and enable the visualization of numerical results. Although researchers often opt for 2D calculations to obtain numerically reliable results within a reasonable computation time frame, it is emphasized that the 3D flow structures generated at the blades and wake zone demand dedicated efforts in terms of calculation procedures and the adoption of 3D CFD modeling. This is particularly crucial when introducing geometric modifications to maximize the power coefficient of Darrieus.

Accurate meshing, meticulous computational grid design, and precise definition of geometric parameters are crucial in analyzing wind turbine systems' aerodynamic behavior and performance. While different modeling approaches (2D, 2.5D, and 3D) offer valuable insights, their limitations and potential sources of uncertainty must be carefully assessed when interpreting the results. Integrating experimental validation and conducting sensitivity analyses further enhance the reliability and robustness of the findings.

The accurate modeling of wind turbine systems necessitates a critical evaluation of the methodologies employed and their associated limitations. Factors such as mesh representation, mesh distribution, computational grid, boundary layer generation, mesh sensitivity analysis, and structured grid configuration must be meticulously examined to ensure precise modeling and optimization.

This critical review emphasizes the importance of assessing the limitations and uncertainties associated with experimental visualization techniques and numerical modeling approaches. Measurement uncertainties, visualization limitations, numerical assumptions, and model limitations must be rigorously evaluated to establish the reliability, accuracy, and significance of the presented results. Such critical assessments contribute to advancing wind turbine design, optimization, and the development of efficient renewable energy systems.

To enhance the aerodynamic design of Darrieus turbines, research teams should conduct a thorough review of existing literature and incorporate advanced CFD simulations. By launching CFD calculations aimed at improving the power coefficient and properly validating turbulence models within CFD solvers, the performance of Darrieus turbines can be significantly enhanced. This approach ensures that design modifications are based on reliable and validated computational results, contributing to optimization efforts.

In conclusion, this critical review highlights the importance of critically evaluating modeling techniques, methodologies, and experimental data to advance wind turbine design and optimization. By addressing limitations and uncertainties, performing CFD simulations, and validating turbulence models, the aerodynamic performance of Darrieus turbines can be significantly improved. Such critical review and computational analysis contribute to the advancement of research in the field and the development of more efficient and effective wind energy systems.

Within the realm of 2D modeling, a wide spectrum of contours has been explored to capture the intricacies of Darrieus flow patterns. Notably, Pellone et al. [66] present vorticity distribution around Darrieus blades, shedding light on their dynamic behavior. Meanwhile, the contours from Wang et al. [49] provide valuable insights into vortex shedding and normalized velocity distribution. These contours serve as windows into the dynamic behavior of the turbine, revealing vortex dynamics, pressure coefficients, and wind deflection angles.

Advancing to the domain of 2.5D modeling, deeper insights into vortical structures emerge. Li et al. [14] illustrate vorticity contours at different span lengths during Darrieus revolution, enriching our understanding of temporal evolution. The visualization of pressure coefficients and the comparison of tip vortices shedding from the blade end winglet, as presented by Balduzzi et al. [16], provide an enriched perspective on flow behavior.

Transitioning to the realm of 3D modeling, the contours portrayed here offer a multidimensional insight into Darrieus aerodynamics. Notably, Rossetti et al. [31] provide a 3D view of dimensional vorticity, highlighting its volumetric structures. Additionally, contours from Joo et al. [34] depict the velocity distribution and streamlines in the 3D model. These visual representations provide a comprehensive view of aerodynamic efficiency, encompassing pressure distribution, velocity profiles, and wake dynamics.

6. Conclusions

This critical analysis, part of the "critical review issue," has extensively explored the capability of CFD modeling to simulate and replicate the intricate flow patterns observed in Darrieus VAWTs. By investigating diverse CFD approaches—2D, 2.5D, and 3D simulations—within a specific velocity ratio (λ) range, a comprehensive comprehension of the challenges associated with accurately capturing the aerodynamic behavior and vortex structures of Darrieus turbines has been achieved. The key results are listed as follows:

1. An in-depth analysis of over 250 published articles has yielded a comprehensive compilation of findings, shedding light on the evolution of crucial parameters, including the power coefficient (C_p), moment coefficient (C_m), lift coefficient (C_l), drag coeffi-

cient (C_d), velocity distributions, pressure fields, vorticity patterns, turbulent kinetic energy production and dissipation, streamlines, and Q-criterion of the vorticity;
2. Notably, the influence of commonly used turbulence models on the power coefficient (C_p) values has been elucidated, substantiated by illustrative figures that underscore the impact of modeling choices;
3. Addressing the prevalent use of 2D CFD modeling, primarily due to its computational efficiency, this review recognizes its advantages while highlighting the limitations inherent in capturing turbulent flow structures resulting from geometric alterations. Comprehensive CFD contour plots underscore these limitations, emphasizing the need for enhanced modeling approaches.

In conclusion, this comprehensive examination advances our understanding of CFD modeling for Darrieus turbines, offering a concise yet comparative evaluation of distinct modeling strategies. By unveiling the disparities among 2D, 2.5D, and 3D simulations and their ramifications for flow prediction and replication, this study aids in selecting the most appropriate modeling approach to ensure accurate and dependable predictions in the realm of VAWTs. The discerned insights hold significant value for researchers and practitioners, empowering them in the pursuit of refined design and optimal performance enhancement of Darrieus VAWTs.

Author Contributions: S.e.-D.F.: Conceptualization, Methodology, Validation, Investigation, Resources, Writing—Original Draft, Writing—Review and Editing, Visualization and Project administration. T.B.: Methodology, Validation, Investigation, Resources. A.K.: Methodology, Validation, Investigation, Resources. A.S.: Methodology, Validation, Investigation, Resources. I.K.: Methodology, Validation, Investigation, Resources. E.B.: Methodology, Validation, Investigation, Resources, Writing—Original Draft—Review. All authors have read and agreed to the published version of the manuscript.

Funding: This research received no external funding.

Data Availability Statement: In this study, no new data were generated. The analysis was conducted based on existing publicly available data and previously published research. Researchers interested in accessing the data used in this study can refer to the sources and references cited in the manuscript.

Acknowledgments: The corresponding author, Saïf ed-Dîn Fertahi, would like to take this opportunity to express sincere gratitude to the editor and reviewers for their valuable remarks that have greatly enhanced the quality of this CFD modeling study.

Conflicts of Interest: The authors declare that they have no known competing financial interests or personal relationships that could have influenced the work reported in this paper.

Nomenclature

A	Area Swept by Darrieus (m^2)
R	Darrieus Radius (m)
Re	Reynolds Number
C_d	Drag Coefficient
C_l	Lift Coefficient
C_m	Moment Coefficient
C_p	Power Coefficient
g	Acceleration Due to Gravity Vector (m/s^2)
P	Pressure (Pa)
T_m	Torque on the Wind Turbine (N.m)
U_∞	Upstream Flow Velocity (m/s)
V_c	Tangential Component of Relative Velocity (m/s)
V_n	Normal Component of Relative Velocity (m/s)
y^+	Dimensionless Wall Distance

Greek symbols	
μ	Dynamic Viscosity (kg/(m.s))
ω	Darrieus Angular Velocity (rad/s)
ρ	Fluid Density (kg/m^3)
α	Angle of Attack (degrees)
λ	Tip Speed Ratio
k-ω-SST	Shear Stress Transport turbulence model
k-ε	Turbulence Model
Abbreviations	
2.5D	Two-and-a-Half Dimensional
2D	Two-Dimensional
3D	Three-Dimensional
6DOF	Six Degrees of Freedom
CAD	Computer-Aided Design
CFD	Computational Fluid Dynamics
CPU	Central Processing Unit
DES	Detached-Eddy Simulation
FEVDTM	Free Vortex Element (FVE) Doublet Theory Model
FSI	Fluid–Structure Interaction
LES	Large Eddy Simulation
LES-SGS	Large Eddy Simulation with Sub-Grid Scale model
MRF	Multiple Reference Frame
NACA	National Advisory Committee for Aeronautics
PIV	Particle Image Velocimetry
RAM	Random Access Memory
RPM	Rotations Per Minute
SGS	Sub-Grid Scale model
SST	Shear Stress Transport
TSR	Tip Speed Ratio
TSST	Turbulent Scale Simulation Transport model
VAWT	Vertical Axis Wind Turbine

References

1. Bianchini, A.; Balduzzi, F.; Di Rosa, D.; Ferrara, G. On the use of Gurney Flaps for the aerodynamic performance augmentation of Darrieus wind turbines. *Energy Convers. Manag.* **2019**, *184*, 402–415. [CrossRef]
2. Scopus Preview. Welcome to Scopus Preview. 2023. Available online: https://www.scopus.com/home.uri (accessed on 6 August 2023).
3. Rosander, M. SeaTwirl: The Future of Offshore Floating Wind Turbines. ANSYS Case Study 2018. Available online: https://www.ansys.com/-/media/ansys/corporate/resourcelibrary/casestudy/seatwirl.pdf (accessed on 20 March 2020).
4. Li, S.; Li, Y.; Yang, C.; Wang, Q.; Zhao, B.; Li, D.; Zhao, R.; Ren, T.; Zheng, X.; Gao, Z.; et al. Experimental investigation of solidity and other characteristics on dual vertical axis wind turbines in an urban environment. *Energy Convers. Manag.* **2021**, *229*, 113689. [CrossRef]
5. Rezaeiha, A.; Montazeri, H.; Blocken, B. A framework for preliminary large-scale urban wind energy potential assessment: Roof-mounted wind turbines. *Energy Convers. Manag.* **2020**, *214*, 112770. [CrossRef]
6. Sun, C.; Lam, W.H.; Cui, Y.; Zhang, T.; Jiang, J.; Guo, J.; Ma, Y.; Wang, S.; Tan, T.H.; Chuah, J.H.; et al. Empirical model for Darrieus-type tidal current turbine induced seabed scour. *Energy Convers. Manag.* **2018**, *171*, 478–490. [CrossRef]
7. Erkan, O.; Özkan, M.; Karakoç, T.H.; Garrett, S.J.; Thomas, P.J. Investigation of aerodynamic performance characteristics of a wind-turbine-blade profile using the finite-volume method. *Renew. Energy* **2020**, *161*, 1359–1367. [CrossRef]
8. Daróczy, L.; Janiga, G.; Petrasch, K.; Webner, M.; Thévenin, D. Comparative analysis of turbulence models for the aerodynamic simulation of H-Darrieus rotors. *Energy* **2015**, *90*, 680–690. [CrossRef]
9. Nguyen, M.T.; Balduzzi, F.; Bianchini, A.; Ferrara, G.; Goude, A. Evaluation of the unsteady aerodynamic forces acting on a vertical-axis turbine by means of numerical simulations and open site experiments. *J. Wind. Eng. Ind. Aerodyn.* **2020**, *198*, 104093. [CrossRef]
10. Trivellato, F.; Castelli, M.R. Appraisal of Strouhal number in wind turbine engineering. *Renew. Sustain. Energy Rev.* **2015**, *49*, 795–804. [CrossRef]
11. Tian, W.; Song, B.; Mao, Z. Numerical investigation of wind turbines and turbine arrays on highways. *Renew. Energy* **2020**, *147*, 384–398. [CrossRef]
12. Gavalda, J.; Massons, J.; Diaz, F. Experimental study on a self-adapting Darrieus—Savonius wind machine. *Sol. Wind. Technol.* **1990**, *7*, 457–461. [CrossRef]

13. Ponta, F.L.; Jacovkis, P.M. A vortex model for Darrieus turbine using finite element techniques. *Renew. Energy* **2001**, *24*, 1–18. [CrossRef]
14. Li, C.; Zhu, S.; Xu, Y.-l.; Xiao, Y. 2.5D large eddy simulation of vertical axis wind turbine in consideration of high angle of attack flow. *Renew. Energy* **2013**, *51*, 317–330. [CrossRef]
15. Saini, G.; Saini, R. A computational investigation to analyze the effects of different rotor parameters on hybrid hydrokinetic turbine performance. *Ocean Eng.* **2020**, *199*, 107019. [CrossRef]
16. Balduzzi, F.; Drofelnik, J.; Bianchini, A.; Ferrara, G.; Ferrari, L.; Campobasso, M.S. Darrieus wind turbine blade unsteady aerodynamics: A three-dimensional Navier-Stokes CFD assessment. *Energy* **2017**, *128*, 550–563. [CrossRef]
17. Ghosh, A.; Biswas, A.; Sharma, K.; Gupta, R. Computational analysis of flow physics of a combined three bladed Darrieus Savonius wind rotor. *J. Energy Inst.* **2015**, *88*, 425–437. [CrossRef]
18. Karimian, S.; Abdolahifar, A. Performance investigation of a new Darrieus vertical axis wind turbine. *Energy* **2020**, *191*, 116551. [CrossRef]
19. Bianchini, A.; Ferrara, G.; Ferrari, L. Design guidelines for H-Darrieus wind turbines: Optimization of the annual energy yield. *Energy Convers. Manag.* **2015**, *89*, 690–707. [CrossRef]
20. Patil, R.; Daróczy, L.; Janiga, G.; Thévenin, D. Large eddy simulation of an H-Darrieus rotor. *Energy* **2018**, *160*, 388–398. [CrossRef]
21. Wang, Y.; Sun, X.; Dong, X.; Zhu, B.; Huang, D.; Zheng, Z. Numerical investigation on aerodynamic performance of a novel vertical axis wind turbine with adaptive blades. *Energy Convers. Manag.* **2016**, *108*, 275–286. [CrossRef]
22. Balduzzi, F.; Bianchini, A.; Maleci, R.; Ferrara, G.; Ferrari, L. Critical issues in the CFD simulation of Darrieus wind turbines. *Renew. Energy* **2016**, *85*, 419–435. [CrossRef]
23. Hand, B.; Kelly, G.; Cashman, A. Numerical simulation of a vertical axis wind turbine airfoil experiencing dynamic stall at high Reynolds numbers. *Comput. Fluids* **2017**, *149*, 12–30. [CrossRef]
24. Arab, A.; Javadi, M.; Anbarsooz, M.; Moghiman, M. A numerical study on the aerodynamic performance and the self-starting characteristics of a Darrieus wind turbine considering its moment of inertia. *Renew. Energy* **2017**, *107*, 298–311. [CrossRef]
25. Celik, Y.; Ma, L.; Ingham, D.; Pourkashanian, M. Aerodynamic investigation of the start-up process of H-type vertical axis wind turbines using CFD. *J. Wind. Eng. Ind. Aerodyn.* **2020**, *204*, 104252. [CrossRef]
26. Bakhumbsh, S.A.; Mohamed, M.H. Effect of micro-cylinder as a passive controller on the Darrieus wind turbine performance. *Ocean Eng.* **2022**, *266*, 113118. [CrossRef]
27. Tigabu, M.T.; Khalid, M.S.U.; Wood, D.; Admasu, B.T. Some effects of turbine inertia on the starting performance of vertical-axis hydrokinetic turbine. *Ocean Eng.* **2022**, *252*, 111143. [CrossRef]
28. Nguyen, M.T.; Balduzzi, F.; Goude, A. Effect of pitch angle on power and hydrodynamics of a vertical axis turbine. *Ocean Eng.* **2021**, *238*, 109335. [CrossRef]
29. Rezaeiha, A.; Montazeri, H.; Blocken, B. CFD analysis of dynamic stall on vertical axis wind turbines using Scale-Adaptive Simulation (SAS): Comparison against URANS and hybrid RANS/LES. *Energy Convers. Manag.* **2019**, *196*, 1282–1298. [CrossRef]
30. He, J.; Jin, X.; Xie, S.; Cao, L.; Wang, Y.; Lin, Y.; Wang, N. CFD modeling of varying complexity for aerodynamic analysis of H-vertical axis wind turbines. *Renew. Energy* **2020**, *145*, 2658–2670. [CrossRef]
31. Rossetti, A.; Pavesi, G. Comparison of different numerical approaches to the study of the H-Darrieus turbines start-up. *Renew. Energy* **2013**, *50*, 7–19. [CrossRef]
32. Castelli, M.R.; Dal Monte, A.; Quaresimin, M.; Benini, E. Numerical evaluation of aerodynamic and inertial contributions to Darrieus wind turbine blade deformation. *Renew. Energy* **2013**, *51*, 101–112. [CrossRef]
33. Le, T.Q.; Lee, K.-S.; Park, J.-S.; Ko, J.H. Flow-driven rotor simulation of vertical axis tidal turbines: A comparison of helical and straight blades. *Int. J. Nav. Archit. Ocean Eng.* **2014**, *6*, 257–268. [CrossRef]
34. Joo, S.; Choi, H.; Lee, J. Aerodynamic characteristics of two-bladed H-Darrieus at various solidities and rotating speeds. *Energy* **2015**, *90*, 439–451. [CrossRef]
35. Dessoky, A.; Bangga, G.; Lutz, T.; Krämer, E. Aerodynamic and aeroacoustic performance assessment of H-rotor Darrieus VAWT equipped with wind-lens technology. *Energy* **2019**, *175*, 76–97. [CrossRef]
36. Dessoky, A.; Lutz, T.; Bangga, G.; Krämer, E. Computational studies on Darrieus VAWT noise mechanisms employing a high order DDES model. *Renew. Energy* **2019**, *143*, 404–425. [CrossRef]
37. Clary, V.; Oudart, T.; Larroudé, P.; Sommeria, J.; Maître, T. An optimally-controlled RANS actuator force model for efficient computations of tidal turbine arrays. *Ocean Eng.* **2020**, *212*, 107677. [CrossRef]
38. Su, J.; Chen, Y.; Han, Z.; Zhou, D.; Bao, Y.; Zhao, Y. Investigation of V-shaped blade for the performance improvement of vertical axis wind turbines. *Appl. Energy* **2020**, *260*, 114326. [CrossRef]
39. Ahmad, M.; Shahzad, A.; Akram, F.; Ahmad, F.; Shah, S.I.A. Design optimization of Double-Darrieus hybrid vertical axis wind turbine. *Ocean Eng.* **2022**, *254*, 111171. [CrossRef]
40. Hashem, I.; Mohamed, M. Aerodynamic performance enhancements of H-rotor Darrieus wind turbine. *Energy* **2018**, *142*, 531–545. [CrossRef]
41. Asr, M.T.; Nezhad, E.Z.; Mustapha, F.; Wiriadidjaja, S. Study on start-up characteristics of H-Darrieus vertical axis wind turbines comprising NACA 4-digit series blade airfoils. *Energy* **2016**, *112*, 528–537. [CrossRef]
42. Mohamed, M.; Dessoky, A.; Alqurashi, F. Blade shape effect on the behavior of the H-rotor Darrieus wind turbine: Performance investigation and force analysis. *Energy* **2019**, *179*, 1217–1234. [CrossRef]

43. Bedon, G.; Betta, S.D.; Benini, E. Performance-optimized airfoil for Darrieus wind turbines. *Renew. Energy* **2016**, *94*, 328–340. [CrossRef]
44. Malipeddi, A.; Chatterjee, D. Influence of duct geometry on the performance of Darrieus hydroturbine. *Renew. Energy* **2012**, *43*, 292–300. [CrossRef]
45. Torresi, M.; Fortunato, B.; Camporeale, S.M. Numerical investigation of a Darrieus rotor for low-head hydropower generation. *Procedia Comput. Sci.* **2013**, *19*, 728–735. [CrossRef]
46. Mohamed, M. Performance investigation of H-rotor Darrieus turbine with new airfoil shapes. *Energy* **2012**, *47*, 522–530. [CrossRef]
47. Mohamed, M.; Ali, A.; Hafiz, A. CFD analysis for H-rotor Darrieus turbine as a low speed wind energy converter. *Eng. Sci. Technol. Int. J.* **2015**, *18*, 1–13. [CrossRef]
48. Balduzzi, F.; Bianchini, A.; Ferrara, G.; Ferrari, L. Dimensionless numbers for the assessment of mesh and timestep requirements in CFD simulations of Darrieus wind turbines. *Energy* **2016**, *97*, 246–261. [CrossRef]
49. Wang, Y.; Sun, X.; Zhu, B.; Zhang, H.; Huang, D. Effect of blade vortex interaction on performance of Darrieus-type cross flow marine current turbine. *Renew. Energy* **2016**, *86*, 316–323. [CrossRef]
50. Chen, J.; Yang, H.; Yang, M.; Xu, H. The effect of the opening ratio and location on the performance of a novel vertical axis Darrieus turbine. *Energy* **2015**, *89*, 819–834. [CrossRef]
51. Sengupta, A.; Biswas, A.; Gupta, R. Studies of some high solidity symmetrical and unsymmetrical blade H-Darrieus rotors with respect to starting characteristics, dynamic performances and flow physics in low wind streams. *Renew. Energy* **2016**, *93*, 536–547. [CrossRef]
52. Mabrouk, I.B.; Hami, A.E.; Walha, L.; Zghal, B.; Haddar, M. Dynamic vibrations in wind energy systems: Application to vertical axis wind turbine. *Mech. Syst. Signal Process.* **2017**, *85*, 396–414. [CrossRef]
53. Zamani, M.; Maghrebi, M.J.; Varedi, S.R. Starting torque improvement using J-shaped straight-bladed Darrieus vertical axis wind turbine by means of numerical simulation. *Renew. Energy* **2016**, *95*, 109–126. [CrossRef]
54. Zamani, M.; Nazari, S.; Moshizi, S.A.; Maghrebi, M.J. Three dimensional simulation of J-shaped Darrieus vertical axis wind turbine. *Energy* **2016**, *116*, 1243–1255. [CrossRef]
55. Qamar, S.B.; Janajreh, I. A comprehensive analysis of solidity for cambered Darrieus VAWTs. *Int. J. Hydrog. Energy* **2017**, *42*, 19420–19431. [CrossRef]
56. Bianchini, A.; Balduzzi, F.; Bachant, P.; Ferrara, G.; Ferrari, L. Effectiveness of two-dimensional CFD simulations for Darrieus VAWTs: A combined numerical and experimental assessment. *Energy Convers. Manag.* **2017**, *136*, 318–328. [CrossRef]
57. Shaheen, M.; Abdallah, S. Efficient clusters and patterned farms for Darrieus wind turbines. *Sustain. Energy Technol. Assess.* **2017**, *19*, 125–135. [CrossRef]
58. Sobhani, E.; Ghaffari, M.; Maghrebi, M.J. Numerical investigation of dimple effects on Darrieus vertical axis wind turbine. *Energy* **2017**, *133*, 231–241. [CrossRef]
59. Shaaban, S.; Albatal, A.; Mohamed, M. Optimization of H-Rotor Darrieus turbines' mutual interaction in staggered arrangements. *Renew. Energy* **2018**, *125*, 87–99. [CrossRef]
60. Ali, S.; Lee, S.-M.; Jang, C.-M. Effects of instantaneous tangential velocity on the aerodynamic performance of an H-Darrieus wind turbine. *Energy Convers. Manag.* **2018**, *171*, 1322–1338. [CrossRef]
61. Patel, V.; Eldho, T.; Prabhu, S. Experimental investigations on Darrieus straight blade turbine for tidal current application and parametric optimization for hydro farm arrangement. *Int. J. Mar. Energy* **2017**, *17*, 110–135. [CrossRef]
62. Barnes, A.; Hughes, B. Determining the impact of VAWT farm configurations on power output. *Renew. Energy* **2019**, *143*, 1111–1120. [CrossRef]
63. Xu, W.; Li, G.; Wang, F.; Li, Y. High-resolution numerical investigation into the effects of winglet on the aerodynamic performance for a three-dimensional vertical axis wind turbine. *Energy Convers. Manag.* **2020**, *205*, 112333. [CrossRef]
64. Saha, U.K.; Jain, S. On the influence of blade thickness-to-chord ratio on dynamic stall phenomenon in H-type Darrieus wind rotors. *Energy Convers. Manag.* **2020**, *218*, 113024. [CrossRef]
65. Castelli, M.R.; Englaro, A.; Benini, E. The Darrieus wind turbine: Proposal for a new performance prediction model based on CFD. *Energy* **2011**, *36*, 4919–4934. [CrossRef]
66. Maître, T.; Amet, E.; Pellone, C. Modeling of the flow in a Darrieus water turbine: Wall grid refinement analysis and comparison with experiments. *Renew. Energy* **2013**, *51*, 497–512. [CrossRef]
67. Jacob, J.; Chatterjee, D. Design methodology of hybrid turbine towards better extraction of wind energy. *Renew. Energy* **2019**, *131*, 625–643. [CrossRef]
68. Atlaschian, O.; Metzger, M. Numerical model of vertical axis wind turbine performance in realistic gusty wind conditions. *Renew. Energy* **2021**, *165*, 211–223. [CrossRef]
69. Fluent, A. *Ansys Fluent Theory Guide*; ANSYS Inc.: Canonsburg, PA, USA, 2020; pp. 1–1026.
70. Ansys Inc. Ansys Fluent 2020 R1 User's Guide, 11.6.5. Six DOF Solver Settings, 11.6.5.1. Setting Rigid Body Motion Attributes for the Six DOF Solver. 2020. Available online: https://www.ansys.com/ (accessed on 14 February 2020).
71. Ni, L.; Miao, W.; Li, C.; Liu, Q. Impacts of Gurney flap and solidity on the aerodynamic performance of vertical axis wind turbines in array configurations. *Energy* **2021**, *215*, 118915. [CrossRef]
72. Hosseini, A.; Goudarzi, N. Design and CFD study of a hybrid vertical-axis wind turbine by employing a combined Bach-type and H-Darrieus rotor systems. *Energy Convers. Manag.* **2019**, *189*, 49–59. [CrossRef]

73. Siddiqui, M.S.; Khalid, M.H.; Zahoor, R.; Butt, F.S.; Saeed, M.; Badar, A.W. A numerical investigation to analyze effect of turbulence and ground clearance on the performance of a roof top vertical–axis wind turbine. *Renew. Energy* **2021**, *164*, 978–989. [CrossRef]
74. Mohamed, M. Criticism study of J-Shaped Darrieus wind turbine: Performance evaluation and noise generation assessment. *Energy* **2019**, *177*, 367–385. [CrossRef]
75. Mabrouk, I.B.; El Hami, A. Dynamic response analysis of Darrieus wind turbine geared transmission system with unsteady wind inflow. *Renew. Energy* **2019**, *131*, 482–493. [CrossRef]
76. Sengupta, A.; Biswas, A.; Gupta, R. Comparison of low wind speed aerodynamics of unsymmetrical blade H Darrieus rotors blade camber and curvature signatures for performance improvement. *Renew. Energy* **2019**, *139*, 1412–1427. [CrossRef]
77. Claessens, M.C. The Design and Testing of Airfoils for Application in Small Vertical Axis Wind Turbines. Master's Thesis, Delft University of Technology, Delft, The Netherlands, 2006.
78. Sun, X.; Wang, Y.; An Q.; Cao, Y.; Wu, G.; Huang, D. Aerodynamic performance and characteristic of vortex structures for Darrieus wind turbine. I. Numerical method and aerodynamic performance. *J. Renew. Sustain. Energy* **2014**, *6*. [CrossRef]
79. Raciti Castelli, M.; Ardizzon, G.; Battisti, L.; Benini, E.; Pavesi, G. Modeling strategy and numerical validation for a Darrieus vertical axis micro-wind turbine. *ASME Int. Mech. Eng. Congr. Expo.* **2010**, *44441*, 409–418.
80. Jain, P.; Abhishek, A. Performance prediction and fundamental understanding of small scale vertical axis wind turbine with variable amplitude blade pitching. *Renew. Energy* **2016**, *97*, 97–113. [CrossRef]
81. Elkhoury, M.; Kiwata, T.; Aoun, E. Experimental and numerical investigation of a three-dimensional vertical-axis wind turbine with variable-pitch. *J. Wind. Eng. Ind. Aerodyn.* **2015**, *139*, 111–123. [CrossRef]
82. Tunio, I.A.; Shah, M.A.; Hussain, T.; Harijan, K.; Mirjat, N.H.; Memon, A.H. Investigation of duct augmented system effect on the overall performance of straight blade Darrieus hydrokinetic turbine. *Renew. Energy* **2020**, *153*, 143–154. [CrossRef]
83. Scungio, M.; Arpino, F.; Focanti, V.; Profili, M.; Rotondi, M. Wind tunnel testing of scaled models of a newly developed Darrieus-style vertical axis wind turbine with auxiliary straight blades. *Energy Convers. Manag.* **2016**, *130*, 60–70. [CrossRef]
84. Jeong, H.; Lee, S.; Kwon, S.-D. Blockage corrections for wind tunnel tests conducted on a Darrieus wind turbine. *J. Wind. Eng. Ind. Aerodyn.* **2018**, *179*, 229–239. [CrossRef]
85. Pereira, T.; Batista, N.; Fonseca, A.; Cardeira, C.; Oliveira, P.; Melicio, R. Darrieus wind turbine prototype: Dynamic modeling parameter identification and control analysis. *Energy* **2018**, *159*, 961–976. [CrossRef]
86. Batista, N.; Melício, R.; Mendes, V.; Calderón, M.; Ramiro, A. On a self-start Darrieus wind turbine: Blade design and field tests. *Renew. Sustain. Energy Rev.* **2015**, *52*, 508–522. [CrossRef]
87. Tescione, G.; Ragni, D.; He, C.; Ferreira, C.S.; Van Bussel, G. Near wake flow analysis of a vertical axis wind turbine by stereoscopic particle image velocimetry. *Renew. Energy* **2014**, *70*, 47–61. [CrossRef]

fluids MDPI

Article

CFD Simulation of SCR Systems Using a Mass-Fraction-Based Impingement Model

Max Quissek *, Uladzimir Budziankou, Sebastian Pollak and Thomas Lauer

Institute of Powertrains and Automotive Technology, TU Wien, Getreidemarkt 9, Object 1, 1060 Vienna, Austria; sebastian.pollak@tuwien.ac.at (S.P.); thomas.lauer@tuwien.ac.at (T.L.)
* Correspondence: max.quissek@tuwien.ac.at; Tel.: +43-1-58801-31572

Abstract: Computational fluid dynamics (CFD) are an essential tool for the development of diesel engine aftertreatment systems using selective catalytic reduction (SCR) to reduce nitrous oxides (NO_x). In urea-based SCR, liquid urea–water solution (UWS) is injected into the hot exhaust gas, where it transforms into gaseous ammonia. This ammonia serves as a reducing agent for NO_x. CFD simulations are used to predict the ammonia distribution in the exhaust gas at the catalyst inlet. The goal is to achieve the highest possible uniformity to realize homogeneous NO_x reduction across the catalyst cross section. The current work focuses on the interaction of UWS droplets with the hot walls of the exhaust system. This is a crucial part of the preparation of gaseous ammonia from the injected liquid UWS. Following experimental investigations, a new impingement model is described based on the superposition of four basic impingement behaviors, each featuring individual secondary droplet characteristics. The droplet–wall heat transfer, depending on surface temperature and impingement behavior, is also calculated using a newly parameterized model. Applying the presented approach, the cooling of a steel plate from intermittent spray impingement is simulated and compared to measurements. The second validation case is the distribution of ammonia at the catalyst inlet of an automotive SCR system. Both applications show good agreement and demonstrate the quality of the new model.

Keywords: SCR; impingement; spray cooling; CFD; ammonia uniformity

Citation: Quissek, M.; Budziankou, U.; Pollak, S.; Lauer, T. CFD Simulation of SCR Systems Using a Mass-Fraction-Based Impingement Model. *Fluids* **2023**, *8*, 216. https://doi.org/10.3390/fluids8080216

Academic Editors: Markus Klein, Ernesto Benini and Francesco De Vanna

Received: 13 June 2023
Revised: 20 July 2023
Accepted: 22 July 2023
Published: 25 July 2023

1. Introduction

1.1. Motivation

Selective catalytic reduction (SCR) systems are a key technology for diesel engines to reduce the concentration of nitrous oxides (NO_x) in the exhaust gas. While combustion-based methods (e.g., exhaust gas recirculation, air staging) are applied in automotive and industrial applications [1], an additional aftertreatment is necessary for vehicles to comply with increasingly strict emission limits. The preparation and distribution of gaseous ammonia upstream of the SCR catalyst are crucial for an effective and efficient NO_x conversion. For the entire catalyst to participate in the reduction process, the ammonia has to be equally distributed across the catalytic brick's cross section. Otherwise, hotspots of over-stoichiometric ammonia concentration may cause ammonia slip, and in particularly lean areas, insufficient NO_x conversion can be expected. The transformation from a liquid urea–water solution (UWS) to gaseous ammonia comprises the evaporation of water and the decomposition of urea to ammonia via thermo- and hydrolysis [2]. Part of this preparation is the interaction of UWS droplets with the exhaust system's hot walls and the following vaporization. The impingement process was already shown to be strongly dependent on the droplet's kinetic energy, and the surface temperature [3,4]. As the spray impingement may cause a cooling of the wall [5,6], the impingement behavior of subsequent droplets can, in turn, be influenced. Additionally, the ammonia formation following this evaporation influences the ammonia distribution in the system. Consequently, apart from generating secondary droplets, the droplet–wall heat transfer must also be captured to predict impingement behavior accurately.

In summary, modeling the impingement behavior and the secondary droplet characteristics, on the one hand, and the associated heat transfer and wall film formation, on the other hand, are crucial aspects of SCR simulation. Observed problems with these two aspects of droplet–wall interaction and possible solutions will be the article's content.

1.2. Impingement Modeling

The first part of modeling droplet–wall interaction describes the droplet's fate after contact with the wall. The description of this interaction mainly relies on empirical models. Following experimental observations, the outcome of an impingement process is described based on numerous influencing parameters. The main parameters are the surface temperature and the droplet impact kinetics [3,6–8]. Additionally, the influence of surface parameters (e.g., roughness, porosity) or fluid parameters (e.g., viscosity, surface tension) have been reported [9–12].

The influence of the droplet's impact conditions and its stability is mostly described based on dimensionless numbers, such as Weber-number (We) or K-number [3,4,13–15]. The impingement model presented in the current paper is based on an experimental characterization using the We-number [11]. It is defined in Equation (1) with the wall-normal component of the velocity v_n, droplet diameter d, density ρ, and surface tension σ.

$$We = \frac{\rho {v_n}^2 d}{\sigma} \tag{1}$$

The temperature influence has been expressed using the absolute surface temperature or a nondimensionalized number T* based on surface temperature and liquid saturation temperature [3,4,6,13]. In the current work, absolute values are used.

Good simulation results were achieved in the past, using the established impingement models, e.g., those of Bai and Gosman, or Birkhold [2,3,16] and model extensions [17,18]. However, validation against experimental data revealed shortcomings in certain areas or under specific conditions. The thermal-induced breakup (TBU) behavior was shown to be an essential impingement regime for SCR simulation. Still, it was only described in a first approximation [18]. The necessity for further detailed analysis of this phenomenon (e.g., secondary droplets, temperature boundaries) could be derived from these results. Corresponding experimental investigations were carried out and reported [11,19]. Apart from the further analysis of TBU, experiments indicated a gradual transition between the various impingement regimes instead of step-like changes [11,17,19].

Following the observations, the available impingement models in the literature were analyzed to find a suitable representation of the results. The earlier works, especially, confined themselves to droplet–wall interaction at room temperature and focused on the formulation of a kinetic threshold between film and secondary droplet formation [14,20–23]. Another group of impingement models was designed for elevated temperatures but still below the boiling point of the droplets [24–27]. Some works describe the impingement behavior across a wider temperature range in principle very well, but without detailed information on boundaries or secondary droplets [3,15]. The Bai–Onera model [16] just as the Bai–Gosman model [3] restrict the detailed modeling to the sub-boiling temperature. Research dedicated to higher temperatures focused on the droplet behavior above the Leidenfrost point (LFP) [28]. Naber analyzed secondary droplets of single impingement events at three temperatures (at saturation point, above LFP, and at a transition temperature in between), but no regime boundaries concerning temperature are specified [29].

Kuhnke presents a modeling approach that includes the thermal effect on impingement behavior [4]. Three impingement regimes comprise the model: adhesion (representing film formation), rebound (drop size and wall film remain unchanged), and splash. The splash regime is designed to cover thermal-induced and kinetic-induced secondary atomization. The differences between the two physical effects are described to be represented by smooth transitions in secondary droplet properties [4]. However, experiments showed a clear difference in physical behavior between the two phenomena [19]. The directions of secondary droplets are different due to the different driving factors, i.e., kinetic energy, on the one hand, and intensive boiling on the other hand. It was observed that, in temperature ranges where mechanical breakup (MBU) and thermal-induced breakup (TBU) coexist, the MBU droplets have very flat ejection angles, and the TBU droplets yield very steep angles [11]. This is in contrast to the described gradual transition of droplet properties for all secondary droplets only with the impact of We-number [4]. Also, no temperature influence on secondary droplets is represented above the deposition limit. The Kuhnke model, therefore, only knows two temperature ranges, one below a dimensionless threshold temperature (T* = 1.1) and one above. The LFP temperature is not represented in the model. Comparable is the model by Senda, which uses the saturation temperature as a threshold [30].

Birkhold's droplet–wall interaction model is designed to model the impingement of UWS droplets [2]. With the Kuhnke model as a basis, also a single temperature threshold in the regime classification is assumed between a cold and a hot wall. For the application of the model to UWS, this value was changed to T* = 1.19. Compared to Kuhnke, two regimes were added: partial rebound and partial breakup. These regimes are designed as a transition with a certain mass fraction assigned to deposition. While these transition regimes are an improvement in modeling the gradual change reported in the experiments, other regime boundaries are still step-like threshold values. In the visualization of spray impingement, a fine aerosol is reported for UWS at 508 K [2]. This is in good agreement with the effect of thermal-induced breakup at this temperature shown in [11]. The reported underestimation of secondary droplets' spray penetration in the axial direction from the wall in the simulation results could be caused by the missing specific TBU regime in the model. The boundaries of Birkhold's regime map were confirmed by Börnhorst, but no characteristics of secondary droplets are given [9].

To the best of the authors' knowledge, no existing impingement model represents the thermal-induced breakup regime as observed and described in previous experimental investigations [11]. Especially the detailed maps about the gradual increase and decrease of TBU influence with changing surface temperature are not yet represented.

1.3. Droplet–Wall Heat Transfer

The second important aspect of the droplet–wall interaction is calculating heat transfer during the contact time. Depending on the surface temperature, the boiling behavior, and the energy transfer vary significantly [3,6,31]; see Figure 1.

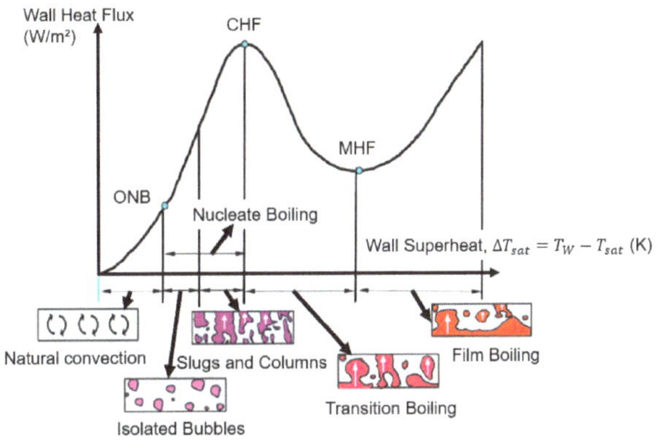

Figure 1. Nukiyama pool boiling curve for water at atmospheric pressure. The onset of nucleate boiling (ONB), critical heat flux (CHF), and minimum heat flux (MHF) are plotted. Reprinted with permission from Elsevier [32]; permission conveyed through Copyright Clearence Center, Inc..

Three significant temperatures are essential for the description of the boiling behavior. First, the fluid's saturation temperature T_{sat}, above which boiling phenomena start to become relevant, and the onset of nucleate boiling (ONB) can be observed. Second, the critical heat flux (CHF) temperature denotes the temperature at which the most vigorous heat transfer between solid and liquid is observed due to the intense formation of vapor bubbles [32]. Third, at the Leidenfrost point (LFP), the vapor film between solid and liquid closes. Consequently, there is no more direct liquid–solid contact and heat transfer is strongly reduced to the minimum heat flux (MHF) by the insulating vapor layer, as shown in Figure 1 [6,32].

Due to this substantial heat flux reduction, the insulation caused by the vapor film is sometimes assumed to be ideal, and the heat transfer above LFP is neglected [3]. However, investigations have shown that, especially for small droplets, this heat transfer is significant [5]. Two phases can be distinguished in this process: a stage of direct liquid–solid contact and another one with reduced heat transfer through a vapor layer [5]. More details are given in Section 2.2.1. In contrast to most impingement models that do not directly cover the heat transfer [3,4,16], a respective modeling approach is discussed and presented in [2]. The proposed model is partly based on investigations by Wruck [5]. For temperatures below the LFP, the Nukiyama approach is used by the author. However, it is stated that the Nukiyama pool boiling correlation is, in the strict sense, only valid for quasi-stationary droplet–wall contact and not for a dynamic impingement scenario [2]. The energy transferred from the wall to the droplet increases the droplet's fluid temperature. In the Wruck heat transfer model, the energy directly evaporates the liquid contact layer of the contacting droplet. Considering the significant influence of heat transfer on the droplet–wall interaction behavior, it seems appropriate to include a heat transfer model directly in the impingement model. Especially at high load points in diesel engines, a temperature in the Leidenfrost range can coincide with high UWS injection mass flows. These conditions can cause significant cooling of the components of the SCR mixing section [33] and, consequently, a sudden increase in film mass in the system when the temperature falls below the LFP. As a possible precursor for deposit formation [34], this phenomenon is crucial in the development of SCR systems. To enable an analysis of the deposition risk, this temperature behavior needs to be captured correctly in the complete temperature range of SCR systems. An established impingement heat transfer model from Wruck was already available. However, experiments showed that the underlying droplet–wall contact changes with the impingement regime. The measured contact time for lower We-numbers allocated to the rebound regime did not match the correlation used by Wruck. The contact area, derived by Wruck from geometrical considerations, can be replaced by well-validated measurement results. A respective adaptation and parameterization of the Wruck approach to UWS and SCR conditions seems a promising approach to obtain an appropriate model.

The scope of the current paper is to present a new impingement model specifically focused on the simulation of urea-based SCR systems. Some experiments that will serve as a basis and part of the resulting data were already presented in [11]. Additional analysis regarding secondary droplets and parameters for a heat transfer calculation is shown in the present report. Finally, with the help of two validation cases, the model's capability will be demonstrated.

2. Materials and Methods

Modeling impingement comprises two major parts. On the one hand, the outcome of the impingement event for the droplet must be described, i.e., the formation of a fluid film and the generation of secondary droplets with velocity, direction, and diameter. On the other hand, heat transfer during the droplet–wall interaction is essential. Consequently, in the first section, the distribution of the droplet mass among the impingement regimes is described using a mass-fraction-based classification. The following section deals with heat transfer modeling, while the third part of this chapter describes the resulting secondary droplets. The commercial code Siemens Star-CCM+ v 17.02.008 is used to implement and test the presented impingement model. However, the general findings and approaches are not limited to a specific code.

The last part will present experiments that serve as validation cases for the developed modeling approach. First, a case demonstrating the heat transfer simulation will be described. Then, the model capabilities to predict ammonia uniformity at the catalyst will be shown for an automotive SCR system.

2.1. *Mass-Fraction-Based Impingement Modeling*

Experimental investigations have shown that the impingement behavior of droplets on hot surfaces can be described using four basic impingement behaviors, shown schematically in Figure 2. Each impingement event either corresponds to one of these regimes or can be described as a mass-fraction-weighted superposition [11].

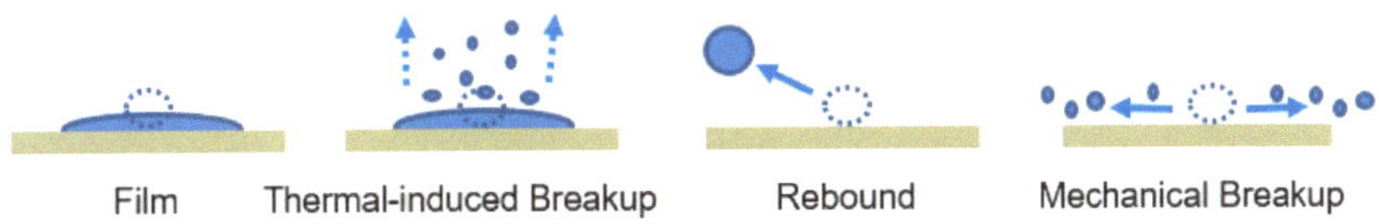

Figure 2. Four basic regimes of impingement behavior with characteristic secondary droplets. The dotted circle represents the impinging droplet at the time of first contact with the wall. The arrows show the direction of the secondary droplets [11].

A detailed description of the impingement regimes was given in a recently published paper on experimental investigations [11]. Nonetheless, a short recapitulation will be provided here to facilitate the comprehension of the following modeling approach. After coming into contact with a surface, a droplet may stay attached to form a liquid film (film regime). If the surface temperature is sufficiently high, the film starts to boil almost immediately, and the formation and bursting of vapor bubbles release a fine spray. The secondary droplets generated from this thermal-induced breakup (TBU regime) are ejected from the film almost perpendicular. In case of a rebound, the spread-out droplet retracts again and lifts off the surface like a solid elastic ball (rebound regime). This behavior is enabled by the insulating vapor cushion, which forms between liquid and wall due to very high surface temperatures. With increasing impact energy, the forces spreading the liquid on the surface overcome its cohesive forces, and the droplet disintegrates in a mechanical breakup (MBU regime). The resulting secondary droplets are ejected in all directions from the impact location with a slight elevation angle.

The allocation of droplet mass to the four regimes is based on We-number and surface temperature T_w. Therefore, when a droplet contacts the wall, the relevant data, e.g., temperatures, diameter, density, velocity, liquid composition from urea and water fraction, and surface tension of the impinging droplet, are extracted and stored. The We-number of the droplet is calculated based on the impingement data.

In CFD simulations of SCR systems, sprays comprising a large number of droplets have to be modeled. A Lagrange parcel approach is used to reduce the numerical effort. This method uses parcels without a numerically resolved diameter, each representing multiple droplets with the same properties (e.g., diameter, density, temperature). The interaction with the surroundings (e.g., drag, gravity, evaporation) is modeled based on these parameters. On the one hand, this reduces the number of droplets to be calculated significantly. On the other hand, it fortifies the demand to correctly represent also mixed impingement behavior, as each parcel represents more mass than a single droplet, thus, increasing its influence on the general result. A set of four impingement maps (e.g., Figure 3) for a given surface roughness is stored in the code as tables. Using the extracted impingement parameters, the relevant mass fraction x_{i_map} for each regime i is then read from the tables. A set of impingement maps for a surface roughness of $R_z = 7$ µm is given in Figure 3 [11].

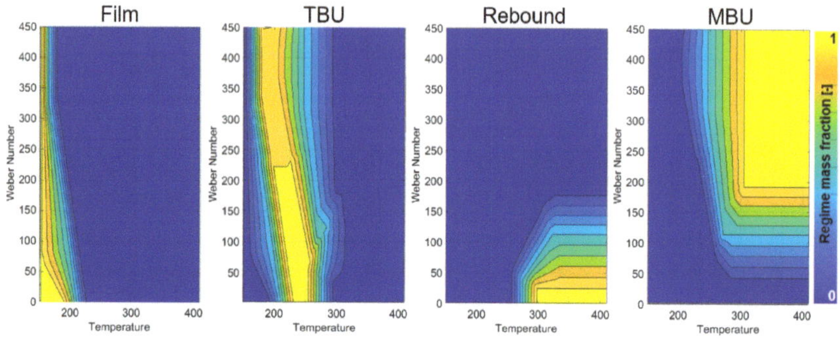

Figure 3. Impingement maps for $R_z = 7$ μm, showing the mass fractions in the four basic regimes over surface temperature and We-number. (**Left**): regime film. (**Middle left**): regime thermal-induced breakup (TBU). (**Middle right**): regime rebound. (**Right**): regime mechanical breakup (MBU) [11].

At low temperatures, an impinging droplet forms a wall film. From 150 °C, an increase in temperature causes the film regime to fade out and the mass fraction of TBU to increase (Figure 3: left and middle left). With a further rise in temperature towards 300 °C, the Leidenfrost effect gains influence, reducing the intensity of boiling and TBU (Figure 3: middle left). At high temperatures, rebound and MBU become the dominant regimes (Figure 3: middle right and right). The mass contained in each parcel is divided among those four impingement behaviors according to the mass fractions shown in the plots. For example, at 200 °C and We-number 50, approximately half of the parcel mass belongs to the film regime, while the other half is assigned to the TBU regime.

The given maps represent the interaction between droplets and a dry surface. Consequently, some extensions are necessary in case of impingement on pre-wetted surfaces. Although not studied in detail during earlier experiments [11], observations were made regarding the effect of a wet surface on the rebound probability. A droplet impinging at a sufficiently high temperature with a low We-number typically undergoes rebound. If two droplets hit the exact location within a short time interval, the second impinges on the spread-out first droplet, i.e., a wet wall. In this case, the droplet that would usually rebound undergoes TBU. So no rebound could be observed from wet surfaces, and, consequently, if a wet wall is detected on impingement (film thickness > 1 μm), any mass fraction assigned to the rebound is shifted to the TBU regime instead.

In case a droplet impinges on an already existing thicker film, an additional mass fraction is shifted toward the formation of secondary droplets. This film splash is modeled using a correlation introduced by Bai and Gosman [3]. As the threshold for a minimal film thickness above which the film splash is active, the dimensionless film thickness is implemented as given in Equation (2) with absolute film thickness h and primary droplet diameter d. Above a threshold of $h^* = 0.1$, film-splash is calculated, i.e., when the film is thicker than 10% of the incoming droplet diameter.

$$h^* = \frac{h}{d} \tag{2}$$

Additionally, the droplet needs sufficient kinetic energy to form secondary droplets and, therefore, has to fulfill the criterion given in Equation (3) [3].

$$We > We_{crit} = 1320 \cdot La^{-0.18} \tag{3}$$

The Laplace-number La is based on droplet density ρ, surface tension σ, and dynamic viscosity μ according to Equation (4).

$$La = \frac{\rho \cdot \sigma \cdot d}{\mu^2} \tag{4}$$

In case all criteria are met, the film-splash mass fraction x_{fS} is calculated as in Equation (5).

$$x_{fS} = 0.2 + rand(0.8) \tag{5}$$

In the original formulation, the random share can reach up to 0.9, which, in total, would yield a mass fraction above 1, meaning that an incoming droplet can strike more liquid out of the existing film than its own mass. As the presented impingement model implementation cannot reduce the existing film mass, this random share was reduced to the given 0.8. This way, no more mass can form secondary droplets than impacted with the primary one. This ensures mass conservation. The calculated mass fraction is directly assigned to the MBU regime. The remaining mass of the impinging droplet enters the calculation procedure for dry surfaces.

As a part of the droplet evaporates during contact with the wall, this part of the mass must be deducted before calculating secondary droplets. The calculation of the heat transfer leading to this evaporation will be described in the next section. To deduct the respective mass, the transferred energy has to be translated into a specific mass. Therefore, the enthalpy of vaporization is needed. As described in [2], the enthalpy of vaporization and the reaction enthalpy from urea to ammonia can be combined in a single-step process. This lumped enthalpy is used to calculate the vapor mass fraction x_{i_vapor} of the droplet in regime i. The resulting mass is then deducted from the secondary droplet mass and added to the film fraction. This approach forces the respective liquid mass to form a liquid film on the hot surface. There it boils and evaporates, generating the desired vapor and cooling the surface. The resulting regime mass fractions are defined as given in Equations (6)–(9).

$$x_{film_{final}} = \left[x_{film_{map}}\left(1 - x_{fs}\right)\right] + \sum x_{i_{vapor}} \tag{6}$$

$$x_{rebound_{final}} = \left[x_{rebound_{map}}\left(1 - x_{fs}\right)\right]\left(1 - x_{rebound_{vapor}}\right) \tag{7}$$

$$x_{MBU_{final}} = \left[x_{MBU_{map}}\left(1 - x_{fs}\right) + x_{fs}\right]\left(1 - x_{MBU_{vapor}}\right) \tag{8}$$

$$x_{TBU_{final}} = \left[x_{TBU_{map}}\left(1 - x_{fs}\right)\right]\left(1 - x_{TBU_{vapor}}\right) \tag{9}$$

2.2. *Droplet–Wall Heat Transfer*

After the regime mass fractions are assessed, the amount of energy transferred between the droplet and the wall during the contact time needs to be calculated. As the authors conducted no direct heat transfer measurements, approaches found in the literature are used as a basis. However, as far as possible, impingement parameters used in the resulting model are taken from or validated with their own experimental data. The authors' data was gathered during impingement investigations reported in [11]. Additional analysis was performed in the scope of the current work to increase the database for modeling. The contact time was evaluated based on high-speed imaging from the first recorded frame with droplet-surface contact to that of fluid–solid separation. The time is then calculated from the number of frames and the frame duration. Regarding the contact area, the maximum length of the contact patch is evaluated. As the images are recorded in side view, only the length of the contact patch is known and not its width. Especially for very flat impingement angles, it can be expected that the form of the contact patch is not circular but elongated. As the width cannot be extracted from the images, a margin of uncertainty remains about the exact contact area between the droplet and the wall.

2.2.1. Modeling Approach

Extensive investigations of heat transfer during droplet–wall interaction were performed by Wruck [5]. These insights are especially beneficial as they cover heat exchange under Leidenfrost conditions. Measurements of the contact temperature during impact on the impingement surface were performed with a very high temporal resolution. A modeling approach is proposed, divided into two heat transfer phases. Figure 4 shows the basic idea with a short phase of direct contact between liquid and solid (Figure 4 (a): 't_2', (b): 'Sudden contact') with a very steep temperature gradient (Figure 4b: 'Boiling retardation'). The liquid contact layer is abruptly evaporated from the energy transferred during this direct contact phase and forms a vapor layer between the droplet and the wall (Figure 4 (a): 't_3'). In the second phase of vapor contact, the surface cools down further until the contact ends with the droplet lifting off (Figure 4 (a): 't_4', (b): 'End of collision'). The reported results show that the smaller the droplet, the more crucial the first direct contact phase is for the overall heat transfer [5].

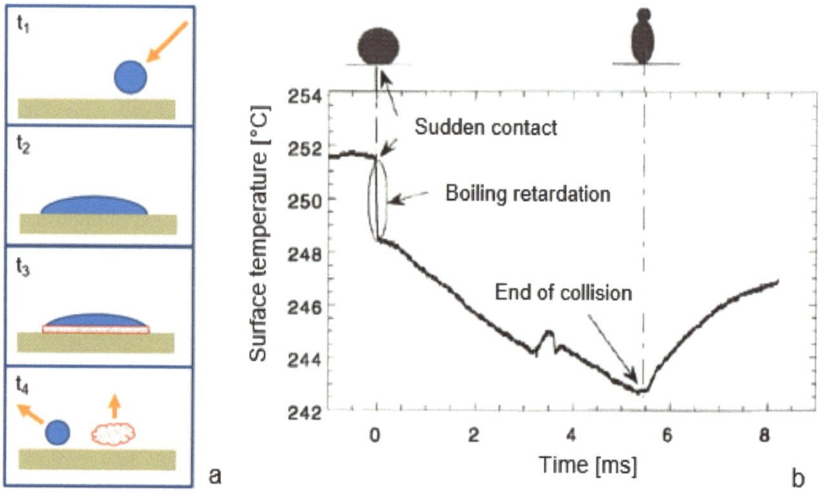

Figure 4. Phases of heat transfer during droplet–wall interaction: (**a**) Schematic of phases. (**b**) Evolution of contact temperature during impingement (adapted from [5]).

During the direct contact phase, the heat transfer is calculated based on the heat equation and the contact temperature T_c, as described in detail by Wruck [5] and summarized in the following section starting with Equation (10).

$$T_c = T_w + \frac{b_f}{b_f + b_w}\left(T_f - T_w\right) \tag{10}$$

Here, T_w and T_f are the temperatures of the wall and the fluid before contact, and b_f and b_w are the respective thermal effusivities. The thermal effusivity of a material i can be calculated based on density ρ, thermal conductivity λ, and heat capacity c, using Equation (11).

$$b_i = \sqrt{\lambda_i c_i \rho_i} \tag{11}$$

Finally, with the resulting contact temperature, the heat transfer Q_{dc} during the direct contact phase is calculated following Equation (12) [5].

$$Q_{dc} = 2A_{dc}\lambda_w(T_w - T_c)\frac{{t_{dc}}^{0.5}}{(\pi\alpha_w)^{0.5}} \tag{12}$$

Here, A_{dc} is the contact area between fluid and solid, the thermal diffusivity of the wall material is α_w, and the duration of direct contact is t_{dc}. The length of the direct contact $t_{dc} = 22$ µs was assessed in [5] experimentally from the duration of the very steep initial gradient of the contact temperature shown in Figure 4b.

Another important finding from these experiments by Wruck regards the calculation of the fluid effusivity from steady-state thermodynamic data as given in Equation (11) [5]. This approach does not yield correct results for the highly overheated liquid resulting from a droplet–wall interaction at high surface temperatures. The liquid effusivity was calculated based on the contact temperature measurements, and the graph in Figure 5 was presented.

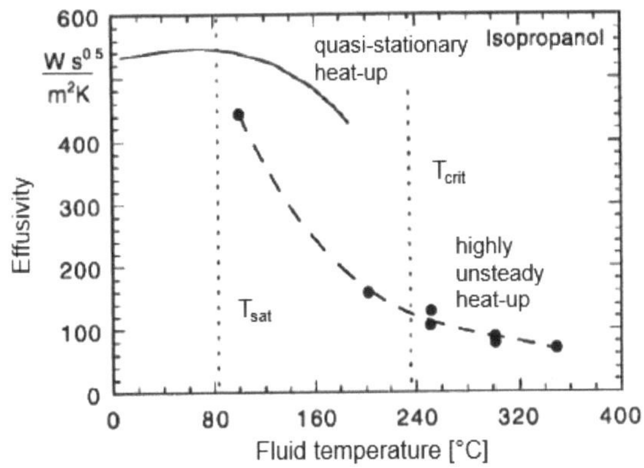

Figure 5. Influence of the overheated state of the liquid on the thermal effusivity during the interaction with a hot surface (adapted from [5]).

For the highly unsteady heat-up observed during the droplet–wall interaction, the effusivity is by a factor lower than in the stationary case. The presented data were measured using isopropanol. Unfortunately, no comparable data could be found for other liquids. The stationary effusivity $b_{UWS_{stat}}$ can be calculated from available UWS fluid properties, Equation (11). On the contrary, the exact behavior in over-heated conditions can only be guessed and leaves a degree of uncertainty. Hence, only a comparable trend was implemented for the unsteady effusivity of UWS. Starting 100 °C above saturation temperature T_{sat}, a gradual reduction of the stationary value up to a factor of 4 is implemented to determine $b_{UWS_{unsteady}}$ as shown in Equation (13).

$$b_{UWS_{unsteady}} = \begin{cases} b_{UWS_{\text{stat}}} & \text{for } T_f < T_{\text{sat}} + 100 \\ b_{UWS_{\text{stat}}}\left(1 - \left(\frac{0.75}{100}\right)\left(T_f - (T_{\text{sat}} + 100)\right)\right) & \text{for } T_{\text{sat}} + 100 < T_f < T_{\text{sat}} + 200 \\ 0.25 b_{UWS_{\text{stat}}} & \text{for } T_{\text{sat}} + 200 < T_f \end{cases} \quad (13)$$

In the vapor contact phase, the vapor layer's thickness influences the heat flux between droplet and wall, and the transferred heat is calculated using Equation (14) [5].

$$Q_{vc} = \int_{t_{dc}}^{t_{vc}} A_{vc}\lambda_v(T_w - T_{sat})S_v{}^{-1}dt \quad (14)$$

Here, A_{vc} is the contact area, t_{vc} is the end of the vapor contact phase, and S_v is the vapor film thickness. The vapor contact heat and the direct contact heat yield the total amount of transferred energy.

For the heat transfer calculation, contact area and contact time are necessary. The parameters were analyzed individually for droplet–wall interaction in each regime.

2.2.2. Analysis of Droplet–Wall Contact

Contact area

The contact area between the droplet and the surface is governed by the influence of the droplet kinetics during impingement. It was shown in several studies [35,36] that for different liquids and ranges of droplet diameters, the spreading factor of a droplet on the surface can be described based on the We-number as given in Equation (15) [35].

$$\frac{d_{max}}{d_0} = 1 + 0.23We^{0.5} \tag{15}$$

Although the data from the current investigation are limited to the contact patch length l_{max}, a good correlation with the given formula can be shown in Figure 6 for droplets in the TBU regime. While the data are scattered, the linearization of each data set matches very well with the calculated reference data. The experimental results for the other regimes can be found in the appendix in Figure A1 and are equally well described by the given correlation.

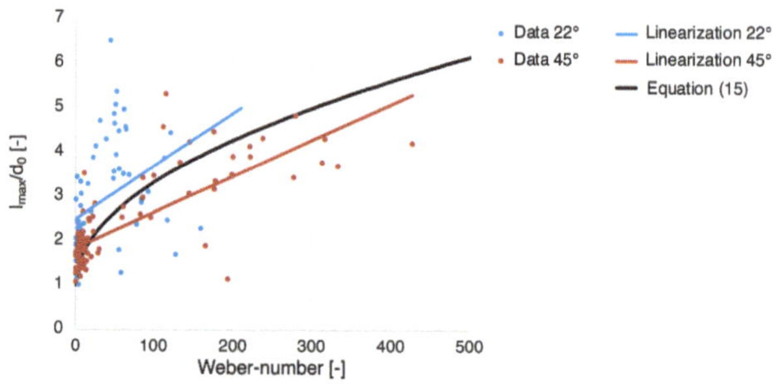

Figure 6. Ratio of contact patch length l_{max} to droplet diameter d_0 over Weber-number. Impingement data from droplets in the TBU regime with 22° and 45° impingement angle and their linearization. The function from Equation (15) [35] is added as a reference.

Contact time

The contact times between the droplet and the wall differ depending on the impingement regime. According to earlier investigations, the contact time of a droplet during rebound is strongly dependent on droplet size [13,15,37]. It can be described by the Rayleigh time t_{Ray}, which is derived from the period of a freely oscillating drop and given in Equation (16) [5,6].

$$t_{Ray} = \frac{\pi}{4}\sqrt{\frac{d^3 \rho_l}{\sigma_l}} \tag{16}$$

However, data found in the current study did not show this clear dependence, as seen in Figure 7a. In spite of this, a correlation given below in Equation (17) was applied, which was initially intended for droplets undergoing a breakup, and considers the influence of the We-number. A much better agreement could be found, as shown in Figure 7b. A droplet diameter of 125 µm was used in the calculation to compare experimental data and the given formula. This droplet size corresponds to a Rayleigh time of 150 µs and yields the reference data depicted as a dashed black line in Figure 7b. For an estimation of the influence of the diameter on the comparison, the contact time was also calculated for droplets with 60 µm (Rayleigh time 50 µs, solid black line) and 200 µm (Rayleigh time 300 µs, dotted black line). This range of diameters was chosen as it represents the experimental conditions quite well.

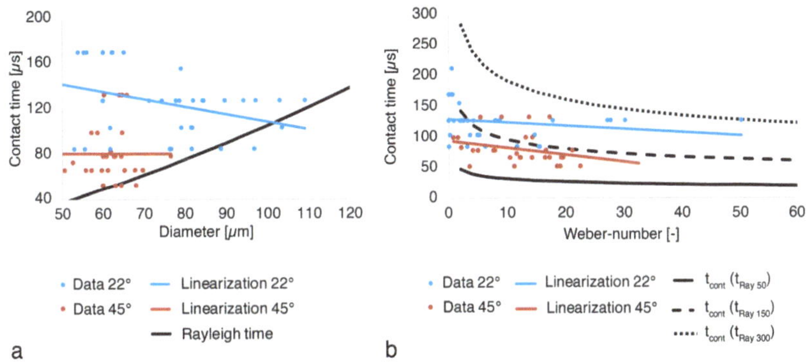

Figure 7. Contact time for the rebound regime: (**a**) As a function of diameter with Rayleigh time (Equation (16)) [5] not matching the measurement data. (**b**) As a function of We-number with Equation (17) for Rayleigh times 50, 150, and 300 μs [5] and much better agreement.

The contact time in the case of MBU is limited by the ripping of the fluid lamella and, thus, is influenced by the kinetic parameters of the droplet. In the literature [5], a correlation was proposed based on the Rayleigh time but extended by a kinetic factor. This correlation is given in Equation (17).

$$t_{cont_{MBU}} = t_{Ray}\frac{2}{\sqrt{\pi}}We^{-0.25} \tag{17}$$

The plot description for Figure 8a is equivalent to that given for Figure 7b. Droplet diameters and corresponding Rayleigh times are larger to represent better the droplets in the MBU regime: 150 μm with a Rayleigh time of 200 μs (dashed line) and 280 μm with a Rayleigh time 500 μs (dotted line).

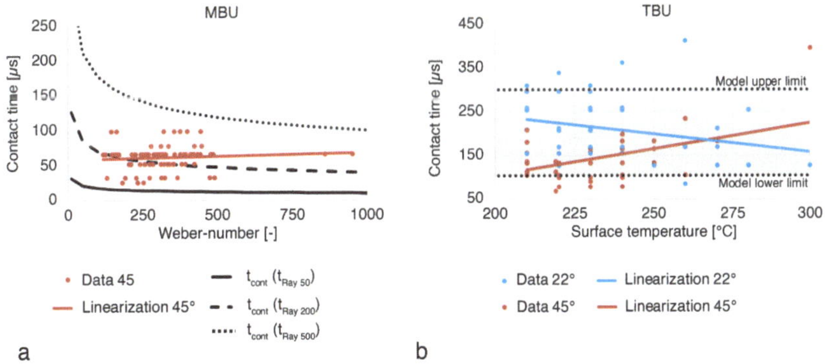

Figure 8. Contact time: (**a**) MBU regime with Equation (17) for Rayleigh times 50, 200 and 500 μs [5]. (**b**) TBU regime.

As a droplet associated with the TBU regime sticks to the surface, the period of a freely oscillating droplet is not regarded as a characterizing parameter. The boiling behavior and, consequently, the surface temperature are assumed to influence the contact time for these droplets. Hence, the evaluation of contact time is shown based on surface temperature in Figure 8b. No apparent influence of the surface temperature or any significant trend could be identified in the data. However, most evaluated data points could be found between 100 μs and 300 μs. More details on the implementation can be found in the following section.

2.2.3. Droplet–Wall Contact Model

Rebound: The total contact time for rebound is calculated using Equation (17). As the spreading of the droplet on the surface is a transient process, the contact area changes during the droplet–wall interaction. For a rebounding droplet, the maximum contact area is reached after half of the contact time, as the droplet then starts to retract again. This maximum area is calculated from the droplet diameter and the spreading factor according to Equation (15). As no boundary is given in the literature for this correlation, an upper limit has to be implemented to prevent unrealistically high values. This limit was chosen so that the film thickness resulting from the calculated area cannot fall below 1 μm. From half of the total contact time, when the maximum diameter is reached, the diameter increase per time is calculated. Hence, the contact diameter and area at every given point in time during the spreading phase of the impinging droplet can be calculated.

MBU: The contact area in the MBU regime can be calculated similarly, with the only difference being that the maximum diameter is reached at the end of the contact time when the lamella starts to rip apart at its peak diameter.

TBU: Following the experimental results, the contact time for TBU is chosen randomly in a range between 100 μs and 300 μs. If the mass fraction of rebound and MBU combined are over 0.9, then the contact time for TBU is shortened. This approach is derived from experimental observations that for high temperatures and small mass fractions of TBU, only very short bursts of secondary TBU droplets appeared. Consequently, also the time for heat transfer must be limited. A possible explanation is the abrupt evaporation of the contact layer during direct contact and the corresponding substantial increase in volume. This expanding vapor was reported in the literature [35] to create a high-velocity vapor flow from beneath the liquid at the rim of the lamella. For very high shares of MBU and rebound, i.e., at temperatures above 300 °C, this vapor expansion is assumed to stop all direct interaction between liquid and wall. Consequently, also TBU ends after the boiling retardation of 22 μs. The contact area for TBU is calculated from the spreading factor and droplet diameter. As the film remains on the surface and the initial spreading phase is short compared to the total contact time, the initial transient phase is neglected, and the peak area is used for calculating the heat transfer.

With these modeled contact parameters, the transferred heat is calculated, and the vaporized mass fraction of the droplet can be obtained.

2.3. *Secondary Droplet Characteristics*

After the regime mass fractions are assessed and reduced by the vapor fraction, the parameters of the resulting secondary droplets need to be calculated.

Crucial parameters of the secondary droplets were evaluated for use in the impingement model during experiments reported in [11]. The data incorporate droplet velocities and directions. As the resulting post-impingement droplets in the experiments are often very deformed, very small, and large in number, the evaluation is prone to a more significant experimental error than in the case of the primary droplets. Otherwise, the droplet detection script follows the same approach described for primary droplets [11]. Droplets from TBU could not all be evaluated individually. Instead, an approximate elevation angle of the 'spray cone' of resulting droplets and a range of secondary velocities from a few selected analyzed droplets was assessed. For rebounding droplets, the post-impingement velocities could be evaluated best. The restitution coefficients of pre- and post-impingement velocities in wall-normal and tangential directions were derived from the gathered information. Droplets resulting from MBU could also only be analyzed for some individual cases, gathering information about elevation angles and estimating energy loss during interaction with the wall.

Rebound

The post-impact velocities of a rebounding droplet can be characterized by using restitution coefficients, as was shown earlier [3,38]. The correlation proposed in [38] and given in Equation (18) for the restitution coefficient in the wall-normal direction e_{R-n} is used as a reference in Figure 9a. A good agreement can be observed.

$$e_{R-n} = 1.0We^{-0.5} \tag{18}$$

While in the literature constant values for the restitution coefficient of the wall-tangential velocity component can be found [3,39], derived from the approach for the wall-normal coefficient, the following correlation is proposed in Equation (19). A good agreement between experimental data and the newly proposed formula for the restitution coefficient in wall-tangential direction e_{R-t} can be seen in Figure 9b.

$$e_{R-t} = 1 - 1.0We^{-0.5} \tag{19}$$

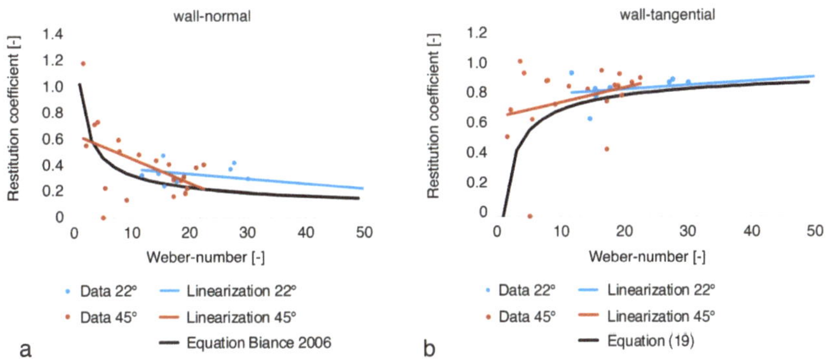

Figure 9. Restitution coefficients for a rebound: (**a**) Wall-normal direction with Equation (18) [38]. (**b**) Wall-tangential direction with proposed Equation (19).

MBU

The direction of secondary droplets from a breakup can be described using azimuthal angles α_{azi} and elevation angles α_{ele}. For the elevation angle, a dependency on the We-number is described [40]. A comparison of data gathered in the current experiments and the literature correlation (green) is shown in Figure 10a. While most data points can be found between 0.1 rad (6°) and 0.5 rad (28°), the reported We-number influence could not be confirmed under the current experimental conditions.

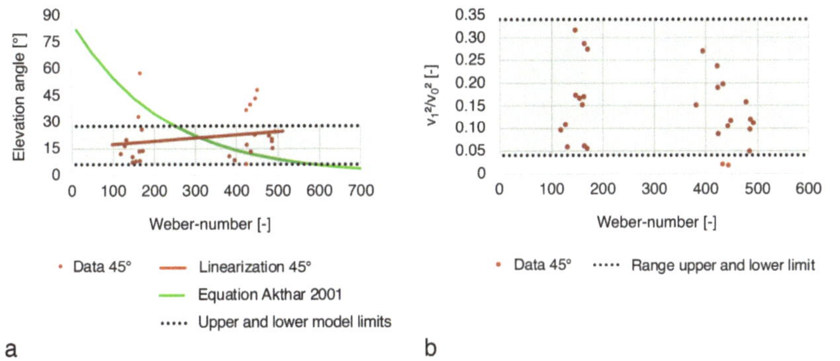

Figure 10. Secondary droplet parameters in the MBU regime: (**a**) Elevation angle with literature proposal from [40] (green). (**b**) Square of pre- and post-impingement velocity as a measure for energy loss.

The ratio of the squared secondary velocity over the squared primary velocity was analyzed regarding the secondary velocity magnitude after MBU. As shown in Figure 10b, a substantial reduction can be observed. Assuming that the overall mass before and after impingement is constant, the given data would suggest a loss of kinetic energy between 68% and 98%. In reality, a small mass fraction will be lost due to evaporation. As this can be assumed to be in the range of only a few percent, the influence is insignificant in this energy estimation. A part of the energy is not lost but transferred to the increased surface energy of the higher number of tiny droplets with a larger surface-to-volume ratio. Additionally, the secondary droplets are tiny and prone to drag-induced velocity reduction. As the velocity measurement takes a few frames, part of this drag might already be factored into the secondary velocities shown in the graph. Hence, a loss of energy from oscillations and friction between 40% and 70% is assumed. More details on the implementation and the used literature correlations can be found in the following section.

Secondary Droplet Model Implementation

All secondary droplets need a diameter, velocity components, density, temperature, mass fractions of urea and water, and an injection point within the simulation domain. As Lagrange parcels are used in the modeling, the number of droplets within each parcel must also be defined. The density is assumed to stay constant before and after the impingement. The mass fractions of urea and water are also directly carried over from the primary impinging droplet. The temperature of the secondary droplets is also assumed to match that of the incoming droplet for rebound and MBU, as heat exchange is short and, therefore, limited to the contact layer that uses all energy for evaporation. Only in the case of TBU, the secondary droplets are assumed to be at saturation temperature as they are formed from the thin walls of vapor-filled exploding bubbles. The remaining parameters are calculated individually for each regime as follows.

Film

Diameter: Film droplets keep their original diameter for the impulse transfer from droplet to fluid film to be calculated correctly.

Particle count: To meet the assigned film regime mass fraction of the droplet, the number of droplets in the parcel is reduced accordingly.

Velocity and direction: For correct input parameters in calculating momentum transfer to the resulting fluid film, velocity in magnitude and direction are taken over from the primary droplet.

Injection location: For every droplet assigned to form film, an injection location is calculated close above the surface. It is found by trailing the path of the primary droplet back a short distance from the impact location to ensure that the injected film droplet impacts within one timestep.

TBU

Diameter: The diameter of secondary droplets from TBU d_{TBU} is modeled using a correlation found experimentally by Cossali et al. [8]. It is based on the surface temperature and given in Equation (20). The surface temperature T_w is in °C, and a lower limit of 1 µm is added for stability reasons.

$$d_{TBU} = d_0 \left(0.306 - \frac{0.306}{1 + e^{\frac{T_w - 162.5}{11.5}}} \right) \tag{20}$$

Particle count: The number of droplets in the particle is calculated to meet the assigned mass fraction of TBU.

Velocity and direction: The velocity magnitude is chosen randomly between 0.4 and 8 m/s. This value is mainly driven thermally, presumably by bubble nucleation density, bubble growth rate, evaporation rate, and possibly fluid parameters that influence the thickness and stability of the lamella. Hence, a more detailed calculation would be supposedly numerically costly, and a more thorough investigation was not possible within the current project. From observations, however, the azimuthal angle of secondary droplets was identified to be randomly distributed around the impact location, and the elevation angle varied randomly between 60° and 90°.

Injection location: The injection location of TBU droplets is the impact location.

Rebound

Diameter: A rebounding droplet is a single droplet. If a part of the mass is lost, its diameter is reduced. Consequently, the diameter is calculated to meet the assigned rebound regime mass fraction.

Particle count: With the mass fraction already considered in the diameter, the number of droplets is kept as in the primary droplet.

Velocity and direction: Looking at the rebound as an elastic interaction between two bodies, the velocity can be calculated using restitution coefficients. This computation is carried out separately for the wall-normal and the tangential direction. The restitution coefficient in the normal direction is calculated using Equation (18). For the restitution coefficient in the tangential direction, the proposed correlation is implemented as given in Equation (19).

Injection location: The injection location is the impingement location.

MBU

In the case of MBU, two secondary Lagrange parcels are created from every impingement event to account for the large variety of possible droplet pathways.

Diameter: The diameter of both droplets is chosen to match half of the mass fraction of the primary droplet assigned to the MBU regime and also considers the increase in particle count due to the breakup.

Particle count: The number of secondary droplets created in the breakup is calculated following an experimentally found correlation [40], which is given in Equation (21).

$$n_{MBUsum} = (0.0427We + 10.46)x_{MBU_{final}} \tag{21}$$

As the correlation was found analyzing We-numbers of up to 750, an upper limit of 50 droplets was implemented to ensure the calculation does not yield exaggerated values for higher We-numbers. With x_{MBU_final}, the mass fraction assigned to MBU and the vaporized portion of the mass are already considered. The particle count is divided among parcel one and parcel two randomly between 30% and 70% each. Finally, for the injection of the parcels, the calculated value is multiplied by the particle count of the primary droplet to meet the conservation of mass.

Velocity and direction: The velocity magnitude of the MBU droplets is found using an energy balance. First, the kinetic and surface energy of the primary droplet is calculated. According to the MBU mass fraction, an equal fraction of the energy is assigned to the MBU droplets and divided equally among the two parcels. Then, using the already available particle counts and diameters, the surface energy of each secondary parcel is calculated and subtracted from the parcel energy. From the remaining part of the energy, a random share of 40% to 70% is assumed to be lost in oscillations and friction, as described in the experimental section. The rest is regarded as the kinetic energy of the parcel. Finally, from the available kinetic energy, the velocity is computed. If there is not enough energy for a positive velocity of the secondary droplets, i.e., if the remaining energy is zero or negative, then the mass is shifted to the film regime to remain on the surface.

Directions of the MBU parcels are defined using azimuthal and elevation angles. The azimuthal angles α_{azi} take the impingement angle α_{ele_0} into account and are calculated according to Equation (22).

$$\alpha_{azi} = (-4 + rand(8))\alpha_{ele\_0} \tag{22}$$

Defined as such, the MBU droplets can take every direction if their impingement angle is above 45°. For very small impingement angles, the secondary droplets tend to move in the direction of the primary droplet. This implementation represents the observation that for those sharp angles, a tangential velocity component of the primary droplet is carried over to secondary droplets. The statement is also confirmed by similar reports found in the literature [4].

Following the experimental data, a random distribution between 0.1 radiant (6°) and 0.5 radiant (28°) was implemented for the elevation angle α_{ele}.

Injection location: The injection location for MBU is the impingement location.

2.4. *Validation Case Surface Cooling*

2.4.1. Experimental Setup

A holistic approach is necessary to validate the heat transfer properties, as the heat transfer of single droplets was not measured directly. Consequently, surface cooling due to intermittent spray injection over several injection cycles was used as a reference. The experimental results used for comparison were recorded during an earlier project published in [41]. In the exhaust line of a heavy-duty diesel engine, an optically accessible box was integrated. The box has a quadratic cross section of 200 × 200 mm and a length of 600 mm. At the inlet and the outlet of the box, an uncoated monolith is fitted to straighten the airflow through the box and achieve fine-scale isotropic turbulence, as shown in [42]

A model of the box is visible in Figure 11 on the right-hand side. The exhaust gas enters the box from the left and streams along a stainless-steel plate positioned inside the box. On top of the box, an injector is mounted vertically targeting the hot steel plate. Starting from a steady-state temperature, intermittent injection of UWS onto the impingement surface was activated using a constant injection frequency of 1 Hz. The plate's temperature was measured at the black-varnished bottom side through zinc-selenide (ZnSe) windows using infrared (IR) thermography. More details regarding the test bench setup can be found in [41]. The recorded IR images, as shown in Figure 11 at the bottom left, are the experimental reference data. For better orientation, the impingement plate was divided into small sections, named CP3 to CP5. The average temperature values of these areas were recorded, as well as the minimum temperature observed in the impingement areas of the different spray cones. Where specified in a legend, the 'left' spray cone means left in the flow direction, i.e., in the upper right corner of the IR picture. The dark blue areas towards the edges of the images are caused by the steel surface of the box holding the round ZnSe windows. They are only displayed to be colder due to the differing radiation properties.

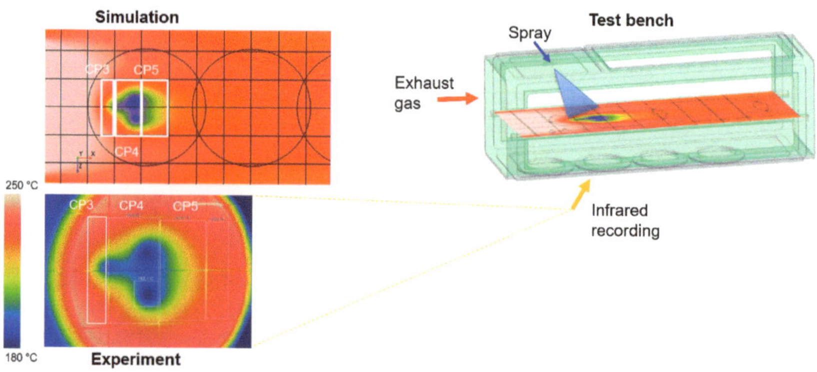

Figure 11. The validation test case for surface cooling from spray impingement. Hot exhaust gas flows into a box (**upper right**) to heat an impingement plate exposed to intermittent UWS spray. The surface temperature on the bottom of the plate was recorded with IR thermography (**lower left**) and simulated (**upper left**).

The experiments were conducted with two injectors with different Sauter mean diameter (SMD) values, cone angles, and static mass flows, as detailed in Table 1.

Table 1. Parameters of the injectors used in the validation cases [41].

Parameter	Unit	Injector 1	Injector 2
SMD	µm	84	178
Static mass flow	kg/h	3.18	7.32
Number of holes	–	3	1
Cone angle	°	6	11.5

Additionally, two different operating points of the diesel engine were used to generate the surrounding exhaust gas flow. Consequently, the influence of the gas temperature on the cooling process could be investigated. Engine conditions and corresponding UWS injection parameters are summarized in Table 2.

Table 2. Engine and injector operating parameters [41].

Parameter	Unit	Operating Point 2	Operating Point 3
Exhaust gas mass flow	kg/h	1000	1200
Exhaust gas temperature	°C	275	350
Injection mass flow injector 1	mg/s	$35.3(t_{inj} = 40$ ms)	$132.5(t_{inj} = 150$ ms)
Injection mass flow injector 2	mg/s	$10.2(t_{inj} = 5$ ms)	$20.3(t_{inj} = 10$ ms)

2.4.2. Simulation Setup

The whole experimental setup was recreated within the simulation domain using Siemens Star-CCM+ v 17.02.008. A polyhedral mesh with a base size of 2 mm and a total number of 5.2 million cells was used. The boundary layer was resolved with seven prism layers to enable a low-y^+-approach and capture the heat transfer correctly. A mesh sensitivity analysis was performed earlier as reported in [42] to select the appropriate mesh. More details on the mesh, including figures, can be found in [42].

First, the setup was simulated with an inlet cone in front of the first monolith. Here, a mass flow inlet was adopted as a boundary condition. The values for mass flow and temperature can be found in Table 2. At the outlet of the second monolith, a pressure condition is applied with ambient pressure and a temperature of 30 °C. At the outside of the box, convection is modeled with the ambient temperature and a heat transfer coefficient of 15 W/m^{2}K [43]. For the radiation heat transfer within and outside of the box, the gray thermal radiation model was adopted together with the surface-to-surface radiation model for calculating the heat exchange within the box [43]. The radiation properties of the materials are given in Table 3. Heat transfer between solid and fluid is modeled with conjugate heat transfer.

Table 3. Radiation properties of the surfaces in the box [43].

Surface	Emissivity	Reflectivity	Transmissivity
Impingement plate steel top	0.1	0.9	0
Impingement plate varnished bottom	0.96	0.04	0
Liquid film	0.85	0.15	0
Catalyst	0.9	0.1	0
ZnSe windows	0.13	0.13	0.74
Borosilicate windows	0.9	0.1	0

With this setup, the steady-state flow and temperature conditions were simulated, comparable to the experimental engine tests. These results were used as initial conditions for the following transient simulation with intermittent UWS injection. The cone was removed to reduce the mesh size and save numerical cost. The recorded temperature and velocity field in front of the first monolith was applied as an inlet condition.

The simulation setup for modeling the physical processes in an SCR system is based on the work by Fischer [44]. The turbulent gas flow is modeled with Reynolds-averaged Navier–Stokes (RANS) equations. The V2F k-ϵ turbulence-model [45,46] is used with an all y^+ treatment. Additionally to the equations for turbulent kinetic energy and turbulent dissipation, the V2F model solves transport equations for the wall-normal stress component and the elliptic relaxation parameter. This improves the prediction of heat transfer in the boundary layers [45,46]. The injected droplet size distribution was defined using a cumulative distribution function measured with laser diffraction analysis [34]. The droplet–wall interaction, including heat transfer, is simulated using the presented modeling approach. For simulation of the spray propagation, the Schiller–Naumann drag coefficient with Cunningham correction factor was adopted [34]. A liquid film model, including the surface tension effect, was selected to simulate film propagation. Evaporation and boiling were considered as well, based on a Nukiyama boiling correlation.

As a validation, the temperature–time diagram for several areas of the impingement plate is compared between simulation and experimental data, as well as the size and outline of the cooled-down area.

2.5. Validation Case Ammonia Uniformity SCR System

The homogeneous distribution of gaseous ammonia at the inlet of the SCR catalyst is crucial for efficient system performance. The ammonia uniformity index (UI) can be used to quantify this homogeneity and, thus, is an indicator for the quality of the ammonia mixing. It is defined as given in Equation (23) with the local concentration c_i, evaluated area A_i, the total area A, and the average concentration $\bar{c}$ [47]. A value of one shows a perfectly homogeneous distribution.

$$UI = 1 - \frac{\sum_{i=1}^{N} A_i |c_i - \bar{c}|}{2A\bar{c}} \tag{23}$$

2.5.1. Experimental Setup

In earlier experimental investigations, the ammonia uniformity of a heavy-duty SCR system was analyzed on an engine test bench [42]. The mixing section and the catalyst are shown in Figure 12.

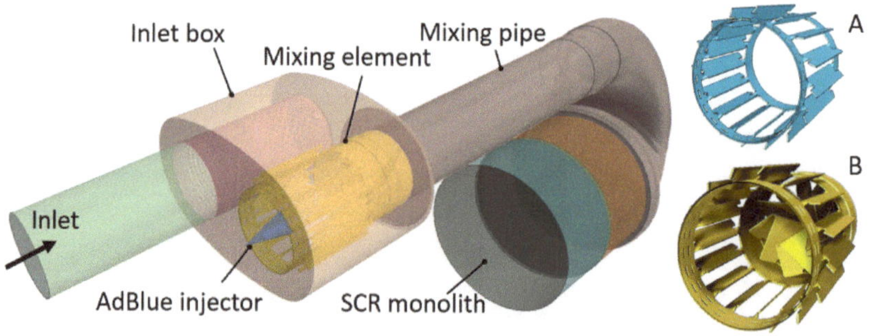

Figure 12. Geometry of the SCR system with swirl element A and additional mixing element B.

The ammonia distribution across the outlet of the catalytic brick was measured on the engine test bench using chemoluminescence detection (CLD). More details on the measurement procedure can be found in [18]. The investigation was performed at an exhaust gas mass flow of 900 kg/h and a temperature of 425 °C. A single cone injector with a spray angle of 48° was used with a measured SMD of 57.8 µm. The UWS injection rate was 694 mg/s with an injection frequency of 1 Hz. This high-load engine operating point was chosen to assess the impingement model quality, as previous results were considered unsatisfactory.

The exhaust system was analyzed with and without an additional swirl mixing element installed in the mixing section, as shown in Figure 12. This extra part is also a primary impingement surface for the spray. Consequently, impingement behavior is regarded as a crucial influencing factor.

2.5.2. Simulation Setup

The mesh consists of 7.7 million polyhedral cells with a base size of 4 mm and is depicted in Figure 13. The boundary layer is resolved with eleven prism layers to enable a low y+ approach. Pipes and mixing elements are modeled as solid parts. The cells for the catalyst are extruded from its inlet and treated as a porous medium.

At the inlet of the exhaust system, a mass flow boundary condition is applied with the mass flow from the engine test bench of 900 kg/h and an exhaust gas temperature of 425 °C. At the outlet of the catalyst, a pressure outlet boundary is applied with ambient pressure and a temperature of 30 °C. As the exhaust system on the test bench was very well insulated, the outside of the system components is modeled as adiabatic. The UWS spray is initialized using a tabled cumulative distribution function (CDF) for the droplet size and the measured values for injection mass flow, spray opening angle, and initial droplet velocity.

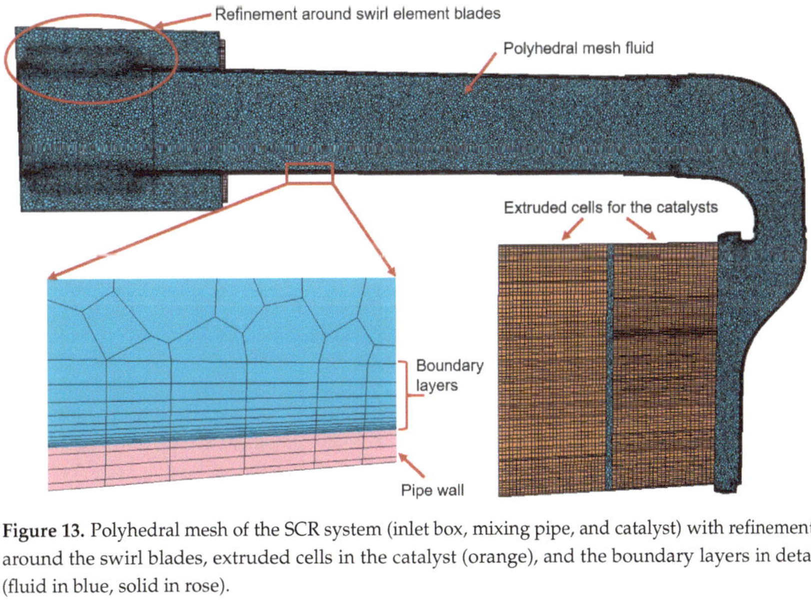

Figure 13. Polyhedral mesh of the SCR system (inlet box, mixing pipe, and catalyst) with refinements around the swirl blades, extruded cells in the catalyst (orange), and the boundary layers in detail (fluid in blue, solid in rose).

The simulation setup is, in principle, the same as described for the first validation case above. The most significant difference is the turbulence model. As the presented exhaust system has a relatively complex flow field (swirl flow in combination with a 180° pipe bend), the lag elliptic blending k-ϵ model is used [46,48]. It is recommended for complex flow cases with streamline curvature and rotation, as it takes the influence of a misalignment of the principal axes of the strain rate and the anisotropy tensor on the turbulent kinetic energy production into account [46,48]. Further, it had been successfully applied to the current geometry previously [42]. The second alteration is eliminating the radiation heat transfer model to save numerical effort. This is enabled by the already-mentioned excellent thermal insulation on the test bench.

Apart from the flow field, impingement is a main influencing factor for ammonia uniformity. Consequently, a high-quality impingement model is crucial for a correct uniformity prediction. CFD simulations using a standard impingement model (Bai–Gosman) underestimated the UI for both mixer geometries. The operating temperature of 425 °C is above the LFP and, therefore, most impingement models assume no evaporation during droplet–wall contact. This seems to cause an overestimation of droplet mass reaching the catalyst in liquid form. The consequence of its evaporation on the catalytic brick is an increased intensity of ammonia-rich areas. Especially for the mixer geometry B, which shows an excellent distribution on the test bench, the error in the simulation result is significant. In this geometry, the interaction between spray and mixer is more intense, so the effect of droplet–wall interaction and its modeling is more dominant. Thus, the presented simulation case is an essential test for the validity of the proposed modeling approach.

3. Results and Discussion

The capabilities of the droplet impingement model for simulation of the interaction between UWS droplets and hot surfaces will be shown using the two validation cases. First, the surface cooling from intermittent spray injection is shown. Second, the ammonia distribution at the cross section of the catalyst of an automotive SCR system is simulated.

3.1. Validation Surface Cooling

The plate's temperature at the beginning of injection was simulated in a steady-state simulation. The temperature distribution could be well-replicated with a maximum deviation between experimental and simulation data in the worst point of 1.6 °C (case 1), 1.1 °C (case 2), and 3.1 °C (case 3).

Case 1: operating point 2/injector 1

The distribution and location of the cooled-down area could be replicated very well, as is visible in the plots in Figure 14. The agreement of the absolute temperature drop in simulation and experiment is also acceptable for most parts of the temperature–time diagram, especially the gradient during the first 30 s. For the second half of the experiment, the simulated temperature drop underestimates the experimental values slightly. Possibly, this is because the evaporation in the simulation is slightly retarded as the evaporating mass has to be injected onto the surface first. In reality, it evaporates instantly, as shown in [5]. Consequently, film thickness might be overestimated somewhat, causing the influence of the film splash to gain traction at some degrees Celsius higher than in reality. After 77 s the minimum temperature is 185.5 °C in the simulation case compared to 175.8 °C in the experiment.

Case 2: operating point 2/injector 2

Using injector 2 with a single hole and larger SMD, again, an excellent agreement for the cooled area and location can be achieved. Figure 15 shows how the initial gradient of the temperature is also captured well. In this case, however, the value at which the cooling eases off to reach a steady value is lower than expected from experiments. The threshold value for minimal film thickness, above which film splash is active, is a ratio to the droplet diameter. Consequently, for the larger SMD of injector 2, the threshold thickness tends to be larger than for the smaller SMD of injector 1. Hence, an influence can be expected only above a larger film thickness that might take longer to build up. Possibly, the correlation for the film splash threshold needs more thorough investigation in the future. After 57 s of injection, the resulting minimum temperature is 188.5 °C compared to 201.0 °C, yielding an error of 12.5 °C.

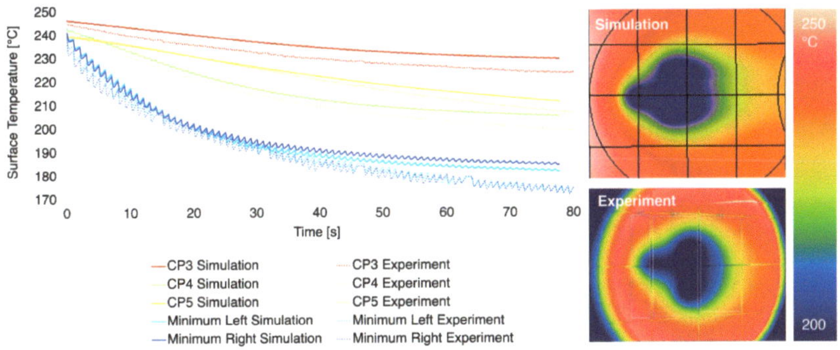

Figure 14. Temperature–time diagram of the surface cooling at engine operating point 2 with injector 1 and resulting cooled areas in simulation and experiment after 60 s.

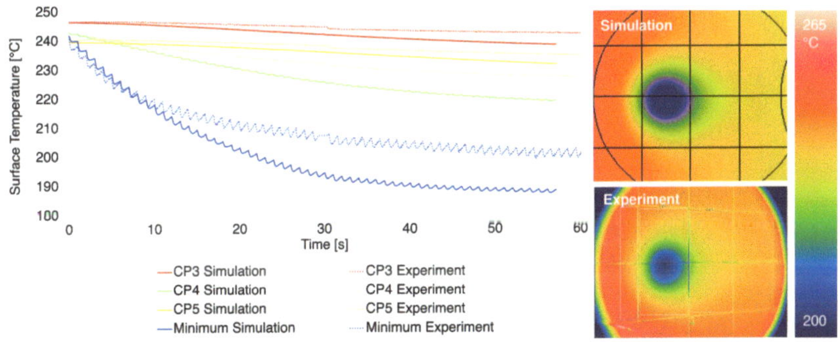

Figure 15. Temperature–time diagram of the surface cooling at engine operating point 2 with injector 2 and resulting cooled areas in simulation and experiment after 50 s.

Case 3: operating point 3/injector 1

The last presented case depicted in Figure 16 is an example of the operating point 3 with a starting temperature of the surface above 300 °C. Here, the much slower cooling at the high-temperature regimes MBU and rebound at the beginning compared to the cooling effect from TBU can be observed. The resulting flat temperature gradient during the first 25 s is predicted accurately. A peculiarity of this experimental data set is the substantial deviation between the cooling from the left and the right-hand side injector spray cone. The plot shows that the shift from slow to fast cooling occurs much earlier in the right impingement area than on the other side. A more detailed analysis of the injector during the investigations in [34] revealed a difference in mass flow through the individual injector holes. Although this mass flow difference was considered in the modeling process, the strong influence could not be reproduced. Apart from the mass flow, other injector parameters may have been altered for the respective injector hole, e.g., the SMD, due to a partial blockage of the orifice. However, these parameters were not investigated at the time, and more detailed data on the injector are unavailable. A possible influence can only be assumed but not represented in the simulation model. Although not as abrupt as in the recordings, the shift from the slow cooling at high temperatures to the steeper temperature gradient from TBU could be replicated in the simulation. Finally, the transition from fast back to slower cooling also agrees well with the experimental data. 75 s after the start of injection, the temperature minimum in the left spray cone (light blue) in the experiment is 212.5 °C and in the simulation 191.1 °C. However, the relatively large error of 21.4 °C is primarily caused by the later cool-down in the experiment. For the impingement area of the right spray cone, the surface temperatures agree well with 189.8 °C compared to 191.5 °C, yielding an error of only 1.7 °C.

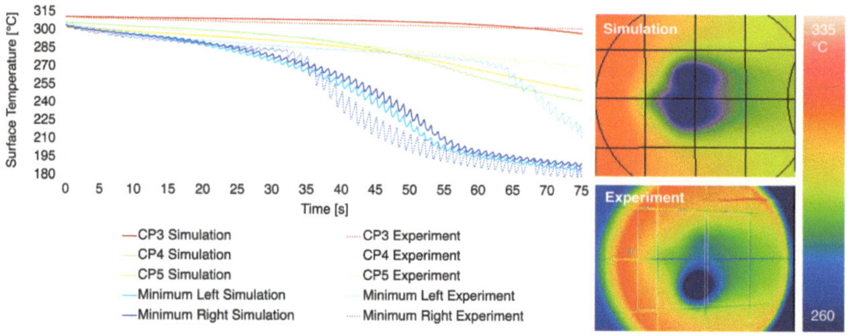

Figure 16. Temperature–time diagram of the surface cooling at engine operating point 3 with injector 1 and resulting cooled areas in simulation and experiment after 50 s.

3.2. *Validation Ammonia Uniformity*

Figure 17 shows the measured ammonia distribution at the SCR catalyst outlet on the engine test bench. On the left, the result with only the swirl element (Figure 12: Mixer A) shows a UI of 93.8%. On the right, the excellent distribution with the additional mixing element (Figure 12: Mixer B) can be seen, yielding a UI of 98.9%.

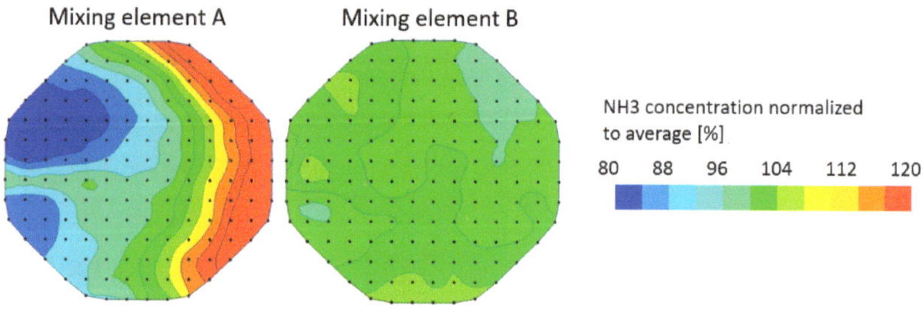

Figure 17. Measured ammonia uniformity at the catalyst outlet on the engine test bench. **Left**: UI = 93.8%. **Right**: UI = 98.9%.

The measurement results for mixing element A show a horizontal line of average ammonia concentration (green) between the two lean areas (blue). It is caused by droplets sliding along the backside of the mixing pipe bend. They detach from the wall due to the pipe geometry, which acts like a trailing edge where the pipe opens towards the larger diameter catalyst housing after the bend, Figure 18. Superimposed on this effect, the gas flow in the inlet box encounters the spray from the side. The spray is pushed asymmetrically toward the right side of the mixing pipe. Then, following the swirling gas flow, the ammonia distribution is also not symmetrically on the catalyst inlet but shows a concentration increase from one side to another. In the case of geometry B, most droplets interact with the mixer blades before the flow can push them toward the side of the mixing pipe. Consequently, the asymmetry in the result is gone, and the uniformity improved.

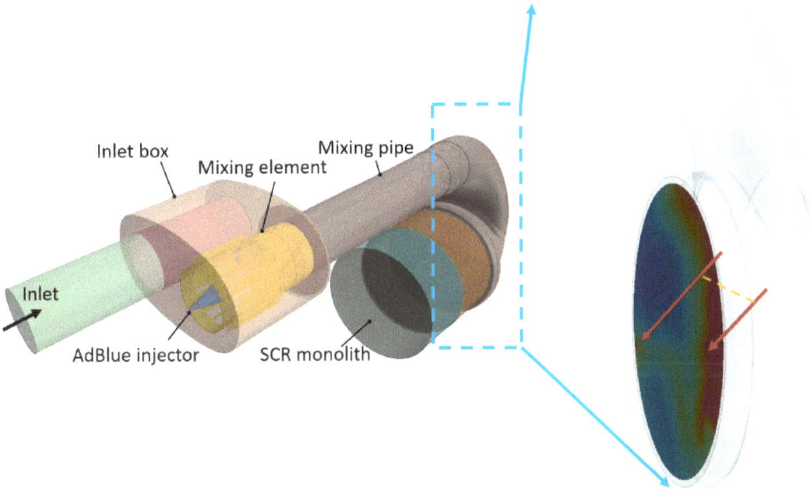

Figure 18. The ammonia-rich horizontal line across the middle of the catalyst is caused by droplets sliding along the pipe wall before impacting and evaporating on the catalyst. The pipe's opening towards the catalyst causes a trailing edge (dashed orange line) from where droplets separate (red arrows).

In Figure 19, the results of the CFD simulation with the presented impingement model are given. For variant A with swirl element, the ammonia distribution almost perfectly matches the measurement results in Figure 17. The high-concentration area on the right is captured correctly. The two lean regions on the left, with a more extensive and leaner upper section and a smaller, less intense lower area, are also well-replicated. The simulated UI of 93.6% is only 0.2%p off the measured value. The simulation result for variant B shows slight deviations from the measurements. An ammonia hot spot not seen on the test bench can be detected towards the left. Consequently, the UI of 97.4% is 1.5%p below the measurement results. Further analysis of the simulation results revealed that the hot spot is caused by UWS droplets impinging and evaporating on the catalyst inlet. The source of this ammonia-rich area is the same as described for the measurement results of mixing element A, with droplets sliding along the pipe wall before impacting the SCR brick, as depicted in Figure 18.

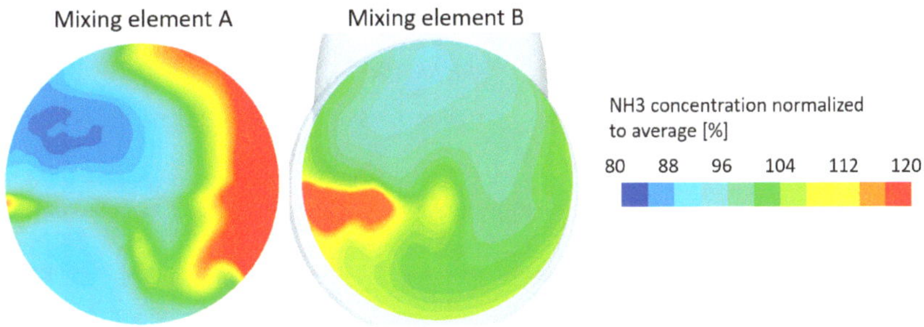

Figure 19. Simulated ammonia uniformity at the catalyst outlet. **Left**: UI = 93.6%. **Right**: UI = 97.4%.

It is assumed that the droplets slide along the pipe wall on a vapor cushion. While moving relative to the surface, the vapor cushion has to be built up constantly, causing ongoing evaporation. This effect could be enhanced by a curvature of the wall, as observed in a pipe bend, as it pushes the droplet against the wall. In the simulation, this droplet sliding consists of multiple individual impingement events to redirect the droplet path constantly. As the impingement angle of these events is very flat, the wall-normal velocity of the repeated rebounds is minimal and yields very small We-numbers. While in reality the sliding is an ongoing contact, i.e., a very long contact time, the current model results in tiny values of contact time and area. Figure A1 shows that in the region of single-digit We-numbers, the observed contact area might be significantly larger than calculated for impingement angles of 22°. For near-wall-parallel droplet tracks, this effect could be even more substantial. Consequently, the evaporation of these sliding droplets in the simulation may be underestimated.

The underestimation of these droplets' evaporation can be seen for both geometries. For mixer A, a little area with overestimated ammonia concentration can be found on the left side of the horizontal center line. This area contributes to the reported minimal deviation in UI in this case. An ammonia hot spot can also be observed in the exact location for geometry B. However, mixer B pushes the droplets much stronger towards the pipe wall, causing more droplet mass to slide along the mixing pipe wall for a more considerable distance. In the measurement, this most likely causes complete evaporation of the droplet mass. Consequently, the described underestimation of this sliding evaporation in the simulation causes a more significant error for geometry B.

Although the result is not as excellent as for variant A, the differences between the two variants of mixing elements observed on the test bench are still captured to a large extent by the simulation results. The remaining deviation between the experiment and the simulation is most likely caused by the aforementioned sliding droplets. Here, further investigations with a specific focus might be needed to enable an extension of the model for this case.

The simulation results are improved by working on the heat transfer and the impingement outcome (regime classification and secondary droplet characteristics). To investigate the weighting of each effect on the overall results, simulations were carried out with the presented impingement model without evaporation from heat transfer. The results were very similar regarding the general ammonia distribution. Only the UI was decreased for both geometries by a few tenth-%p (mixer A: UI=93.1%, mixer B: UI=97.1%). The more significant improvement is, therefore, the impingement outcome modeling. For comparison, the results using a Bai–Gosman impingement model with best-practice settings [18] are shown in Appendix A.2, Figure A2. A significant improvement in accuracy with the new impingement model can be observed in the ammonia distribution and the resulting UI. The UI-deviations from the measurement results with the new model (0.2%p and 1.5%p) are significantly smaller than previously (6.9%p and 5.3%p).

4. Conclusions

The article engages in modeling droplet–wall interaction. A particular focus is on the case of urea–water solution (UWS) droplets injected into the hot pipes of a selective catalytic reduction (SCR) system.

1. A new impingement model was presented based on superpositioned mass fractions of four basic impingement behaviors. Thus, the number of regimes can be reduced for simplification while the replication of observed transitional behavior is enabled. Measurements of secondary droplet characteristics were displayed and included in the model.
2. The heat transfer during droplet–wall interaction was implemented in the model following the approach presented in the literature by Wruck and adapted to the use with urea-water solution. This ensures that the effect of surface cooling from droplet–wall contact on the following impingement events is considered. For the model parameterization, additionally measured data of droplet–wall interaction were used.

The model was implemented in commercial CFD code Siemens Star-CCM+ v 17.02.008 for validation against measurement results:

3. The performance of the presented model in predicting the spray cooling behavior was shown on a flat impingement plate in the exhaust line of a diesel engine. The outline of the cooled-down areas could be simulated very well for different sprays and temperatures. The temperature–time diagrams also showed satisfactory agreement for multiple control areas. Especially the temperature-dependent change between fast- and slow-cooling impingement regimes could be replicated.
4. The potential of the impingement model to predict the ammonia distribution in a heavy-duty SCR aftertreatment system was illustrated. An almost perfect agreement between the experimental and the simulation results could be demonstrated for one geometry with a swirl element regarding ammonia distribution in the exhaust gas and uniformity index (UI). The simulation of a second mixer geometry showed slight deviations from the experimental results with an additional ammonia-rich area. Simulation analysis revealed droplets sliding along the hot pipe wall and then impinging on the catalyst as the source of the error. This effect of underestimated evaporation of sliding droplets is more significant with the additional mixing element B, which is the reason for the more substantial deviation from the experiments.

Applying the presented impingement model, a significant step forward in the simulation of droplet–wall interaction could be achieved. This is evidenced by the results of the validation cases. Nonetheless, the results also reveal possibilities for further improvement:

i The particular case of sliding droplets on a hot wall seems to require further experimental investigations on a specific test bench setup and a derived modeling effort. Thus, slight remaining deviations between the simulation and experiment might be eliminated.

ii The roughness influence on ammonia uniformity should be investigated, and a respective modeling approach presented in [11] assessed.

iii An uncertainty regarding the effusivity of UWS in the highly transient case of droplet impingement was revealed in the current work. Appropriate experiments should be conducted to close this knowledge gap.

Author Contributions: Conceptualization, M.Q. and T.L.; Investigation, M.Q.; Data curation, M.Q. and S.P.; Funding acquisition, T.L.; Investigation, M.Q.; Methodology, M.Q. and U.B.; Validation, M.Q.; Formal analysis, M.Q.; Project administration, T.L.; Software, M.Q.; Supervision, T.L.; Writing—original draft, M.Q.; Writing—review and editing, M.Q. and T.L. All authors have read and agreed to the published version of the manuscript.

Funding: This work was funded by the Federal Ministry for Economic Affairs and Climate Action (BMWK) through the German Federation of Industrial Research Associations eV (AiF) based on a decision taken by the German Bundestag and the Research Association for Combustion Engines eV (FVV) (funding no. 601400). The work at the TU Wien was funded by the Ministry for Climate Action, Environment, Energy, Mobility, Innovation and Technology (BMK) through the Austrian Research Promotion Agency (FFG), grant number 877645. The research was carried out within the framework of the industrial collective programme (IGF/CORNET no. 270 EN). Open Access Funding by TU Wien.

Data Availability Statement: The data are not publicly available due to confidentiality reasons.

Acknowledgments: Special thanks to Melanie Walter for support with editing the diagrams.

Conflicts of Interest: The authors declare no conflicts of interest. The funders had no role in the design of the study; in the collection, analyses, or interpretation of data; in the writing of the manuscript; or in the decision to publish the results.

Abbreviations

The following abbreviations are used in this manuscript:

CFD	Computational fluid dynamics
CHF	Critical heat flux
IR	Infrared
La	Laplace-number
LFP	Leidenfrost point
MBU	Mechanical breakup
MHF	Minimum heat flux
NO_x	Nitric oxides
ONB	Onset of nucleate boiling
RANS	Reynolds averaged Navier Stokes
Re	Reynolds-number
SCR	Selective catalytic reduction
SMD	Sauter mean diameter
TBU	Thermal-induced breakup
t_{inj}	Injection duration
UI	Uniformity index
UWS	Urea-water solution
We	Weber-number
ZnSe	Zinc-selenide

Appendix A

Appendix A.1. Contact Parameter Analysis

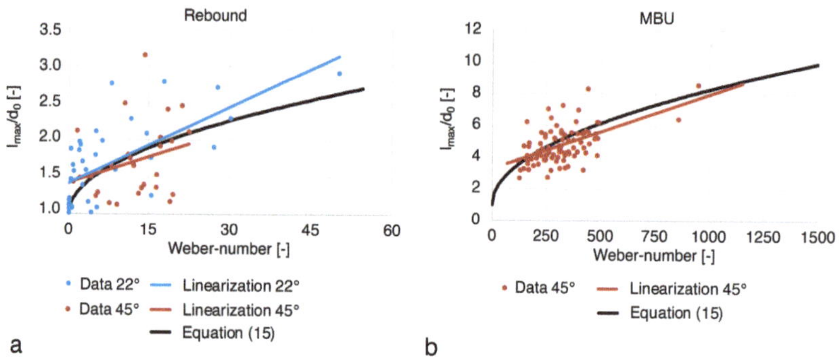

Figure A1. Measured ratio of contact patch length l_{max} to droplet diameter d_0 over Weber-number for impingement data from 22° and 45° impingement angle and their linearization. As a reference, the function from Equation (15) [35] is added: (**a**) Rebound. (**b**) Mechanical breakup (MBU).

Appendix A.2. Results Uniformity Modeling Best Practice Case

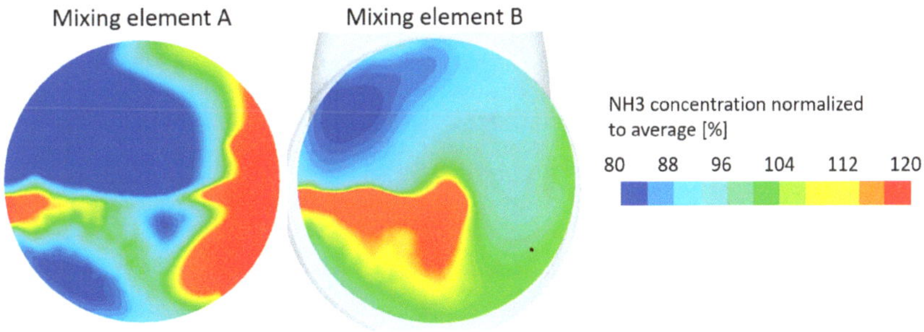

Figure A2. Simulated ammonia uniformity at the catalyst outlet with Bai–Gosman best-practice impingement model [18] without the additional implementation of a thermal-induced breakup. **Left**: UI = 86.9%. **Right**: UI = 93.6%.

References

1. Cravero, C.; de Domenico, D.; Marsano, D. The Use of Uncertainty Quantification and Numerical Optimization to Support the Design and Operation Management of Air-Staging Gas Recirculation Strategies in Glass Furnaces. *Fluids* **2023**, *8*, 76. [CrossRef]
2. Birkhold, F. *Selektive katalytische Reduktion von Stickoxiden in Kraftfahrzeugen: Untersuchung der Einspritzung von Harnstoffwasserlösung*; Berichte aus der Strömungstechnik, Shaker: Aachen, Germany, 2007.
3. Bai, C.; Gosman, A.D. Developtment of Methodology for Spray Impingement Simulation. *SAE Trans.* **1995**, *104*, 550–568.
4. Kuhnke, D. Spray Wall Interaction Modelling by Dimensionless Data Analysis. Ph.D. Thesis, Technische Universität Darmstadt, Darmstadt, Germany, 2004.
5. Wruck, N. *Transientes Sieden von Tropfen beim Wandaufprall*; Berichte aus der Verfahrenstechnik, Shaker: Aachen, Germany, 1999.
6. Liang, G.; Mudawar, I. Review of drop impact on heated walls. *Int. J. Heat Mass Transf.* **2017**, *106*, 103–126. [CrossRef]
7. Moita, A.S.; Moreira, A. Drop impacts onto cold and heated rigid surfaces: Morphological comparisons, disintegration limits and secondary atomization. *Int. J. Heat Fluid Flow* **2007**, *28*, 735–752. [CrossRef]
8. Cossali, G.E.; Marengo, M.; Santini, M. Thermally induced secondary drop atomisation by single drop impact onto heated surfaces. *Int. J. Heat Fluid Flow* **2008**, *29*, 167–177. [CrossRef]
9. Börnhorst, M.; Deutschmann, O. Single droplet impingement of urea water solution on a heated substrate. *Int. Heat Fluid Flow* **2018**, *69*, 55–61. [CrossRef]
10. Kuhn, C.; Schweigert, D.; Kuntz, C.; Börnhorst, M. Single droplet impingement of urea water solution on heated porous surfaces. *Int. J. Heat Mass Transf.* **2021**, *181*, 121836. [CrossRef]
11. Quissek, M.; Lauer, T. Impact of Surface Roughness on the Impingement of Urea–Water Solution Droplets. *Fluids* **2023**, *8*, 152. [CrossRef]
12. Rioboo, R.; Tropea, C.; Marengo, M. Outcomes from a Drop Impact on Solid Surfaces. *At. Sprays* **2001**, *11*, 12. [CrossRef]
13. Tran, T.; Staat, H.J.J.; Prosperetti, A.; Sun, C.; Lohse, D. Drop impact on superheated surfaces. *Phys. Rev. Lett.* **2012**, *108*, 036101. [CrossRef]
14. Mundo, C.; Sommerfeld, M.; Tropea, C. Droplet-wall collisions: Experimental studies of the deformation and breakup process. *Int. J. Multiph. Flow* **1995**, *21*, 151–173. [CrossRef]
15. Wachters, L.; Westerling, N. The heat transfer from a hot wall to impinging water drops in the spheroidal state. *Chem. Eng. Sci.* **1966**, *21*, 1047–1056. [CrossRef]
16. Bai, C.X.; Rusche, H.; Gosman, A.D. Modeling of Gasoline Spray Impingement. *At. Sprays* **2002**, *12*, 1–28. [CrossRef]
17. Smith, H.; Zöchbauer, M.; Lauer, T. *Advanced Spray Impingement Modelling for an Improved Prediction Accuracy of the Ammonia Homogenisation in SCR Systems*; SAE Technical Paper Series; SAE International: Warrendale, PA, USA, 2015. [CrossRef]
18. Quissek, M.; Lauer, T.; García-Afonso, O.; Fowles, S. *Identification of Film Breakup for a Liquid Urea-Water-Solution and Application to CFD*; SAE Technical Paper Series; SAE International: Warrendale, PA, USA, 2019. [CrossRef]
19. Quissek, M.; Budziankou, U.; Lauer, T. *A Novel Approach for the Impingement of AdBlue-Droplets based on Smooth Regime Transitions*; SAE Technical Paper Series; SAE International: Warrendale, PA, USA, 2020. [CrossRef]
20. Mundo, C.; Tropea, C.; Sommerfeld, M. Numerical and experimental investigation of spray characteristics in the vicinity of a rigid Wall. *Exp. Therm. Fluid Sci.* **1997**, *15*, 228–237. [CrossRef]
21. Gavaises, M.; Theodorakakos, A.; Bergeles, G. Modeling wall impaction of diesel sprays. *Int. J. Heat Fluid Flow* **1996**, *17*, 130–138. [CrossRef]
22. Wang, D.M.; Watkins, A.P. Numerical modeling of diesel spray wall impaction phenomena. *Int. J. Heat Fluid Flow* **1993**, *14*, 301–312. [CrossRef]
23. O'Rourke, P.J.; Amsden, A.A. A Spray/Wall Interaction Submodel for the KIVA-3 Wall Film Model. *J. Engines* **2000**, *109*, 281–298.

24. Nagaoka, M.; Kawazoe, H.; Nomura, N. Modeling Fuel Spray Impingement on a Hot Wall for Gasoline Engines. *J. Engines* **1994**, *103*, 878–896.
25. Naitoh, K.; Takagi, Y.; Kokita, H.; Kuwahara, K. Numerical Prediction of Fuel Secondary Atomization Behavior in SI Engine based on the Oval-Parabola Trajectories(OPT) Model. *J. Engines* **1994**, *103*, 897–914.
26. Naber, J.D.; Reitz, R.D. Modeling Engine Spray/Wall Impingement. *J. Engines* **1988**, *97*, 118–140.
27. Stanton, D.W.; Rutland, C.J. Multi-dimensional modeling of thin liquid films and spray-wall interactions resulting from impinging sprays. *Int. J. Heat Mass Transf.* **1998**, *41*, 3037–3054. [CrossRef]
28. Akao, F.; Araki, K.; Mori, S.; Moriyama, A. Deformation Behaviors of a Liquid Droplet Impinging onto Hot Metal Surface. *Trans. Iron Steel Inst. Jpn.* **1980**, *20*, 737–743. [CrossRef]
29. Naber, J.D.; Farrell, P.V. Hydrodynamics of Droplet Impingement on a Heated Surface. *J. Engines* **1993**, *102*, 1346–1361.
30. Senda, J.; Kobayashi, M.; Iwashita, S.; Fujimoto, H. Modeling of Diesel Spray Impingement on a Flat Wall. *J. Engines* **1994**, *103*, 1918–1931.
31. Bernardin, J.D.; Stebbins, C.J.; Mudawar, I. Effects of surface roughness an water droplet impact history and heat transfer regimes. *Int. J. Heat Mass Transf.* **1997**, *40*, 73–88. [CrossRef]
32. Koizumi, Y. Outline of Boiling Phenomena and Heat Transfer Characteristics. In *Boiling*; Elsevier: Amsterdam, The Netherlands, 2017; pp. 1–11. [CrossRef]
33. Budziankou, U.; Lauer, T.; Yu, X.; Schmidt, B.M.; Cho, N. *Modeling Approach for a Wiremesh Substrate in CFD Simulation*; SAE Technical Paper Series; SAE International: Warrendale, PA, USA, 2017. [CrossRef]
34. Budziankou, Uladzimir. Modeling of Deposit Formation in SCR-Systems. Doctoral thesis, Technische Universität Wien, Wien, Austria, 24 March 2022.
35. Castanet, G.; Caballina, O.; Lemoine, F. Drop spreading at the impact in the Leidenfrost boiling. *Phys. Fluids* **2015**, *27*, 063302. [CrossRef]
36. Karl, A.; Anders, K.; Rieber, M.; Frohn, A. Deformation of liquid droplets during collisions with hot walls: Experimental and Numerical Results. *Part. Part. Syst. Charact.* **1996**, *13*, 186–191. [CrossRef]
37. Ueda, T.; Enomoto, T.; Kanetsuki, M. Heat Transfer Characteristics and Dynamic Behavior of Saturated Droplets Impinging on a Heated Vertical Surface. *Bull. JSME* **1979**, *22*, 724–732. [CrossRef]
38. Biance, A.L.; Chevy, F.; Clanet, C.; Lagubeau, G.; Quéré, D. On the elasticity of an inertial liquid shock. *J. Fluid Mech.* **2006**, *554*, 47. [CrossRef]
39. Karl, A.; Frohn, A. Experimental investigation of interaction processes between droplets and hot walls. *Phys. Fluids* **2000**, *12*, 785–796. [CrossRef]
40. Akhtar, S.W.; Yule, A.J. Droplet Impaction on a Heated Surface at High Weber Numbers. In Proceedings of the 17th ILASS-Europe, Zurich, Switzerland, 2–6 September 2001.
41. Budziankou, U.; Quissek, M.; Lauer, T. Deposit Formation in SCR-Systems - Optical Investigations. *SAE Int. J. Adv. Curr. Pract. Mobil.* **2021**, *3*, 501–515. [CrossRef]
42. Budziankou, U.; Quissek, M.; Lauer, T. *A Fast Modeling Approach for the Numerical Prediction of Urea Deposit Formation*; SAE Technical Paper Series; SAE International: Warrendale, PA, USA, 2020. [CrossRef]
43. Börnhorst, M.; Budziankou, U.; Deutschmann, O.; Lauer, T. *Fundamental Experimental and Numerical Investigations on the Deposit Formation and Decomposition from AdBlue in SCR-Systems*; FVV Final Project Report 1262; FVV: Frankfurt am Main, Germany, 2019.
44. Fischer, S. Simulation of the Urea-Water-Solution Preparation and Ammonia- Homogenization with a Validated CFD-Model for the Optimization of Automotive SCR-Systems. Ph.D. Thesis, Technische Universität Wien, Wien, Austria, 2012.
45. Davidson, L.; Nielsen, P.; Sveningsson, A. Modifications of the V2 Model for Computing the Flow in a 3D Wall Jet. In Proceedings of the Proceedings of the International Symposium on Turbulence, Heat and Mass Transfer, Antalya, Turkey, 12–17 October 2003.
46. Siemens. Simcenter STAR-CCM+ User Guide. 2022. Available online: https://docs.sw.siemens.com/en-US/doc/226870983/PL20220315376513299.starccmp_userguide_html?audience=external (accessed on 7 June 2023).
47. Zöchbauer, M.; Fischer, S.; Lauer, T.; Siegmann-Hegerfeld, T.; Harasek, M.; Krenn, C.; Pessl, G. *Validation of Turbulence Models for an Automotive SCR System with Laser Doppler Anemometry Measurements*; SAE Technical Paper Series; SAE International: Warrendale, PA, USA, 2013. [CrossRef]
48. Lardeau, S.; Billard, F. Development of an elliptic-blending lag model for industrial applications. In Proceedings of the 54th AIAA Aerospace Sciences Meeting, San Diego, CA, USA, 4–8 January 2016; American Institute of Aeronautics and Astronautics: Reston, VA, USA, 2016. [CrossRef]

fluids

MDPI

Article

Simulation of Dynamic Rearrangement Events in Wall-Flow Filters Applying Lattice Boltzmann Methods

Nicolas Hafen [1,2,3,*], Jan E. Marquardt [2,3], Achim Dittler [1,3] and Mathias J. Krause [2,3,4]

1 Gas Particle Systems, Karlsruhe Institute of Technology, 76131 Karlsruhe, Germany
2 Lattice Boltzmann Research Group, Karlsruhe Institute of Technology, 76131 Karlsruhe, Germany
3 Institute of Mechanical Process Engineering and Mechanics, Karlsruhe Institute of Technology, 76131 Karlsruhe, Germany
4 Institute for Applied and Numerical Mathematics, Karlsruhe Institute of Technology, 76131 Karlsruhe, Germany
* Correspondence: nicolas.hafen@kit.edu; Tel.: +49-721-46575

Abstract: Wall-flow filters are applied in the exhaust treatment of internal combustion engines for the removal of particulate matter (PM). Over time, the pressure drop inside those filters increases due to the continuously introduced solid material, which forms PM deposition layers on the filter substrate. This leads to the necessity of regenerating the filter. During such a regeneration process, fragments of the PM layers can potentially rearrange inside single filter channels. This may lead to the formation of specific deposition patterns, which affect a filter's pressure drop, its loading capacity and the separation efficiency. The dynamic formation process can still not consistently be attributed to specific influence factors, and appropriate calculation models that enable a quantification of respective factors do not exist. In the present work, the dynamic rearrangement process during the regeneration of a wall-flow filter channel is investigated. As a direct sequel to the investigation of a static deposition layer in a previous work, the present one additionally investigates the dynamic behaviour following the detachment of individual layer fragments as well as the formation of channel plugs. The goal of this work is the extension of the resolved particle methodology used in the previous work via a discrete method to treat particle–particle and particle–wall interactions in order to evaluate the influence of the deposition layer topology, PM properties and operating conditions on dynamic rearrangement events. It can be shown that a simple mean density methodology represents a reproducible way of determining a channel plug's extent and its average density, which agrees well with values reported in literature. The sensitivities of relevant influence factors are revealed and their impact on the rearrangement process is quantified. This work contributes to the formulation of predictions on the formation of specific deposition patterns, which impact engine performance, fuel consumption and service life of wall-flow filters.

Keywords: wall-flow filter; lattice Boltzmann methods; porous media; resolved particle simulations; homogenized lattice Boltzmann method

Citation: Hafen, N.; Marquardt, J.E.; Dittler, A.; Krause, M.J. Simulation of Dynamic Rearrangement Events in Wall-Flow Filters Applying Lattice Boltzmann Methods. *Fluids* **2023**, *8*, 213. https://doi.org/10.3390/fluids8070213

Academic Editors: Ramesh Agarwal and Francesco De Vanna

Received: 19 June 2023
Revised: 12 July 2023
Accepted: 19 July 2023
Published: 21 July 2023

1. Introduction

Modern exhaust treatment systems of internal combustion engines rely on wall-flow filters as one of the key components for enabling compliance with present emission limits [1]. In such filters, a porous structure comprised of oppositely arranged inlet and outlet channels traps the contained particulate matter (PM) with efficiencies of up to 99% [2]. With increased filter loading, deposition layers are formed on the porous walls' surfaces and the filter's pressure drop increases, accompanied by a degradation of engine performance [3–5]. The introduced PM mostly consists of combustible soot and small amounts of inorganic non-combustible ash, which allows the filter's regeneration by continuously or periodically oxidizing its reactive components. The regeneration affects the deposition layer composition over the long term, as it leads to an accumulation of ash, which eventually forms larger agglomerates. The resulting inhomogeneities can then lead to a break-up of the continuous layer into individual PM layer fragments, which potentially detach from the filter surface and relocate inside a channel [3]. This leads to the formation of specific deposition patterns, which affect a single channel's pressure drop contribution, its loading capacity and the separation efficiency.

Various scientific works [4,6–8] contribute to the identification of probable causes and influence factors leading to the individual patterns, but a universal and consistent formulation without partially contradictory statements cannot be found. Calculation models that enable a consistent quantification of relevant influence factors on the individual parts of the regeneration process are missing completely. Those are, however, vital for respective predictions on the formation of specific deposition patterns, which impact engine performance, fuel consumption and service life of wall-flow filters.

In order to close this gap, the authors of the present manuscript approached different relevant aspects in three previous, consecutive works [9–11]: In an initial work [9], a holistic lattice Boltzmann method (LBM) approach was presented, which can be employed both for the simulation of the movement of surface resolved particles and the flow through porous media. With it, the gas flow and the movement of PM layer fragments was simulated with the open source software OpenLB [12,13], while super-linear grid convergence and good agreement with a reference solution [14] could be shown. A literature review addressing relevant findings on the deposition pattern formation, general numeric approaches to wall-flow filter modelling and LBM-related research regarding flow through porous media, as well as surface resolved particle simulations, can be found in this work. The approach was then used in a following work [10] for the investigation of the detachment, acceleration and deceleration of individual PM layer fragments, resulting from the fragmentation of a continuous PM deposition layer during the filter's regeneration. This work includes a pressure drop comparison with investigations conducted on an experiment rig described in Thieringer et al. [15]. In the most recent work [11], a comprehensive quantification of the stability and accuracy of both particle-free and particle-including flow was provided for elevated inflow velocities of up to $60\,\mathrm{m\,s^{-1}}$. By considering static, impermeable PM layer fragments attached to the porous substrate's surface, the detachment likelihood of individual layer fragments, including its dependency on local flow conditions and the layer's fragmentation, was determined.

The present and fourth work represents a direct sequel to Hafen et al. [11] by additionally considering the PM fragments' dynamic behaviour following the layer fragmentation. This includes the investigation of the detachment and transport of the fragments along the channel, as well as the subsequent formation of a channel plug. The latter represents a porous, unordered packing, resulting from the accumulation of PM fragments, which can occupy a significant amount of the available channel volume [6,8,16,17]. This can only be achieved by considering the interaction of moving fragments with each other, of moving fragments with a static fragment accumulation and of fragments with the substrate walls.

One goal of this work is therefore the extension of the resolved particle methodology [9–11,18] by a previously developed and thoroughly validated discrete method to treat particle–particle and particle–wall interactions [19] and its application to dynamic rearrangement events in wall-flow filters. A second goal is the evaluation of the influence of the fragmented layer topology, PM properties and operating conditions on the described process itself and the determination of relevant key quantities. The latter include the final pressure drop, as well as the size and the mean density of resulting channel plugs.

Contrary to previous works in which the simulations' stability, accuracy and validity was accessed with convergence studies, literature findings and experimental results, the present work does not attempt such evaluations, but rather focusses on the derivation of quantitative statements. The challenges inherent to experimental investigations of the transient processes of interest [15], which are the motivation behind the present study in the first place, impede a thorough validation of the simulation results at this point. Single-valued key quantities, such as the plug density, are, however, compared to literature findings.

The remainder of the present work is structured as follows: A brief elaboration on the mathematical modelling and relevant numerical methods can be found in Section 2. Their application to a wall-flow filter is then described in Section 3, followed by the discussion and interpretation of the studies' results in Section 4. A summary of all relevant findings and resulting implications can then be found in Section 5. As the considered rearrangement process consists of multiple individual sub-processes, the present work contains a large number of flow field visualizations accompanying the quantitative evaluations in order to improve overall comprehension.

2. Mathematical Modelling and Numerical Methods

Analogous to the previous work [11], the evolution of conserved fluid quantities is described with the incompressible Navier–Stokes equations (NSE) [20] by using the LBM as a discretization approach in form of a mesoscopic description of gas dynamics. As an alternative to conventional computational fluid dynamics (CFD) methods, this enables the retrieval of a fluid's velocity $\boldsymbol{u}(\boldsymbol{x}, t)$. and the pressure $p(\boldsymbol{x}, t)$ at a three-dimensional position $\boldsymbol{x}$ in time t inside a specified domain.

In order to avoid a sole duplication of the existing modelling description, the respective equations are not carried over from Hafen et al. [11], but rather are referenced here. This includes a description of the LBM principles, an explanation of porous media and surface resolved particle modelling and the quantification of errors and convergence behaviour. As the modelling of discrete particle contacts according to [19] was not part of the previous works, however, it is laid out in the following:

2.1. Discrete Contact Modelling

In contrast to the contact treatment of spherical particles, the respective treatment for arbitrarily shaped ones comes with a significantly increased complexity, as the distance between surfaces cannot simply be deduced from their position and diameter. A model accounting for the interaction of arbitrary geometries including the computation of the contact normal $\boldsymbol{n}_c$ can be found in Nassauer and Kuna [21]. With it, the magnitude of the normal contact force

$$F_{c,n} = 0.75E^* \sqrt{V_c d_c}\left(1 + c\dot{d}_n\right) \tag{1}$$

can be obtained by means of the overlap volume V_c, the indentation depth d_c, the damping factor c and the magnitude of the relative velocity between two objects in contact in the direction of the normal force $\dot{d}_n$. Those parameters are computed following the method described in Marquardt et al. [19]. The effective modulus of elasticity E^* accounts for the material properties of both objects. With

$$E^* = \left(\frac{1-\nu_A^2}{E_A} + \frac{1-\nu_B^2}{E_B}\right)^{-1} \tag{2}$$

the moduli of elasticity E_A and E_B as well as the Poisson's ratios ν_A and ν_B of two colliding objects A and B are combined into a single effective value. According to Carvalho and Martins [22], the damping factor

$$\begin{aligned} c &= 1.5\frac{(1-e)(11-e)}{(1+9e)}\frac{1}{v_0} && v_0 > 0, \\ c &= 0 && v_0 \leq 0 \end{aligned} \tag{3}$$

depends on the initial relative velocity magnitude at contact v_0 and the coefficient of restitution e. The force's three-dimensional vector components can be obtained via

$$\boldsymbol{F}_{c,n} = \boldsymbol{n}_c F_{c,n}. \tag{4}$$

Inside a dense packing, a particle's kinetic energy

$$Y_{p,kin} = Y_{p,trans} + Y_{p,rot} = \frac{m_p \mathbf{u}_p^2}{2} + \frac{\mathbf{I}_p \boldsymbol{\omega}_p^2}{2} \tag{5}$$

can be used as a threshold to avoid a never-ending evaluation of particle contacts. It depends on the particle's moment of inertia $\mathbf{I}_p$, its mass m_p, its velocity $\mathbf{u}_p$ and its rotational velocity $\boldsymbol{\omega}_p$, which represent its translational and rotational energy $Y_{p,trans}$ and $Y_{p,rot}$.

3. Application to a Wall-Flow Filter

The simulation setup of a single wall-flow filter channel used in previous works [9–11,18] is considered in the present one as well. Figure 1 shows a sketch of the model, which represents a hashtag-shaped porous structure with a wall thickness of $l_w = 0.4$ mm, which separates a central inflow channel, four additional quarter-sized inlet channels and four half-sized outflow channels. It features solid walls at the end of all inflow channels and at the beginning of all outflow channels, resulting in the described layout of oppositely arranged inlet and outlet channels.

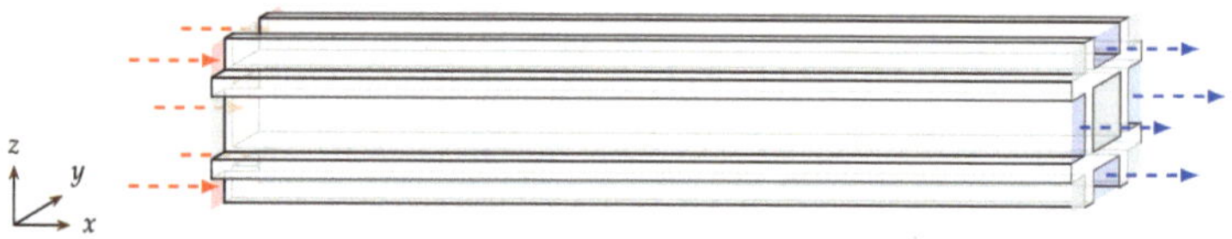

Figure 1. Sketch of wall-flow filter model, consisting of one central inflow (red) channel with fractions of surrounding inflow and outflow (blue) channels separated by porous walls.

Within this model, a channel width of $l_y = 1.6$ mm and a scaled channel length of $l_{x,s} = 24$ mm is used. The latter accounts for a fifth of a real world representative of $l_x = 120$ mm in order to achieve a reduction in computational load that enables the computationally demanding studies in the present work in the first place [11]. All pressure-related quantities in the present work can therefore be related to each other, but should not be considered for a quantitative comparison of their absolute value with those reported in different studies. The model's general capability in recovering the correct pressure drop when considering a full-length channel could, however, be ensured by showing quantitative accordance with literature findings in Hafen et al. [9] and measurements on an experimental test rig in Hafen et al. [10]. The explicit influence of the gas temperature is not considered in this work and ambient conditions are assumed for the whole channel domain.

The model domain is spatially discretized with a resolution N that is defined as the number of cells per channel width l_y. Periodic boundary conditions are then applied on the four surrounding sides, which let it serve as a representation of a real wall-flow filter comprised of hundreds of individual channels. Dirichlet boundary conditions [20] are used to impose a constant velocity at the inlets and a constant pressure at the outlets. No-slip conditions are assumed for the solid walls at the channel end caps on both sides. For the LBM-specific consideration, a simple bounce-back condition [23] is used for no-slip wall, regularized local boundary conditions [24] at the inlets and interpolated boundary conditions [25] at the outlets.

As the present work directly builds upon the results of its predecessor [11], a deposition layer during break-up due to the oxidization of the majority of its soot content [3] is considered as well. The individual layer fragments are modelled as a field of identical cubic discs that cover the porous walls inside an inflow channel. Here, a uniform field of 3×40 fragments is considered as a base configuration on each of the four porous walls of the central inflow channel.

In order to avoid numerical instabilities from initially large gradients, all simulation runs are conducted by smoothly ramping up the inflow velocity to the prescribed one, while the fragments are artificially held in place. Afterwards, convergence of velocity, pressure and hydrodynamic force are evaluated in reoccurring steps in the whole fluid field analogously to [11]. Contrary to the previous work, however, simulations are not concluded when reaching a converged state, defined by the convergence criteria for velocity, pressure and hydrodynamic force. Instead, the fragments are released from their static state, which enables them to potentially detach and move through the channel when sufficient hydrodynamic forces are present. In order to retain reproducibility, no artificial randomness is considered either for the fragmented layer creation or during detachment. Due to the channel's symmetry, this leads to the simultaneous detachment of multiple uniform fragments, which results in an accelerated rearrangement process with respect to a real-world counterpart. As reproducibility represents one of the prime advantages of simulations over experiments, this trade-off is chosen deliberately. A simulation is concluded when the cumulated kinetic energy over all fragments according to (5) falls below a predefined energy threshold. Every simulation run can accordingly be divided into four consecutive parts:

1. Fluid velocity ramp-up and convergence with static fragmented layer;
2. Detachment of fragments;
3. Transport of fragments;
4. Plug formation.

While the first one has been thoroughly investigated in the previous work, the present one focusses specifically on the last one. The conclusion of both periods is denoted *fragmented layer state* and *plug state*, respectively, and will be referenced throughout this work accordingly.

While such a rearrangement process may depend on many influence factors, the ones listed in Table 1 are investigated in this work.

Table 1. Relevant quantities with base value and variation range considered in the respective investigations.

Quantity	Symbol	Base Value	Variation Range (min, max)	Investigation
Fragment height	d_z	170.0 µm	42.5 µm, 212.5 µm	Layer height
Equilateral fragment width	d_{xy}	340.0 µm	170.0 µm, 425.0 µm	Fragment dimensions
Fragment rows along channel	$n_{f,x}$	40	20, 60	Layer structure
Fragment density	ρ_p	500 kg m^{-3}	100 kg m^{-3}, 800 kg m^{-3}	PM density
Fragment's modulus of elasticity	E_p	1×10^{-1} MPa	5×10^{-2} MPa, 5 MPa	Contact properties
Adhesion in normal direction	F_N	0 µN	0 µN, 25 µN	Adhesive forces
Inflow velocity	$\bar{u}_{in}$	10 m s^{-1}	4 m s^{-1}, 14 m s^{-1}	Inflow velocity

The first three factors cause modifications in the initial fragmented layer topology and the converged fluid field, respectively. They thus affect the complete rearrangement process. While the layer height depends on the loading time and PM concentration in the entering exhaust stream, the variation of fragment dimensions and the layer structure provides a simplified approach to account for the impact of the local distribution of ash and soot inside the PM layer. The latter four factors represent additional changes in PM properties and operating conditions, which impact the individual parts of a rearrangement process differently. The choice of base values and variation ranges results from a combined attempt to ensure stability, to keep the analogy to previous studies, to respect physical limitations and to maximize the sensitivity of the individual factors on the process.

In the previous work [11], an inflow velocity of $\bar{u}_{in} = 20\,\mathrm{m\,s^{-1}}$ and a resolution of $N = 96$ was selected for all studies of a deposition layer during break-up. Due to the significantly increased computational demand resulting from the additional consideration of the transient effects, the present work adopts a smaller inflow velocity of $\bar{u}_{in} = 10\,\mathrm{m\,s^{-1}}$, which enables a lower resolution of $N = 64$ in order to avoid exceeding the available computational resources.

The discrete contact model, described in Section 2.1, requires the specification of respective material and mechanical properties. While those are in part subject to some uncertainty, they are kept constant in all studies to nonetheless ensure consistency throughout all investigations. For the porous substrate, a modulus of elasticity of $E_w = 10.0\,\text{GPa}$ and a Poisson's ratio of $\nu_w = 0.2$ can be assumed, following experimental measurements of Cordierite wall-flow filters [26,27]. Due to the lack of respective measurements for the PM under study, in this work, a Poisson's ratio of $\nu_p = 0.4$ is assumed for the PM layer fragments, adopting the properties of saturated clay [28]. It is worth noting at this point, however, that the effective modulus of elasticity E^* in (2) exhibits a minor sensitivity to changes in the Poisson's ratio of the fragments, owing to the large magnitudes associated with the moduli of elasticity in the denominators. Based on its minor stiffness, we assume a fragment's modulus of elasticity of $E_p = 0.1\,\text{MPa}$. Due to the contact model's sensitivity to this parameter and the lack of existing data to substantiate this assumption, its impact is investigated as representative for the mechanical properties of PM as well (cf. Table 1). In order to ensure sufficient dissipation for the comparably coarse temporal resolution in the present work ($\Delta t = 1.32 \times 10^{-7}\,\text{s}$), a coefficient of restitution of $e = 0.3$ is used.

Analogous to the previous works [9–11,18], some relevant quantities are averaged at discrete locations over the cross-section of the central inlet channel and one representative outlet channel (denoted as C_{in} and C_{out}) at recurring intervals and significant points in time. Next to the axial fluid velocity u_x and the pressure p, this work now additionally considers the PM density ρ, eventually providing a velocity, pressure and density profile $\bar{u}_x(x)$, $\bar{p}(x)$ and $\bar{\rho}(x)$ along the channel length both at the converged fragmented layer state and the plug state. Such a density profile can then provide information about:

- Areas of complete detachment;
- Areas of incomplete detachment and re-deposition;
- Volume occupied by the plug;
- Local compactness inside the plug.

In order to enable a consistent quantification of the plug extent and its properties, a mean density profile $\varrho(x)$ is evaluated at the plug state. For this, the raw profile at plug state is smoothed by applying a Savitzky–Golay filter [29] with an order of 2 and a window size of 40 discrete points. The largest gradient in the smoothed profile is then interpreted as the transition point on the x-axis between layer and plug region. The original raw profile is averaged separately in both regions, yielding the mean density profile. This way, the plug starting position and its average density can be obtained as single-valued quantities, available for the comparison with other simulation. The individual steps are laid out at the first occurring of a density profile in Figure 7.

4. Results and Discussion

In the present section, the results from all simulation runs are presented, discussed and related to each other. First, the rearrangement process assuming the base configuration in Table 1 is evaluated in detail with respect to initial flow conditions in the converged state, the transient behaviour during rearrangement events and the final state in Section 4.1. Afterwards, the impact of influence factors that lead to a change in the fragmented layer topology on both the initial flow condition and the rearrangement process are laid out in Section 4.2. Additional PM properties and operating conditions that do not alter the initial state are then investigated in Section 4.3.

4.1. Rearrangement in Base Configuration

The base configuration represents the channel model with a fragmented deposition layer, assuming the exact values listed in Table 1. Its converged fragmented layer state (cf. Section 4.1.1), the following transient behaviour during rearrangement (cf. Section 4.1.2) and the final plug state are investigated separately in the following (cf. Section 4.1.3).

4.1.1. Fragmented Layer State

The flow field inside the channel including the fragmented layer at the converged state is shown in Figure 2.

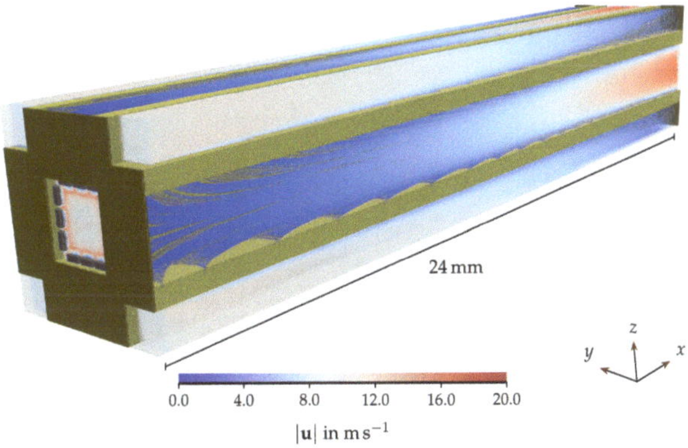

Figure 2. Flow field with uniformly fragmented PM layer in gray. Dark yellow structures represent porous filter substrate, brown structures solid walls. Streamlines exhibit local flow direction. Colour scale indicates local velocity magnitude.

The flow field is represented by streamlines, which indicate the local flow direction and are coloured according to the local velocity magnitude. With those, the decrease in velocity magnitude inside the inlet channels can be identified over the channel length. A continuous magnitude increase can in turn be found in the outflow channels, eventually forming a developed channel profile at the outflow. As the fragmented deposition layer covers a large part of the central inlet channel's cross-section, the area available for the flow is reduced, leading to significantly elevated velocities of up to $|\mathbf{u}| = 16\,\mathrm{m\,s^{-1}}$ with respect to the imposed average inflow velocity of $\bar{u}_{in} = 10\,\mathrm{m\,s^{-1}}$.

The determination of the hydrodynamic forces acting on the particulate structures and the deduction of the local detachment likelihood of individual layer fragments, including their dependency on local flow conditions and the layer's fragmentation, have been the subject of the predecessor work [11]. The corresponding velocity and pressure profiles $\bar{u}_x(x)$ and $\bar{p}(x)$ in the converged fragmented layer state can thus be found there, but have, additionally, been added as a reference to Figure 8.

4.1.2. Transient Behaviour

The transition from a uniformly fragmented deposition layer to an immobile plug configuration consists of detachment and transport of the individual fragments as well as the plug formation. The transient process in the central inlet channel is shown in Figure 3 at four exemplary points in time **T1**, **T2**, **T3** and **T4**.

Time **T1** represents the inner view of Figure 2, with the uniformly fragmented PM layer and a somewhat continuous layer of elevated fluid velocity closely above the fragmented layer over most of the channel length [11]. The fragments are then released (cf. Section 2), leading to a consecutive row-by-row detachment, which is caused by both increased flow exposition of initially shielded fragments and tear-off due to contact with suspended fragments flying by. This way, the number of simultaneously suspended fragments increases continuously until a major portion of the channel's cross-section is occupied by them at **T2**. Due to the restored channel cross-section in the channel's front, the local velocity magnitude decreases significantly here. At time **T3**, re-deposited fragments can be observed in the channel-mid, as fragment–fragment contact rather causes interception than tear-off due to the different flow conditions here. In a previous study on single fragment behaviour [10], fragments were found to always travel to the channel end once detached.

When considering multi-fragment environments, this statement has to be extended with the possibility of interception, respectively. Time **T4** shows the channel's final state, where no fragments are suspended any more and a compact plug has formed at the channel end with a small region of undetachable fragments in front of it. Summing the kinetic energy $Y_{p,kin}$ in (5) over all individual layer fragments results in the cumulated kinetic energy Y_{kin}, which provides a measure of the progress of the rearrangement process. Its temporal development is shown in Figure 4.

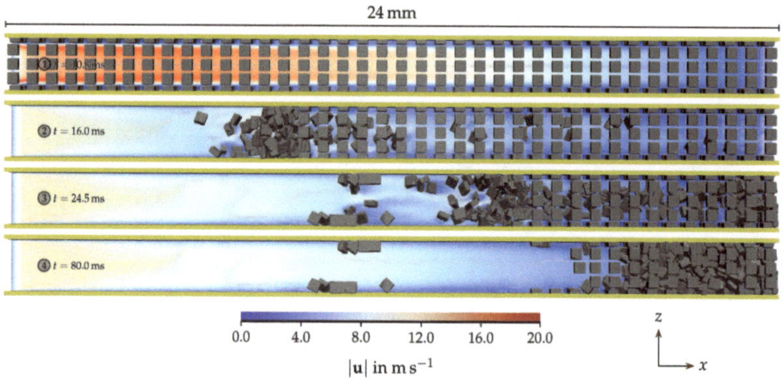

Figure 3. Time series of detachment, transport and plug formation. Dark yellow structures represent porous filter substrate. Streamlines exhibit local flow direction. Colour scale indicates local velocity magnitude.

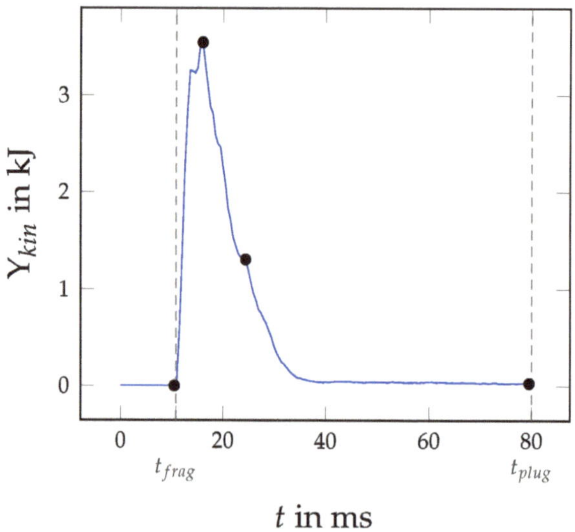

Figure 4. Transient behaviour of fragments' cumulated kinetic energy Y_{kin}. Black dots represent relevant times in Figure 3. Dashed lines indicate time of converged fragmented layer state t_{frag} and final plug state t_{plug}.

The cumulated kinetic energy features a steep increase following the fragment's release after fluid convergence, as fragments continuously detach and resuspend into the flow. A maximum is reached when the number of simultaneously suspended fragments is the highest and detaching fragments approximately equals those joining the plug formation or being intercepted mid-way. The cumulated kinetic energy then starts to decrease until reaching a level at around $t \approx 40$ ms, where it approaches zero with a shallow slope. During this period, fragments steadily arrange inside the newly formed plug, whose compactness continuously increases. The transient development of the cumulated kinetic energy differs in part significantly throughout the different studies considered in this work. In order to ensure the comparability, a maximum simulation time of $t_{plug} = 80$ ms is identified as sufficient to capture all relevant transient effects in all simulation runs. That way, it simultaneously represents the time assumed for the evaluation of the *plug state*.

By averaging the discrete positions of all layer fragments, the average relative position of the total PM mass in the channel $\bar{x}_{rel}$ can be determined, which provides a simple single-valued measure for the spatial development during the rearrangement process. It can also provide some information about a continuing compaction, which reduces the channel volume occupied by the plug and may affect the total pressure drop. The temporal evolution of the relative average PM position $\bar{x}_{rel}$ and the pressure drop Δp is shown in Figure 5.

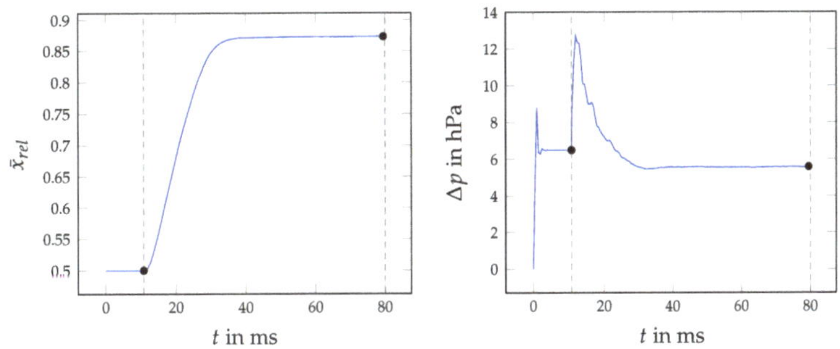

Figure 5. Transient development of relative average PM position $\bar{x}_{rel}$ and pressure drop Δp. Dashed lines indicate time of converged fragmented layer state t_{frag} and final plug state t_{plug}. Black dots represent sampling points at relevant states.

Due to its symmetry around the x-axis, a uniformly fragmented layer leads to an initial relative average PM position of $\bar{x}_{rel} = 0.5$. When the fragments are released after convergence and start to detach (cf. Section 3), the value increases steeply. The apparent smoothness of the course shows that the considered total number of fragments of $n = 480$ can be reasoned to be statistically representative. After the first fragments hit the channel's back wall, the slope decreases and starts to gradually approach a constant value, analogously to the cumulated kinetic energy in Figure 4.

The pressure drop, in turn, settles at a constant value of $\Delta p = 6.50$ hPa right after a start-up-induced fluctuation. When the fragments start to detach, the additional cross-section reduction observable at **T2** in Figure 3 causes more flux to be redirected through the porous walls with an elevated wall-penetrating velocity. Here, the momentum loss increases, causing the pressure drop to rise as well. When the cross-section starts to clear up due to fewer particles detaching, the pressure starts decreasing again and continues to get smaller as more volume becomes available again for the flow. Analogously to the cumulated kinetic energy and the average PM position, the pressure drop then approaches a constant value. The resulting pressure drop of $\Delta p = 5.58$ hPa is smaller than the one at the converged layer state, with a difference of $\Delta p = 92$ Pa. This value is defined by the relation between an increase in the channel's cross-section and the open flow area through the filter substrate on the one hand and a decrease in the available channel volume on the other hand. Both quantities in Figure 5 are sampled at the converged fragmented layer state and the final plug state for the comparison between different simulations in Sections 4.2 and 4.3.

4.1.3. Plug State

The transient behaviour is concluded when reaching the plug state. A detailed view of the resulting plug packing is shown in Figure 6. All surfaces are coloured according to the magnitude of the local hydrodynamic force.

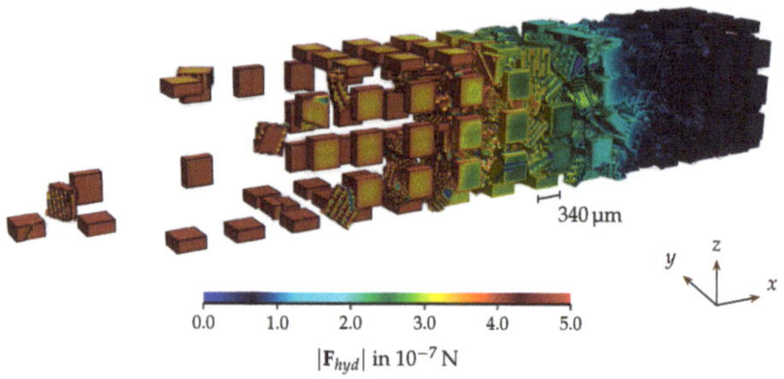

Figure 6. Packing at plug state. Fragment surfaces coloured according to local contribution to hydrodynamic force magnitude $|\mathbf{F}_{hyd}|$. Porous substrate and flow removed for visual clarity.

It can be observed how the transported fragments form a dense packing over multiple fragment rows. Some undetached fragments can be found right in front of it, where the hydrodynamic forces in the normal direction of the substrate's surface are not large enough to cause detachment. It also shows some redeposited particles, which have been intercepted during transport when loosing momentum due to contact with undetached fragments. The hydrodynamic force's magnitude $|\mathbf{F}_{hyd}|$ can be seen to decrease quickly when advancing further into the packing, indicating a continuously decreasing amount of fluid penetrating the plug. As the filter walls are permeable, they are subject to small currents inside them, causing a non-zero force contribution on each fragment's bottom side.

The density profile for both fragmented layer and plug state is shown in Figure 7. The methodology for the evaluation of a mean density profile as described in Section 2 is outlined as well.

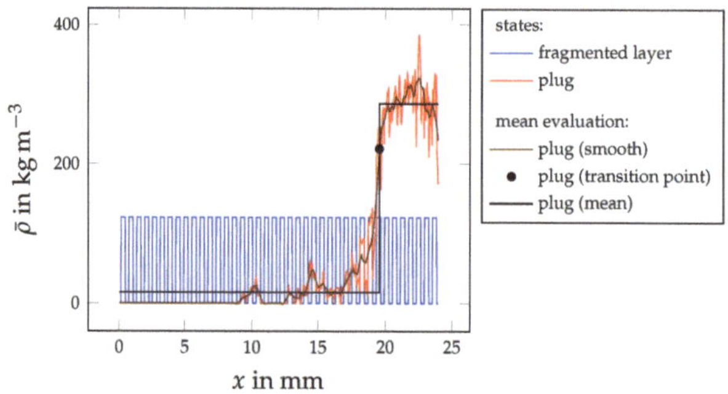

Figure 7. PM density profile $\bar{\rho}(x)$ at fragmented layer and plug state. Additionally, layout of mean evaluation methodology according to Section 2.

It can be seen that the fragmented layer state is characterized by a non-negligible density which exhibits a periodic nature with a constant maximum, directly reflecting the PM layer height. The plug state, in turn, features a distinct profile, which shows no density in the frontal part of the channel, a small one in the channel-mid and a clearly identifiable plug region in the rear part of the channel beginning. The density reaches its maximum in the middle of the plug and decreases slightly both towards its front and the channel back wall.

The transition between layer and plug according to the mean density methodology lies at $x = 19.6$ mm, which reveals that the plug occupies approximately 18% of the central inflow channel's volume at plug state. The resulting mean density profile (denoted $\varrho(x)$ in the following) exhibits an average plug density of $\varrho = 286\,\mathrm{kg\,m^{-3}}$, which lies well in the ranges of $220\,\mathrm{kg\,m^{-3}}$ to $330\,\mathrm{kg\,m^{-3}}$ [6] and $160\,\mathrm{kg\,m^{-3}}$ to $400\,\mathrm{kg\,m^{-3}}$ [16,17] reported in the literature for unsintered lube oil-derived ash.

The respective velocity and pressure profiles $\bar{u}_x(x)$ and $\bar{p}(x)$ in the central inlet channel and the representative outlet channel (cf. Section 3) are shown in Figure 8 for both fragmented layer and plug state.

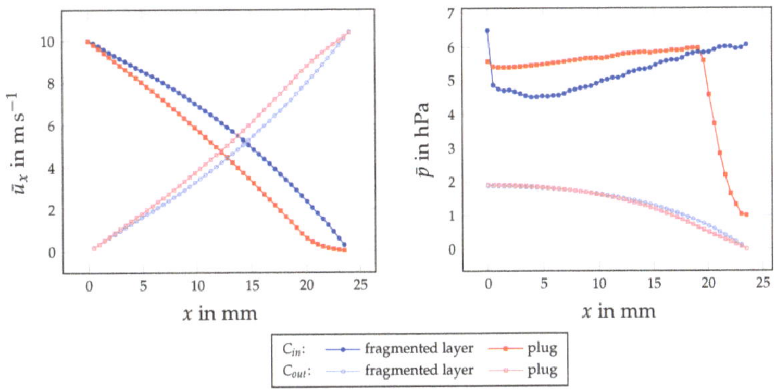

Figure 8. Velocity profile $\bar{u}_x(x)$ and pressure profile $\bar{p}(x)$ in inlet channel domain C_{in} and outlet channel domain C_{out} at fragmented layer and plug state.

In the predecessor work [11], distinct profiles were identified in the fragmented layer state: The velocity is characterized by a transition from the imposed one $\bar{u}_{in}$ to zero in the inflow channel domain $C_{in} = 10\,\mathrm{m\,s^{-1}}$ and exhibits the opposite behaviour in the outflow channel domain C_{out}. It should be noted at this point that all averaging is performed on the full channel cross-section. That way, an increase in the axial velocity due to a reduced area available for the flow does not impact the velocity profile, as fewer but higher velocities lead to the same average. It consequently represents a measure of the x-directed flux at a specific position. The pressure, however, decreases at a reduced available cross-section, as the velocity effectively increases. This results in a sudden pressure drop close to the inlet in the fragmented layer state. The pressure then gradually increases until reaching a local maximum near the back wall. The overall pressure level in the inflow channel results from the momentum loss due to the porous walls between the channels. In the outflow channel, in turn, the pressure decreases continuously with a decreasing slope, leading to a continuously increasing pressure difference between the inflow and outflow channel over its length. At the plug state, the velocity profile in the inflow channel changes significantly, as the rear part of the channel becomes occupied and the effective volume available for the flow decreases. As the flow resistance in the remaining part decreases additionally due to the increase in free substrate surface area, more fluid passes through the walls in the channel front and the mid-section. This becomes evident when considering the significantly decreased axial velocity in the inflow channel and the increased velocity over the whole length of the outflow channel. According to the observations made on Figure 7, the plug is not completely impenetrable, but rather exhibits a quickly decreasing permeability, which results in the observed continuously decreasing hydrodynamic surface force on the fragments. As derived from Figure 5, the pressure in the inlet decreases by $\Delta p = 93.2\,\mathrm{Pa}$ at the plug state. Due to the complete detachment of the first fragment row, the sudden drop in pressure near the inlet vanishes. The profile then exhibits a shallow and nearly linear increase until reaching the plug's beginning. Inside the plug, the pressure drops quickly to a level that lies even below the average outflow channel pressure.

4.2. Influence of Fragmented Layer Topology

In the following, the influence of the layer topology on the complete rearrangement process is investigated by considering differences in the layer height in Section 4.2.1, the fragment dimensions in Section 4.2.2 and the layer structure in Section 4.2.3. Next to quantitative evaluations, flow fields are shown for both the converged layer state and the final plug state.

4.2.1. Influence of Layer Height

A layer as thin as $d_z = 42.5\,\mu m$ does not enable any detachment. The remaining range of considered PM layer heights in Table 1 is enclosed by the borderline cases $d_z = 63.8\,\mu m$ and $d_z = 212.5\,\mu m$, for which the resulting flow field is shown in Figure 9 at the converged fragmented layer stage.

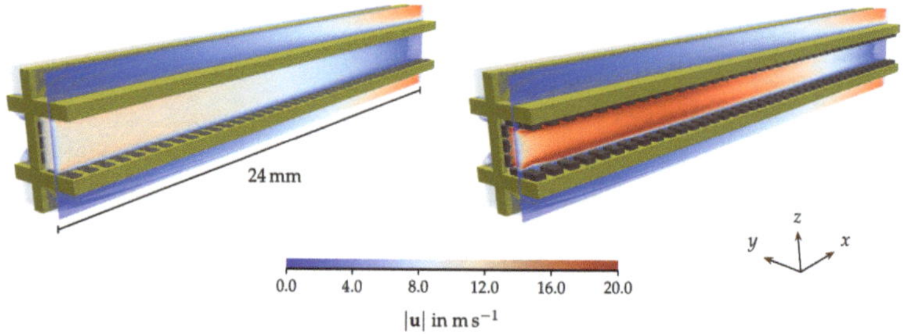

Figure 9. Converged flow fields with uniformly fragmented PM layer in gray for layer heights $d_z = 63.8\,\mu m$ (**left**) and $d_z = 212.5\,\mu m$ (**right**) at fragmented layer state. Dark yellow structures represent porous filter substrate. Streamlines exhibit local flow direction. Colour scale indicates local velocity magnitude.

A direct comparison reveals large differences in the fluid velocity over the major part of the central inlet channel, as a cross-section reduction leads to significantly increased local and averaged axial fluid velocities. Those already hint at more fragment detachment due to an increased detachment likelihood [11] and faster fragment transport. The resulting flow fields at the final plug state are shown in Figure 10.

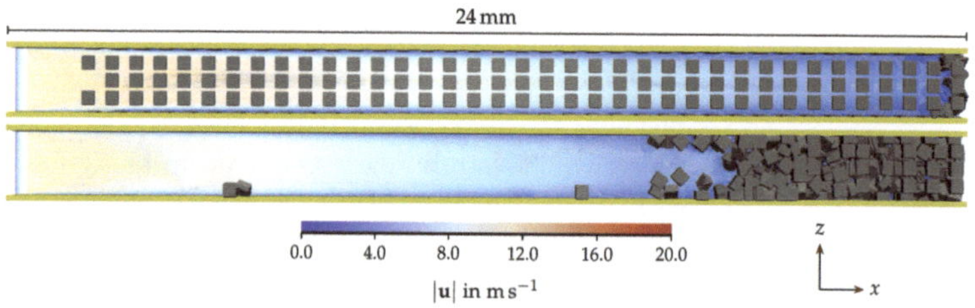

Figure 10. Final flow fields with PM fragments in gray for layer heights $d_z = 63.8\,\mu m$ (**top**) and $d_z = 212.5\,\mu m$ (**bottom**) at plug state. Dark yellow structures represent porous filter substrate. Streamlines exhibit local flow direction. Colour scale indicates local velocity magnitude.

The formulated assumption can be confirmed, as detachment only occurred at the very first fragment rows for a layer height of $d_z = 63.8\,\mu m$, but over the whole channel length for a layer height of $d_z = 212.5\,\mu m$. The PM density $\bar{\rho}(x)$ profile and mean density profile $\varrho(x)$ (as laid out in Figure 7) are shown in Figure 11 for the considered range of layer heights.

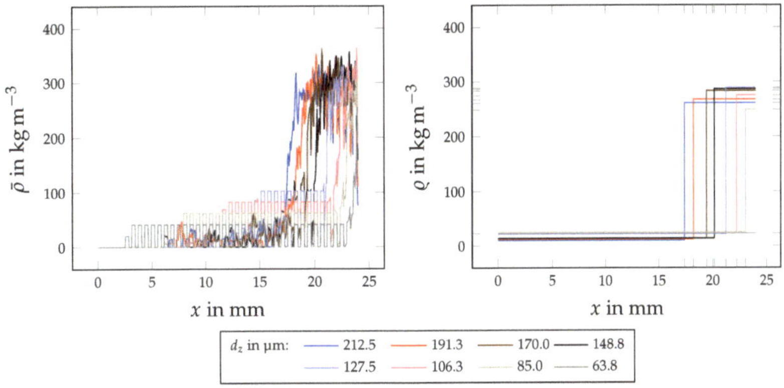

Figure 11. PM density profile $\bar{\rho}(x)$ (**left**) and mean density profile $\varrho(x)$ (**right**) at plug state for different layer heights d_z.

The profile exposes regions of different detachability: while for small layer heights, only the first few fragment rows detach, nearly complete detachment can be found for the largest height. The volume occupied by the resulting plug differs accordingly, as more detachment leads to more PM available for the plug formation, which gradually extends towards the inlet.

The dependency of the relative average PM position $\bar{x}_{rel}$ and the pressure drop Δp on the layer height at both fragmented layer and plug state (cf. Figure 5) is laid out in Figure 12.

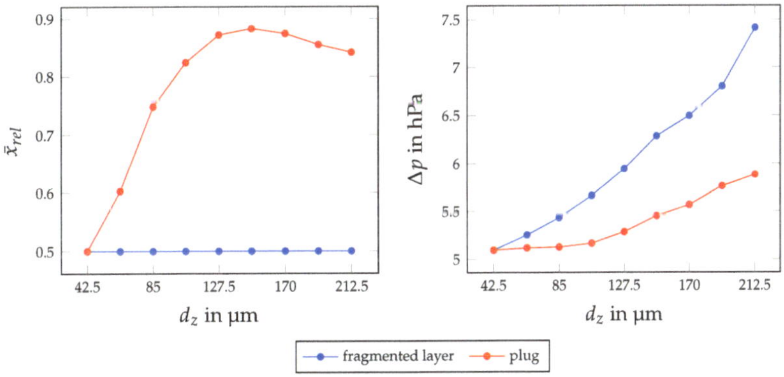

Figure 12. Relative average PM position $\bar{x}_{rel}$ and pressure drop Δp for different layer heights d_z at fragmented layer and plug state.

As no detachment occurs for $d_z = 42.5\,\mu m$, both quantities are identical between the states. With increasing layer height, the relative average PM position then shifts towards the channel back at plug state due to the increasing number of detachable fragments. After reaching a maximum position of $\bar{x}_{rel} = 0.88$ for $d_z = 148.75\,\mu m$, it decreases for larger layer heights as the plug continues to grow in a negative x-direction. Such a local maximum can consequently only be found for quantities that impact the detachment likelihood in a similarly pronounced way. The pressure drop increases continuously for both states, while exposing a much steeper slope at the fragmented state. This way, the difference becomes larger with increasing layer heights, with a maximum of $\Delta p = 153\,\text{Pa}$ for $d_z = 212.5\,\mu m$.

4.2.2. Influence of Fragment Dimensions

The flow fields for the lower and upper end of the considered fragment dimension range in Table 1 is shown in Figure 13.

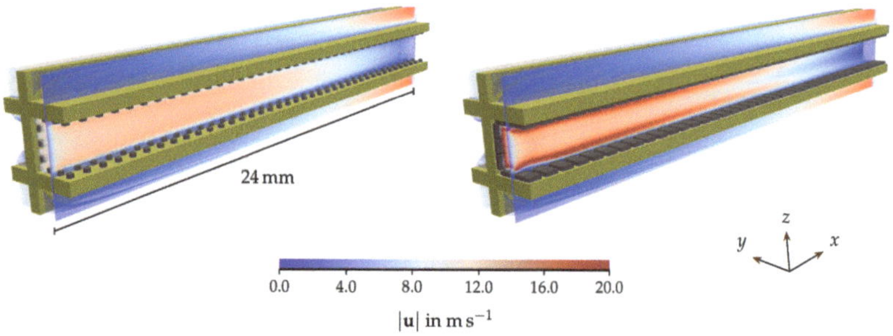

Figure 13. Converged flow fields with uniformly fragmented PM layer in gray for fragment dimensions $d_{xy} = 170\,\mu\text{m}$ (**left**) and $d_{xy} = 425\,\mu\text{m}$ (**right**) at fragmented layer state. Dark yellow structures represent porous filter substrate. Streamlines exhibit local flow direction. Colour scale indicates local velocity magnitude.

Similarly to the relation in Figure 9, the larger fragment dimension leads to higher magnitudes of the fluid velocity due to the reduction in the channel's cross-section. Additionally, the free surface area on the substrate is greatly reduced, which leads to an inhomogeneous velocity distribution of the channel cross-section with a pronounced layer of elevated fluid velocity closely above the fragmented layer. The resulting flow fields at the final plug state are shown in Figure 14.

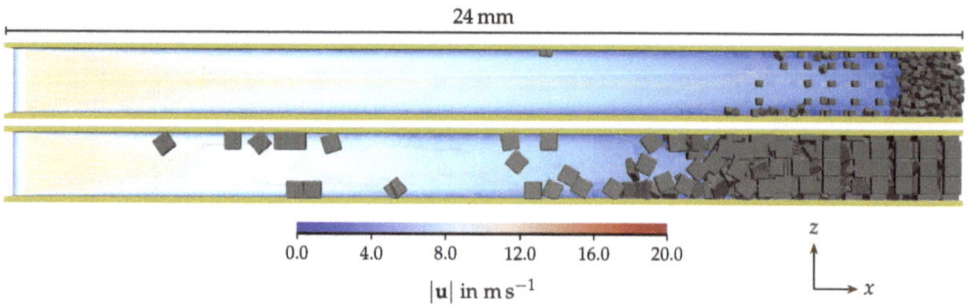

Figure 14. Final flow fields with PM fragments in gray for fragment dimensions $d_{xy} = 170\,\mu\text{m}$ (**top**) and $d_{xy} = 425\,\mu\text{m}$ (**bottom**) at plug state. Dark yellow structures represent porous filter substrate. Streamlines exhibit local flow direction. Colour scale indicates local velocity magnitude.

Due to the large difference in available PM mass, the resulting plug size differs accordingly. For fragment dimensions of $d_{xy} = 170\,\mu\text{m}$, a larger volume remains available for the flow. Complete detachment in the front and mid-section of the channel can be found in both cases, while the smaller fragment size exhibits an undetachable region in the additional flow volume due to insufficient axial flow velocities. The density profile for all considered fragment dimensions is shown in Figure 15.

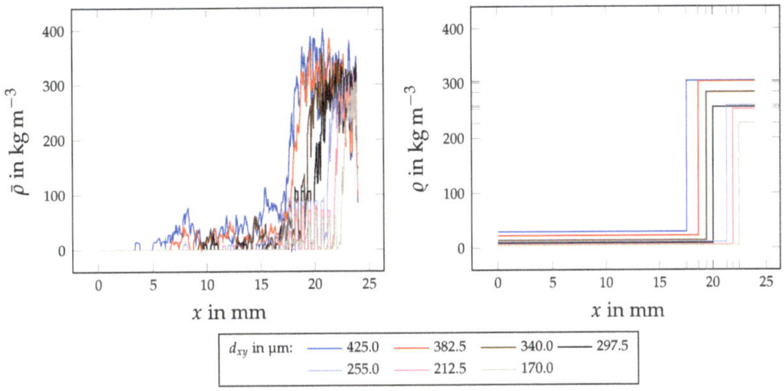

Figure 15. PM density profile $\bar{\rho}(x)$ (**left**) and mean density profile $\varrho(x)$ (**right**) at plug state for different fragment dimensions d_{xy}.

Contrary to the observations made in Figure 11, no regions of undetachable fragments can be identified in the front and mid-section, even for small fragment dimensions. Detachability can consequently be assumed to feature a negligible sensitivity to the fragment dimensions in the considered range. Larger fragments, however, seem to be associated with a higher likelihood for interception during the transport, as non-zero density values reach further towards the inlet at the plug state. Inside the plug, differences are similar to those observed for the layer height, as the cumulated mass, hence the mean density, increases in both cases with larger fragments. The dependency of the relative average PM position $\bar{x}_{rel}$ and the pressure drop Δp on the fragment dimensions at both states is shown in Figure 16.

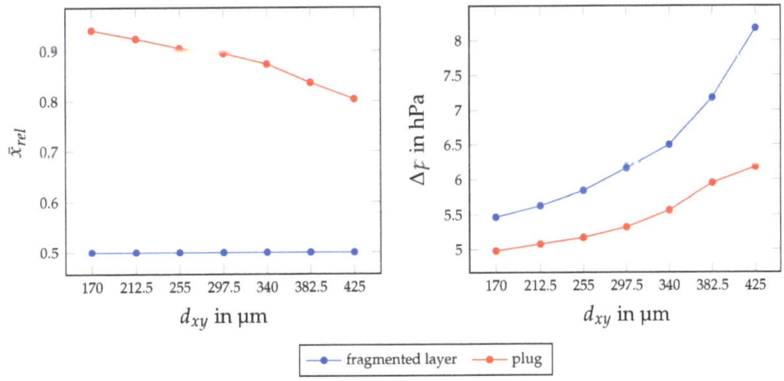

Figure 16. Relative average PM position $\bar{x}_{rel}$ and pressure drop Δp for different fragment dimensions d_{xy} at fragmented layer and plug state.

Contrary to a variation of the layer height in Figure 12 the average PM position continuously decreases with increasing fragment dimension. Due to negligible differences in the detachability, no inflection point can be found here and the average PM position directly reveals how far the plug reaches into the channel. The pressure increases for both states, with the difference between them continuously becoming larger. A fragment size of $d_{xy} = 425\,\mu\text{m}$ eventually results in a difference of $\Delta p = 199.6\,\text{Pa}$.

4.2.3. Influence of Layer Structure

Depending on the amount and nature of inhomogeneities in the local distribution of reactive (soot) and inert (ash) components in a PM layer, a fragmentation due to the oxidation of the reactive components can lead to different amounts of equally sized PM fragments [3]. The borderline cases for the resulting layer structures are displayed in Figure 17.

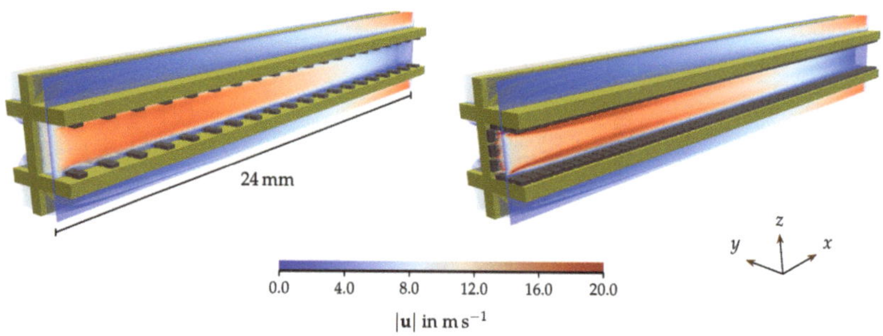

Figure 17. Converged flow fields with uniformly fragmented PM layer in gray for numbers of fragment rows $n_{f,x} = 20$ (**left**) and $n_{f,x} = 60$ (**right**) at fragmented layer state. Dark yellow structures represent porous filter substrate. Streamlines exhibit local flow direction. Colour scale indicates local velocity magnitude.

While the cross-section at the position of fragment rows remains identical, the free surface area on the porous substrate differs significantly. Analogously to the observations in Figure 13, a reduction in free surface area leads to the development of a more inhomogeneous velocity distribution over the channel's cross-section. A layer structure of $n_{f,x} = 20$ shows the trailing fluid structures observed in Hafen et al. [11] behind each fragment row due to the large distance to the next one. When considering $n_{f,x} = 60$, in turn, each fragment row is shielded from the flow by the surrounding ones.

At plug state, the two borderline cases yield the flow fields shown in Figure 18.

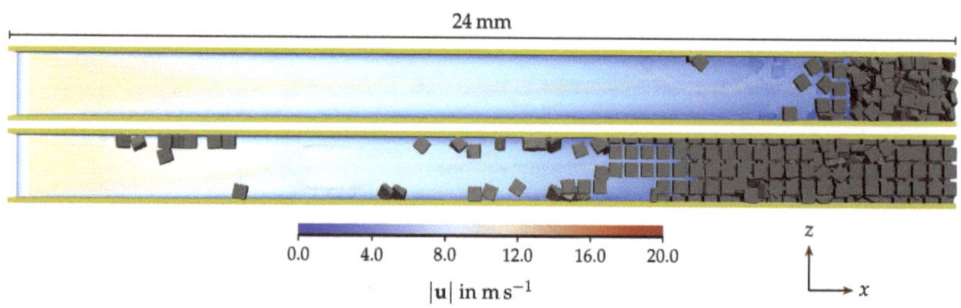

Figure 18. Final flow fields with PM fragments in gray for numbers of fragment rows $n_{f,x} = 20$ (**top**) and $n_{f,x} = 60$ (**bottom**) at plug state. Dark yellow structures represent porous filter substrate. Streamlines exhibit local flow direction. Colour scale indicates local velocity magnitude.

The plug size, again, directly reflects the difference in available PM mass. The influence of the layer structure on the density profiles is shown in Figure 19.

It shows similar tendencies as the fragment dimension in Figure 15: Full detachment can be found for all variations in the channel's front and mid-section. Solely a small section right in front of the plug shows a reduced detachability for a greater number of fragment rows $n_{f,x}$ due to the reduced exposition to the hydrodynamic forces at this position. Contrary to the influence of the fragment dimensions, the plug's mean density exhibits a smaller sensitivity.

The relative average PM position in Figure 20 directly reflects the cumulated PM mass provided at the fragmented layer state, as it depicts plug growth towards the inlet. This growth leads to a nearly linear increase in the pressure drop at plug state. The pressure drop at the fragmented layer state exhibits a steeper slope and rises to larger values when the free substrate surface area becomes smaller. This rise eventually leads to a pressure drop difference of $p = 442\,\text{Pa}$ at $n_{f,x} = 60$ between both states.

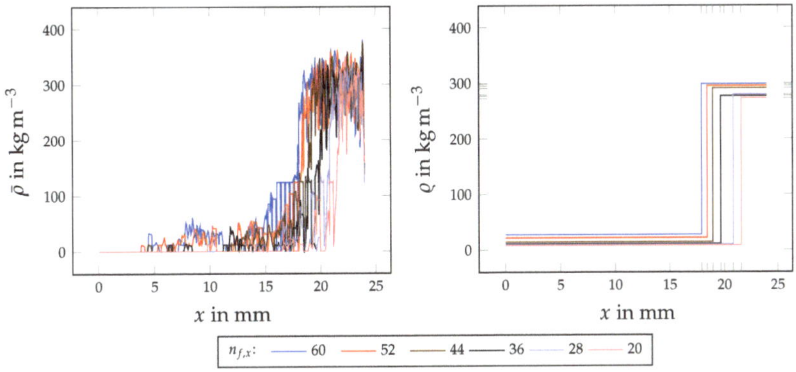

Figure 19. PM density profile $\bar{\rho}(x)$ (**left**) and mean density profile $\varrho(x)$ (**right**) at plug state for different numbers of fragment rows $n_{f,x}$

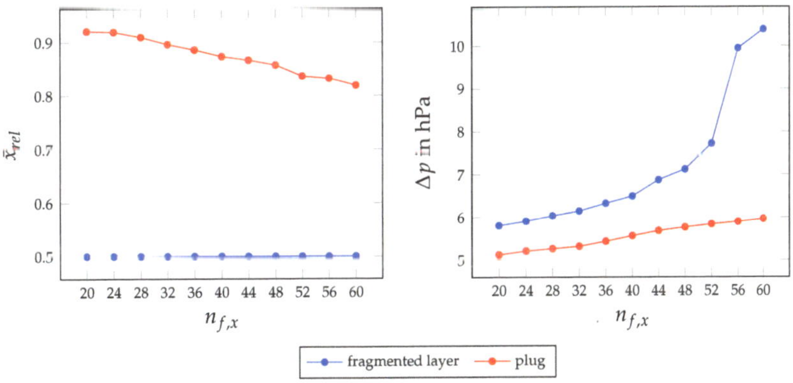

Figure 20. Relative average PM position $\bar{x}_{rel}$ and pressure drop Δp for different numbers of fragment rows $n_{f,x}$ at fragmented layer and plug state.

Some intermediate conclusions can be drawn from the investigation of all three topology variations, which can be attributed to two major causes: Firstly, the total PM mass present in the channel. As no continuing soot oxidization is considered, this mass stays constant throughout each simulation. Secondly, the free substrate surface area available for the flow at fragmented layer state. A reduction in this area leads to an inhomogeneity in the velocity distribution over the channel's cross-section, which includes the development of a layer of elevated velocities closely above the fragments. This potentially leads to larger hydrodynamic forces on the deposited fragments and less propulsion in the channel's centre during transport. In general, it can be stated that the topology at the fragmented layer state affects the final plug structure, while the impact on the pressure drop is smaller at plug state. The pressured dependency at plug state shows a more linear behaviour than the part super-linearly increasing fragmented layer state with potential jumps (cf. Figure 20).

More specifically, larger layer heights, which represent longer loading times or higher PM concentrations, lead to greater pressure drops at plug state and an increased difference between both states. Larger fragment dimensions, associated with a smaller soot-to-ash ratio, lead to the same result. The more fragments remain after the soot oxidization, the larger the final pressure drop, with a pressure jump at $n_{f,x} = 65$ and a substrate surface coverage of approximately 50%. For a given PM mass present in the channel, the transition towards a plug becomes especially beneficial for the pressure drop when the fragmented PM layer only provides a small free substrate surface area available for the flow.

4.3. *Influence of PM Properties and Operating Conditions*

In addition to the investigation of the layer topology's influence, the four remaining factors in Table 1, which do not alter the fragmented layer state, are considered in the following. Those include the PM density in Section 4.3.1, the fragments' mechanical properties in Section 4.3.2, the adhesion between fragments and the porous substrate in Section 4.3.3 and the inflow velocity in Section 4.3.4.

4.3.1. Influence of PM Density

As a variation in the PM density represents a change in the fragment's properties that eventually join the plug formation, the plug's average density directly depends on the prescribed one. The volume occupied by the plug shows a non-negligible sensitivity, as it slightly grows with decreasing PM density (c.f. Figure 21). Due to the fact that the PM density neither impacts the hydrodynamic surface force nor the contact forces in (1), it only affects a fragment's inertia, which is solely relevant for the detachment and transport process. Fragments with higher densities accordingly hit an existing plug structure with more momentum, causing a small compaction. Such a compaction can clearly be identified in the behaviour of the relative average PM position $\bar{x}_{rel}$ in Figure 22.

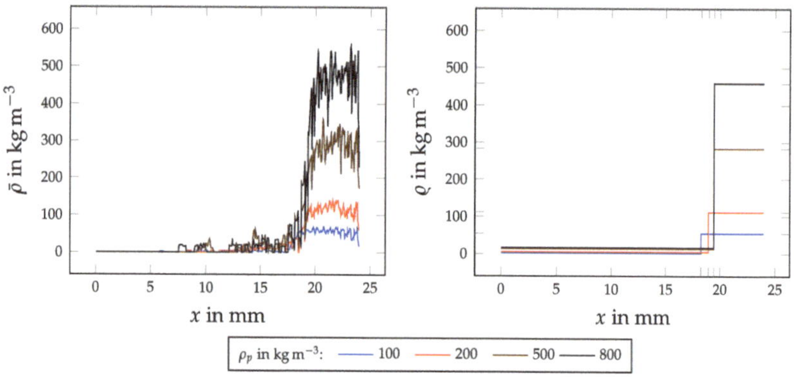

Figure 21. PM density profile $\bar{\rho}(x)$ (**left**) and mean density profile $\varrho(x)$ (**right**) at plug state for different PM densities ρ_p.

The pressure drop at plug state decreases respectively with decreasing plug volume. A larger PM density, therefore, leads to a smaller pressure drop at plug state due to the formation of a more compact plug as a result of higher inertia.

4.3.2. Influence of Mechanical Properties

The fragments' mechanical properties neither affect the fragmented layer state nor the detachment process. While not directly impacting the transport either, any contact with the substrate wall or other fragments on their way can cause a change in trajectory or velocity. The density profile in Figure 23 therefore exhibits no distinct difference in the detachability.

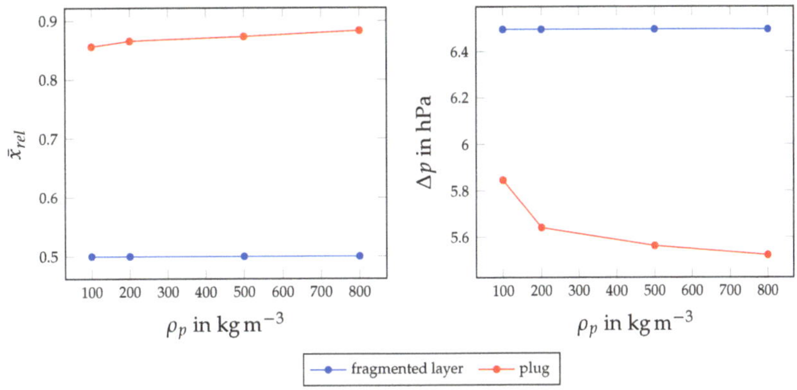

Figure 22. Relative average PM position $\bar{x}_{rel}$ and pressure drop Δp for different PM densities ρ_p at fragmented layer and plug state.

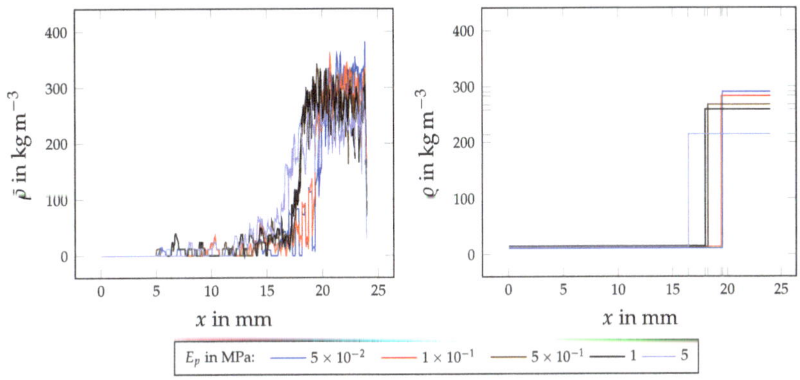

Figure 23. PM density profile $\bar{\rho}(x)$ (**left**) and mean density profile $\varrho(x)$ (**right**) at plug state for different fragment moduli of elasticity E_p.

It can be observed that a large elasticity of $E_p \geq 1\,\text{MPa}$ leads to an increase in the plug volume, as repulsion forces become larger at fragment–fragment contacts according to (1) and counteract further compaction. As the density inside the plug becomes smaller for those as well, the plug's porosity increases additionally. The relative average PM position in Figure 24 directly reflects the volume increase due to the less compact packing.

The pressure drop at plug state initially increases with the plug volume. It then transitions into a nearly constant value of $\Delta p \approx 5.9\,\text{hPa}$ at $E_p = 5 \times 10^{-1}\,\text{MPa}$ due to the continuous increase in the plug's porosity, which compensates for the reduced volume available for the flow. The results suggest that using additives, which decrease the fragments' stiffness, would be beneficial for the reduction in the pressure drop due to a more compact plug.

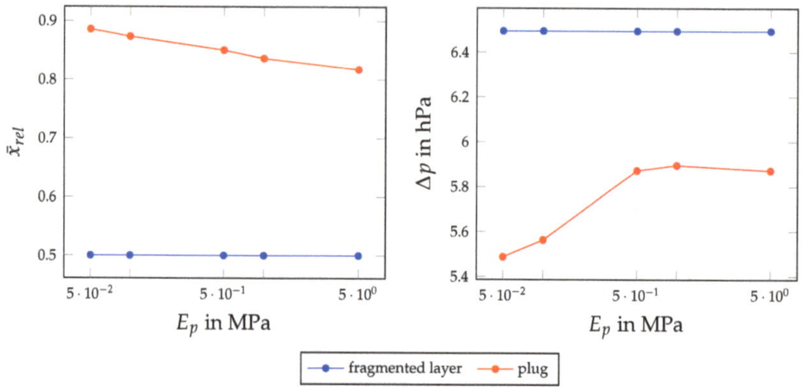

Figure 24. Relative average PM position $\bar{x}_{rel}$ and pressure drop Δp for different fragment moduli of elasticity E_p at fragmented layer and plug state.

4.3.3. Influence of Adhesive Forces

According to the studies conducted in Hafen et al. [11], detachment of the fragments in the first row is assisted by a hydrodynamic force of $F_N = 29.2\,\mu\text{N}$ assuming an inflow velocity of $\bar{u}_{in} = 10\,\text{m}\,\text{s}^{-1}$. As the acting forces are significantly smaller for all other rows, no detachment occurs with adhesion forces in normal direction of the substrate's surface of $F_N = 30\,\mu\text{N}$ or larger present. For the remaining range of considered adhesion forces in Table 1, the density profile is shown in Figure 25.

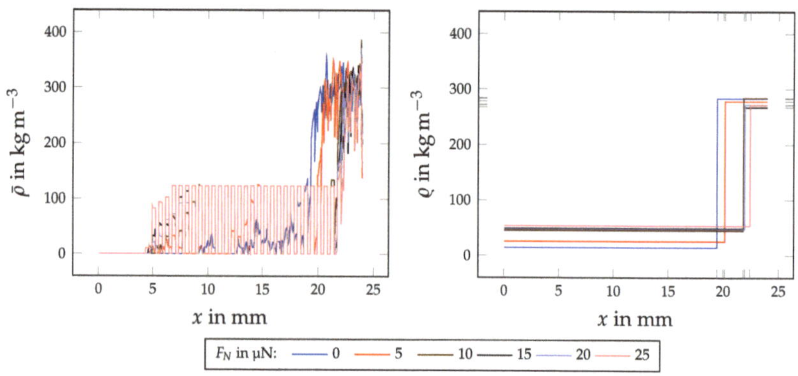

Figure 25. PM density profile $\bar{\rho}(x)$ (**left**) and mean density profile $\varrho(x)$ (**right**) at plug state for different adhesive forces F_N.

As predicted by the detachment likelihood formulated in Hafen et al. [11], the fragment rows closest to the inlet detach first when reducing the adhesion force to $F_N < 29.2\,\mu\text{N}$. With a continuous adhesion reduction, more neighbouring fragment rows become detachable. While the plug volume stays similar for higher values, it increases for $F_N \leq 5\,\mu\text{N}$. Thus, the relative average PM position in Figure 26 can at smaller adhesive forces be attributed to the plug volume only. At larger ones, it then reflects regions of partial detachment and fragment interception, similarly to findings in Figure 12 between $d_z = 42.5\,\mu\text{m}$ and $d_z = 148.75\,\mu\text{m}$.

The pressure drop at the plug state revolves around a small range between 5.40 hPa and 5.59 hPa and stays nearly constant for adhesion forces of $F_N \geq 10.0\,\mu\text{N}$, which corresponds to the observed similar plug volume. For smaller adhesion forces, the pressure drop increases slightly due to the increased plug volume. It becomes clear that the removal of the first few fragment rows already causes the major part of the pressure drop reduction with respect to the fragmented layer state. Additionally, it can be observed that adhesive forces are only relevant around the detachment threshold of the first fragment row, as a further reduction has a negligible influence on the detachment and the pressure drop accordingly.

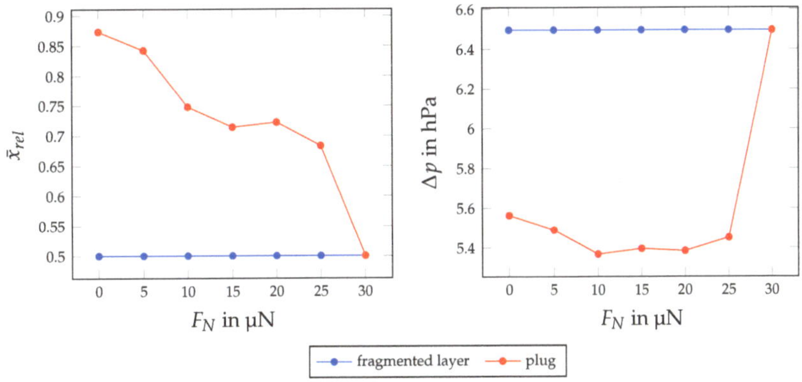

Figure 26. Relative average PM position $\bar{x}_{rel}$ and pressure drop Δp for different adhesive forces F_N at fragmented layer and plug state.

4.3.4. Influence of Inflow Velocity

The inflow velocity $\bar{u}_{in}$ is directly responsible for the magnitude of the hydrodynamic forces present in the channel. It accordingly impacts the fragmented layer state, the detachment process, the pneumatic transport and the plug formation. A velocity of $\bar{u}_{in} \leq 8\,\mathrm{m\,s^{-1}}$ shows a significant number of undetachable fragments in Figure 27.

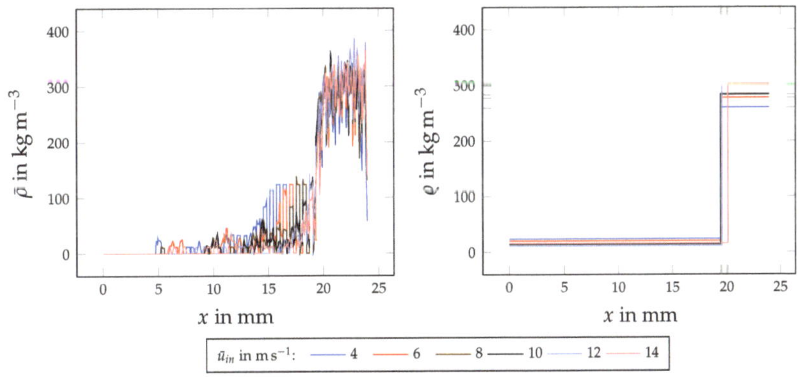

Figure 27. PM density profile $\bar{\rho}(x)$ (**left**) and mean density profile $\varrho(x)$ (**right**) at plug state for different inflow velocities $\bar{u}_{in}$.

For higher inflow velocities, the plug's density profile exhibits a very similar structure, with small differences in the plug's mean density. The relative average PM position in Figure 28 does not show a large sensitivity on the inflow velocity either.

A slight increase is caused by the continuously increasing inertia of fragments hitting the plug and causing a compaction similarly to the observations in Figure 22. At smaller velocities, this effect is superimposed by the increasing number of detachable fragments. As a result of an increased momentum loss due to a higher wall through-put, the pressure drop shows a strong sensitivity to the inflow velocity for both states. While the plug state exhibits a nearly linear profile, the fragmented layer state increases super-linearly, leading to a continuously growing difference between both states until reaching $\Delta p = 214.3\,\mathrm{Pa}$

at $\bar{u}_{in} = 14\,\mathrm{m\,s^{-1}}$. In comparison with the strong relation between gas velocity and momentum loss in the filter substrate, however, the superposed pressure drop dependency at plug state proves to be less relevant.

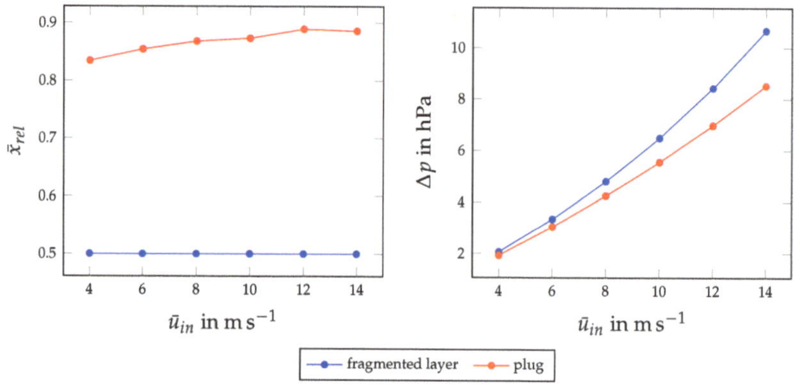

Figure 28. Relative average PM position $\bar{x}_{rel}$ and pressure drop Δp for different inflow velocities $\bar{u}_{in}$ at fragmented layer and plug state.

5. Conclusions

In this work, the dynamic rearrangement process during the regeneration of a wall-flow filter channel was investigated using the open source software OpenLB [12,13]. As a direct sequel to the investigation of the static fragmented layer exposed to elevated fluid velocities in Hafen et al. [11], the present work additionally investigated the dynamic behaviour during fragment detachment, fragment transport and plug formation. First, the complete rearrangement process was evaluated in detail, considering a constant base configuration for the influence factors (cf. Section 4.1). With it, it could be shown that fragments detach row-by-row, confirming the assumption of the predecessor work. Fragment–fragment contact in the channel's front turned out to potentially cause tear-off, while it causes interception in the channel-mid. The inclusion of the discrete contact model proved to enable the successful formation of an end plug occupying a significant amount of the available channel volume. A simple mean density methodology was presented and could be shown to represent a reproducible way of determining the plug extent and its average density, which agrees well with values reported by Kimura et al. [16], Sappok et al. [17] and Dittler [6]. Afterwards, the impact of influence factors that lead to a change in the fragmented layer topology on both the initial flow condition and the rearrangement process were laid out (cf. Section 4.2). While the topology of the fragmented layer could be shown to affect the final plug structure, its pressure drop impact on the initial fragmented layer state was identified to be greater. Larger layer heights, which represent longer loading times or higher PM concentrations, could be shown to lead to greater pressure drops at plug state and an increased difference between both states. Larger fragment dimensions, associated with a smaller soot-to-ash ratio, were shown to lead to the same result. It turned out that a larger number of fragments remaining after the soot oxidization leads to a larger final pressure drop. For a given mass of PM present in the channel, the transition towards a PM plug was shown to be especially beneficial when the fragmented layer only provides a small free substrate surface area available for the flow. Lastly, additional PM properties and operating condition that do not alter the initial fragmented layer topology were investigated (cf. Section 4.3). A larger PM density could be shown to lead to a smaller pressure drop at plug state due to the formation of a more compact plug as a result of higher fragment inertia. It was reasoned that the use of additives, which decrease the fragments' stiffness, would be beneficial for the reduction in the pressure drop due to a more compact plug. It could be shown that the removal of the first few fragment rows already causes the major part of the pressure drop reduction with respect to the fragmented layer state. Additionally, it could be shown that adhesive forces are only relevant around the detachment threshold of the first fragment row, as a further reduction has a negligible influence on the detachment and the pressure drop accordingly. In comparison with the strong relation between gas velocity and momentum loss in the filter substrate, the superposed pressure drop dependency at plug state was identified as less relevant.

While not attempting a systematic comparison of the different influence factors with each other, it can be argued that the inflow velocity (cf. Section 4.3.4) and the layer structure (cf. Section 4.2.3) have the largest impact on the final pressure drop. They can, accordingly, be designated as the most crucial ones for the dynamic rearrangement process in wall-flow filters.

The present work demonstrated the successful extension of the resolved particle methodology via a discrete method to treat particle–particle and particle–wall interactions and its application to the transient rearrangement process in wall-flow filters. Additionally, it reveals the sensitivity of relevant influence factors related to the fragmented layer topology, the PM properties and the operating conditions, and quantifies their impact on the rearrangement process. This way, it contributes to the formulation of respective predictions on the deposition pattern formation, which impact engine performance, fuel consumption and service life of wall-flow filters.

Author Contributions: Conceptualization, N.H., A.D. and M.J.K.; methodology, N.H., J.E.M., A.D. and M.J.K.; software, N.H., J.E.M. and M.J.K.; validation, N.H.; formal analysis, N.H., J.E.M., A.D. and M.J.K.; investigation, N.H.; resources, A.D. and M.J.K.; data curation, N.H.; writing—original draft preparation, N.H.; writing—review and editing, N.H., J.E.M., A.D. and M.J.K.; visualization, N.H.; supervision, A.D. and M.J.K.; project administration, N.H., A.D. and M.J.K.; funding acquisition, N.H., A.D. and M.J.K. All authors have read and agreed to the published version of the manuscript.

Funding: Funded by the Deutsche Forschungsgemeinschaft (DFG, German Research Foundation)—422374351.

Data Availability Statement: Not applicable.

Acknowledgments: We acknowledge support by the KIT-Publication Fund of the Karlsruhe Institute of Technology. This work was performed on the HoreKa supercomputer funded by the Ministry of Science, Research and the Arts Baden-Württemberg and by the Federal Ministry of Education and Research.

Conflicts of Interest: The authors declare no conflict of interest.

Abbreviations

The following abbreviations are used in this manuscript:

CFD	computational fluid dynamics
LBM	lattice Boltzmann method
NSE	Navier–Stokes equation
PM	particulate matter

Nomenclature

The following symbols are used in this manuscript:

u	fluid velocity
p	fluid pressure
x	position
t	time
N	resolution of voxel mesh
u_p	particle velocity
ω_p	particle angular velocity
m_p	particle mass
I_p	particle's momentum of inertia

$Y_{p,kin}$	particle's kinetic energy
l_x	channel length
l_y	channel width
$l_{x,s}$	scaled channel length
$\bar{u}_{in}$	average inflow velocity
ρ_p	particle density
ϱ	mean density (along layer and plug)
d_x	fragment's x-dimension
d_{yz}	fragment's equilateral width
F_N	hydrodynamic normal force
$n_{f,x}$	number of fragment rows over channel length
t_{frag}	time at converged fragmented layer state
t_{plug}	time at final plug state
C_{in}	Inlet channel domain
C_{out}	Outlet channel domain
$\boldsymbol{F}_{c,N}$	normal contact force
$E*$	effective modulus of elasticity
k	contact type constant
V_c	contact overlap volume
d_c	contact indentation depth
c	damping factor
d_n	relative velocity between objects
E_A	modulus of elasticity of object A
ν_A	Poisson's ratio of object A
e	coefficient of restitution
C_{in}	domain of central inlet channel
C_{out}	domain of representative outlet channel

References

1. Wang, Y.; Kamp, C.J.; Wang, Y.; Toops, T.J.; Su, C.; Wang, R.; Gong, J.; Wong, V.W. The origin, transport and evolution of ash in engine particulate filters. *Appl. Energy* **2020**, *263*, 114631. [CrossRef]
2. Gaiser, G. Berechnung von Druckverlust, Ruß- und Ascheverteilung in Partikelfiltern. *MTZ-Mot. Z.* **2005**, *66*, 92–102. [CrossRef]
3. Sappok, A.; Govani, I.; Kamp, C.; Wang, Y.; Wong, V. In-Situ Optical Analysis of Ash Formation and Transport in Diesel Particulate Filters during Active and Passive DPF Regeneration Processes. *SAE Int. J. Fuels Lubr.* **2013**, *6*, 336–349. [CrossRef]
4. Ishizawa, T.; Yamane, H.; Satoh, H.; Sekiguchi, K.; Arai, M.; Yoshimoto, N.; Inoue, T. Investigation into Ash Loading and Its Relationship to DPF Regeneration Method. *SAE Int. J. Commer. Veh.* **2009**, *2*, 164–175. [CrossRef]
5. Aravelli, K.; Heibel, A. Improved Lifetime Pressure Drop Management for Robust Cordierite (RC) Filters with Asymmetric Cell Technology (ACT). In *Proceedings of the SAE World Congress & Exhibition*; SAE International: Warrendale, PA, USA, 2007. [CrossRef]
6. Dittler, A. Ash Transport in Diesel Particle Filters. In *Proceedings of the SAE 2012 International Powertrains, Fuels & Lubricants Meeting*; SAE International: Warrendale, PA, USA, 2012. [CrossRef]
7. Dittler, A. *Abgasnachbehandlung mit Partikelfiltersystemen in Nutzfahrzeugen*, 1st ed.; Wuppertaler Reihe zur Umweltsicherheit; Shaker: Herzogenrath, Germany, 2014.
8. Wang, Y.; Kamp, C. The Effects of Mid-Channel Ash Plug on DPF Pressure Drop. In *Proceedings of the SAE 2016 World Congress and Exhibition*; SAE International: Warrendale, PA, USA, 2016. [CrossRef]
9. Hafen, N.; Dittler, A.; Krause, M.J. Simulation of particulate matter structure detachment from surfaces of wall-flow filters applying lattice Boltzmann methods. *Comput. Fluids* **2022**, *239*, 105381. [CrossRef]
10. Hafen, N.; Thieringer, J.R.; Meyer, J.; Krause, M.J.; Dittler, A. Numerical investigation of detachment and transport of particulate structures in wall-flow filters using lattice Boltzmann methods. *J. Fluid Mech.* **2023**, *956*, A30. [CrossRef]
11. Hafen, N.; Marquardt, J.E.; Dittler, A.; Krause, M.J. Simulation of Particulate Matter Structure Detachment from Surfaces of Wall-Flow Filters for Elevated Velocities Applying Lattice Boltzmann Methods. *Fluids* **2023**, *8*, 99. [CrossRef]
12. Krause, M.J.; Kummerländer, A.; Avis, S.J.; Kusumaatmaja, H.; Dapelo, D.; Klemens, F.; Gaedtke, M.; Hafen, N.; Mink, A.; Trunk, R.; et al. OpenLB—Open source lattice Boltzmann code. *Comput. Math. Appl.* **2021**, *81*, 258–288. [CrossRef]
13. Kummerländer, A.; Avis, S.; Kusumaatmaja, H.; Bukreev, F.; Crocoll, M.; Dapelo, D.; Hafen, N.; Ito, S.; Jeßberger, J.; Marquardt, J.E.; et al. OpenLB Release 1.6: Open Source Lattice Boltzmann Code. 2023. Available online: https://doi.org/10.5281/zenodo.7773497 (accessed on 18 June 2023).

14. Konstandopoulos, A.G.; Skaperdas, E.; Warren, J.; Allansson, R. Optimized Filter Design and Selection Criteria for Continuously Regenerating Diesel Particulate Traps. In *Proceedings of the International Congress & Exposition*; SAE International: Warrendale, PA, USA, 1999. [CrossRef]
15. Thieringer, J.R.D.; Hafen, N.; Meyer, J.; Krause, M.J.; Dittler, A. Investigation of the Rearrangement of Reactive ash; Inert Particulate Structures in a Single Channel of a Wall-Flow Filter. *Separations* **2022**, *9*, 195. [CrossRef]
16. Kimura, K.; Lynskey, M.; Corrigan, E.R.; Hickman, D.L.; Wang, J.; Fang, H.L.; Chatterjee, S. Real World Study of Diesel Particulate Filter Ash Accumulation in Heavy-Duty Diesel Trucks. In *Proceedings of the Powertrain & Fluid Systems Conference and Exhibition*; SAE International: Warrendale, PA, USA, 2006. [CrossRef]
17. Sappok, A.; Santiago, M.; Vianna, T.; Wong, V.W. Controlled Experiments on the Effects of Lubricant/Additive (Low-Ash, Ashless) Characteristics on DPF Degradation. In Proceedings of the Diesel Engine-Efficiency and Emissions Research (DEER) Conference, Dearborn, MI, USA, 4–7 August 2008.
18. Hafen, N.; Krause, M.J.; Dittler, A. Simulation of Particle-Agglomerate Transport in a Particle Filter using Lattice Boltzmann Methods. In *22. Internationales Stuttgarter Symposium*; Bargende, M., Reuss, H.C., Wagner, A., Eds.; Springer: Wiesbaden, Germany, 2022; pp. 292–303.
19. Marquardt, J.E.; Römer, U.J.; Nirschl, H.; Krause, M.J. A discrete contact model for complex arbitrary-shaped convex geometries. *Particuology* **2023**, *80*, 180–191. [CrossRef]
20. Ferziger, J.H.; Perić, M. *Numerische Strömungsmechanik*; Springer eBook Collection; Springer: Berlin/Heidelberg, Germany, 2008. [CrossRef]
21. Nassauer, B.; Kuna, M. Contact forces of polyhedral particles in discrete element method. *Granul. Matter* **2013**, *15*, 349–355. [CrossRef]
22. Carvalho, A.S.; Martins, J.M. Exact restitution and generalizations for the Hunt–Crossley contact model. *Mech. Mach. Theory* **2019**, *139*, 174–194. . [CrossRef]
23. Ladd, A.J.C. Numerical simulations of particulate suspensions via a discretized Boltzmann equation. Part 1. Theoretical foundation. *J. Fluid Mech.* **1994**, *271*, 285–309. [CrossRef]
24. Latt, J.; Chopard, B.; Malaspinas, O.; Deville, M.; Michler, A. Straight velocity boundaries in the lattice Boltzmann method. *Phys. Rev. E* **2008**, *77*, 056703. [CrossRef] [PubMed]
25. Skordos, P.A. Initial and boundary conditions for the lattice Boltzmann method. *Phys. Rev. E* **1993**, *48*, 4823–4842. [CrossRef]
26. Cooper, R.C.; Bruno, G.; Onel, Y.; Lange, A.; Watkins, T.R.; Shyam, A. Young's modulus and Poisson's ratio changes due to machining in porous microcracked cordierite. *J. Mater. Sci.* **2016**, *51*, 9749–9760. [CrossRef]
27. Pandey, A.; Shyam, A.; Watkins, T.R.; Lara-Curzio, E.; Stafford, R.J.; Hemker, K.J. The Uniaxial Tensile Response of Porous and Microcracked Ceramic Materials. *J. Am. Ceram. Soc.* **2014**, *97*, 899–906. [CrossRef]
28. Thota, S.K.; Cao, T.D.; Vahedifard, F. Poissons Ratio Characteristic Curve of Unsaturated Soils. *J. Geotech. Geoenviron. Eng.* **2021**, *147*, 04020149. [CrossRef]
29. Savitzky, A.; Golay, M.J.E. Smoothing and differentiation of data by simplified least squares procedures. *Anal. Chem.* **1964**, *36*, 1627–1639. [CrossRef]

fluids

MDPI

Article

Numerical and Experimental Investigation of Supersonic Binary Fluid Ejector Performance

Nikolay Bukharin [1,*] and Mouhammad El Hassan [2]

1 School of Manufacturing and Automation, Southern Alberta Institute of Technology, Calgary, AB T2M 0L4, Canada

2 Mechanical Engineering Department, Prince Mohammad Bin Fahd University, Al Khobar 34218, Saudi Arabia; melhassan@pmu.edu.sa

* Correspondence: nikolay.bukharin@sait.ca

Abstract: Ejectors are simple mechanical devices with no moving parts which convert the pressure energy of a motive fluid to kinetic energy and generate suction of the secondary fluid. The ability to recover waste heat, to operate using solar power and the ability to use geothermal energy make ejector-based systems attractive in different industrial applications. The main challenge of ejector-based refrigeration systems is their relatively low coefficient of performance (COP). In order to increase the ejector performance, two chemically distinct fluids can be used in the refrigeration cycle. It is suggested that a higher molecular mass be used for the motive fluid to improve the entrainment ratio of the binary fluid ejector (BFE) and thus the system COP. Inert gas combinations of argon–helium and krypton–air are studied in this paper using computational fluid dynamics (CFDs) and experimental measurements. All CFD cases were axisymmetric and the appropriate turbulence model was selected based on experimental validation. Specifically, the entrainment ratio and the static pressure along the ejector wall were measured to validate the CFD predictions. It was found that the molar entrainment ratio was significantly higher in argon–helium compared to krypton–air. The static pressure measurements along the wall, in addition, exhibited good agreement with the results obtained via computational fluid dynamics (CFDs).

Keywords: supersonic ejector; CFDs; binary fluid ejector; mixing; entrainment ratio

Citation: Bukharin, N.; El Hassan, M. Numerical and Experimental Investigation of Supersonic Binary Fluid Ejector Performance. *Fluids* **2023**, *8*, 197. https://doi.org/10.3390/fluids8070197

Academic Editors: Francesco De Vanna and D. Andrew S. Rees

Received: 17 May 2023
Revised: 19 June 2023
Accepted: 27 June 2023
Published: 29 June 2023

1. Introduction

A supersonic steam ejector is a mechanical device that uses steam as the motive fluid to create a vacuum or low pressure that entrain a secondary fluid in a process system [1]. The ejector has no moving parts and is relatively simple in design, making it a cost-effective alternative to traditional vacuum pumps. One of the main advantages of using a steam ejector is its ability to utilize waste heat [2], making it an efficient and cost-effective solution for a variety of applications. Additionally, steam ejectors can be used with renewable energy sources such as solar energy [3], providing a sustainable solution in the refrigeration, chemical and power generation industries.

While steam ejectors offer numerous advantages over traditional refrigeration systems, they also have some disadvantages. One key drawback of steam ejectors is their relatively low coefficient of performance (COP), which refers to the ratio of the energy (cooling, refrigeration, etc.) output to the energy input. The COP of steam ejector systems is generally lower than that of other refrigeration systems, which can make them less attractive for some applications. Additionally, the performance of steam ejectors is highly dependent on the operating conditions and ejector geometry.

Numerous studies have been conducted to explore both the performance of supersonic steam ejectors and methods for enhancing their performance. The effects of the primary steam nozzle distance and mixing chamber throat diameter on steam ejector performance have been studied [4]. It was shown that the distance between the primary steam nozzle and the mixing chamber greatly impacts the steam ejector's entrainment ratio (ER) and thus the ejector performance. Altering the distance between the primary steam nozzle's outlet section and the mixing chamber's inlet section can result in the relocation of the shock wave, which may improve or worsen the steam ejector's entrainment mechanism. As described in [5], the coefficient of performance (COP) for an ejector system is directly related to the mass entrainment ratio. Hence, the entrainment ratio is a standard measure of an ejector's efficiency. In [6], the performance of a small capacity solar thermally driven variable ejector geometry cooling system, using R600a, was experimentally studied under different operating conditions. By adjusting the nozzle exit position (NXP) within the ejector in response to operating conditions, performance enhancement of up to 24% was achieved, in comparison to the performance of an ejector with fixed geometry. A numerical analysis in [7] specifically examined the ability to predict the entrainment ratio using different nozzle geometries. Mach number contours revealed the presence of two sets of oblique shock waves and their propagation inside the ejector. Notably, this research demonstrated that the position of the shock waves inside the mixing chamber and the expansion angle of the primary fluid jet in the mixing chamber were key factors affecting the ejector's performance. Another study [8] examined the impact of supersonic nozzle placement and size on pressure distribution, as well as analyzed the flow dynamics for various geometry parameters in order to identify the mechanisms responsible for enhanced ejector performance. The key finding was that the optimal performance of a supersonic ejector requires the simultaneous consideration of different geometric parameters under specific operational conditions.

To enhance the performance of an ejector, one approach consists of using a binary fluid ejector (BFE), where two distinct chemicals are employed as the primary and secondary fluids. Initially introduced in 1939, the notion that an ejector's entrainment ratio is influenced by the molecular weight of the primary and secondary fluids was demonstrated in [9]. In [10], an ejector heat pump was tested and analyzed. It was shown that the ejector performance was improved by using binary fluids (HFE7500 and water). In [11], CFD simulations were used to explore the impact of thermo-physical characteristics of the primary and secondary fluids on ejector performance. This study revealed that the entrainment ratio is primarily dependent on factors such as molecular mass, specific heat ratio, ejector geometry and operational parameters. It was concluded that the entrainment ratio is not significantly affected by dynamic viscosity, thermal conductivity or mass diffusivity. A supersonic ejector that has low entrainment ratios (0.004–0.065) was studied in [12] for fuel cell applications with air, helium and argon as the secondary fluids, while the primary fluid was air. It was shown that the entrainment ratio of the ejector strongly depends on the molecular weight of the secondary fluid, with helium giving the lowest ER and argon giving the highest ER.

This paper describes a series of experiments using inert gases to validate a CFD model's ability to predict the mass entrainment ratio and thus describe the flow dynamics. Mass flow rates and wall pressure measurements were conducted and compared to CFD predictions. The study aimed to deepen the understanding of the supersonic flow dynamics inside the ejector. It is common practice to utilize RANS-based turbulence models to analyze the performance of ejectors in axisymmetric flows; this provides adequate predictions of the static pressure, the temperature and the entrainment ratio, but it should be noted that the choice of the RANS model for analyzing the flow inside an ejector depends on the operating conditions. Typically, the SST (shear stress transport) k-ω model [13–15] or realizable k-ε model [16,17] are commonly used.

This paper focuses on the analysis of two pairs of inert gases, namely argon and helium, and krypton and air, which exhibit significant variance in their molecular weights. To the best of the authors' knowledge, there has not been any previous study conducted on the behavior of these two pairs of gases within a supersonic ejector. A preliminary study showed that the SST k-ω model was the most suitable choice, as it demonstrated better agreement with experimental data for the inert gas pairs being used. Through an examination of flow dynamics and wall quantities such as wall shear stress and static pressure, new insights into the mixing between primary and secondary fluids were discovered. Such discoveries hold significant interest for both the engineering and research communities.

2. Experimental Setup

Figure 1 illustrates the schematic diagram of the experimental setup. The testing system for inert gases included gas bottles equipped with pressure regulators and needle valves to ensure precise pressure control. Additionally, the system included electric circulation heaters with PID controllers, a configurable ejector with an attached pressure manifold, a vacuum pump (liquid sealed with silencer/separator) and a heat exchanger. All of the flow components and heaters were covered with ceramic and fiberglass insulation.

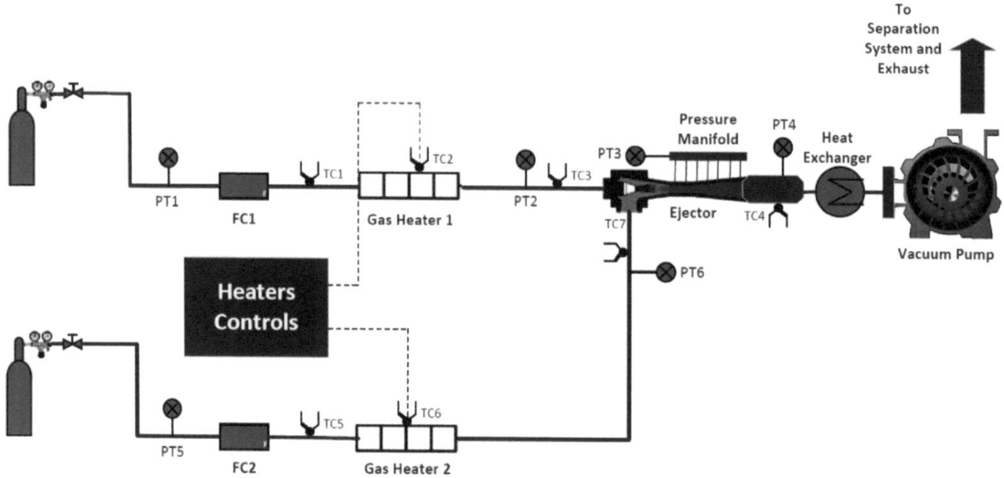

Figure 1. Schematic of the binary fluid ejector experimental setup.

For temperature measurements, a total of seven type K thermocouples (labeled TC1–TC7) were used. Thermocouples were calibrated using a reference probe with a reference junction, a thermocouple calibration furnace and a high-precision temperature scanner. All uncertainties (measurement noise, readout accuracy of reference probe, reference probe calibration uncertainty and long-term drift, furnace axial and radial uniformities) were combined using root sum squares (RSSs). A coverage factor of k = 2 was used in these calculations (which corresponds to a 95% confidence interval). At 500 °C (max. heat temperature), the expanded uncertainty was evaluated as being 0.5 °C. To measure pressure at different points within the system, six absolute static pressure transducers (labeled PT1–PT6) were used. The static error band was utilized as a means of assessing the expected accuracy of a pressure sensor operating under constant temperature conditions. It is expressed as a percentage of the sensor's full range (span), denoted by a ± value, and is relative to the best straight line (BSL). The pressure transducers used were OMEGA PX309-300A5V (range: 0–300 psia, static accuracy (combined effects of linearity, hysteresis and repeatability): ±0.25% BSL, maximum) for PT1 and PT2, OMEGA PX409-005AV (range: 0–5 psia, static accuracy (combined effects of linearity, hysteresis and repeatability): ±0.08% BSL, maximum) for PT3, OMEGA PX309-015A5V (range: 0–15 psia, static accuracy (combined effects of linearity, hysteresis and repeatability): ±0.25% BSL, maximum) for PT4 and OMEGA PX309-005A5V (range: 0–5 psia, static accuracy (combined effects of linearity, hysteresis and repeatability): ±0.25% BSL, maximum) for PT5 and PT6. Two OMEGA FMA5544ST Mass Flow Controllers labeled FC1 and FC2, with a range of 0–500 standard L/min and an accuracy of 0–20% range ±3%FS for the lower range and 20–100% range ±1.5% FS for the upper range, were used for flow rate measurements and control. The programming language used was National Instruments LabVIEW. The thermocouples were positioned 2 inches away from the inlets and outlet of the ejector, while the pressure sensors were located 4 inches away from the ejector's inlets and outlet. To validate the CFD models, a configurable ejector was constructed using the Engineering Sciences Data Unit (ESDU) recommendations: ESDU 86030 [18]. ESDU 86030 provides systematic calculation methods, utilizing performance charts, to facilitate the optimal design of steam/liquid and steam/gas ejectors. Additionally, the document outlines the mechanical design considerations for various components such as the primary nozzle, mixing chamber and contraction or diffuser. The ejector offers adaptable geometry control and the ability to reposition the nozzle within it. The ejector consists of four primary components: the nozzle, mixing chamber, throat and diffuser.

The mixing chamber length was 145 mm, the throat length was 95 mm and the diffuser length was 180 mm. The throat had a diameter of 19 mm, and the diffuser's inlet diameter and outlet diameter were 19 mm and 40 mm, respectively. The diameter of the mixing chamber's inlet diameter was 29 mm. Twenty-nine pressure ports were installed along the ejector body for static pressure measurements. The detailed nozzle geometry is shown in Figure 2. The nozzle exit position (NXP) is defined as the distance between the nozzle exit plane and the mixing chamber inlet plane, and it was fixed at 35 mm. The primary objective of this paper was to maintain this configuration, along with fixed inlet and outlet pressure, and investigate the resulting ejector operation and flow dynamics. The only variable being altered was the inert gas, with different gases utilized for primary and secondary flow. The experimental setup is shown in Figure 3.

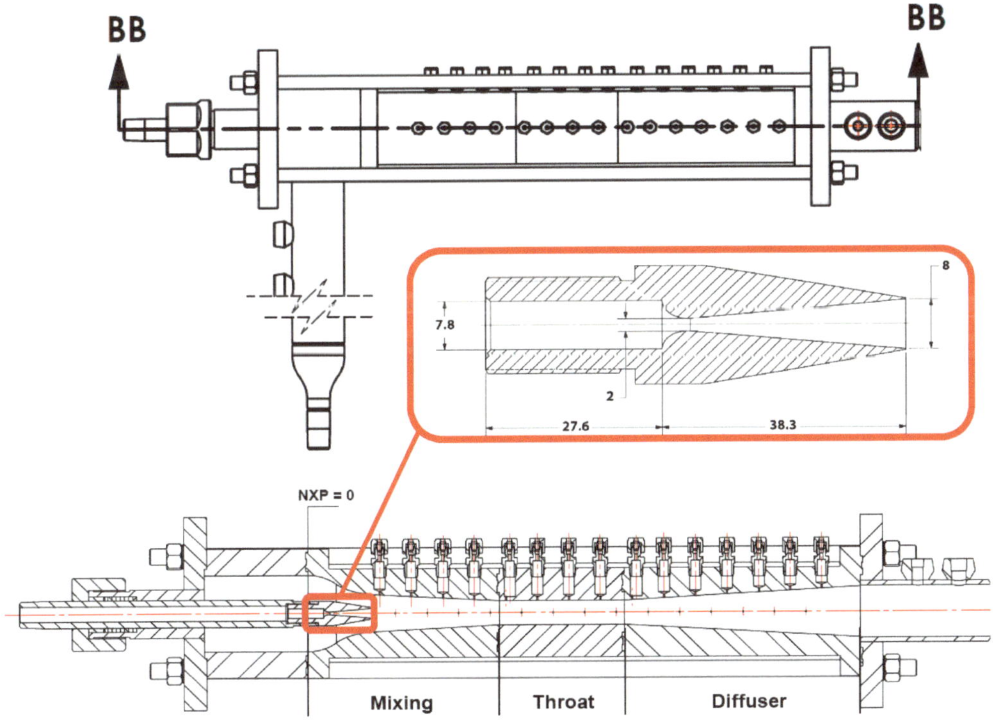

Figure 2. Configurable ejector (all dimensions are in mm).

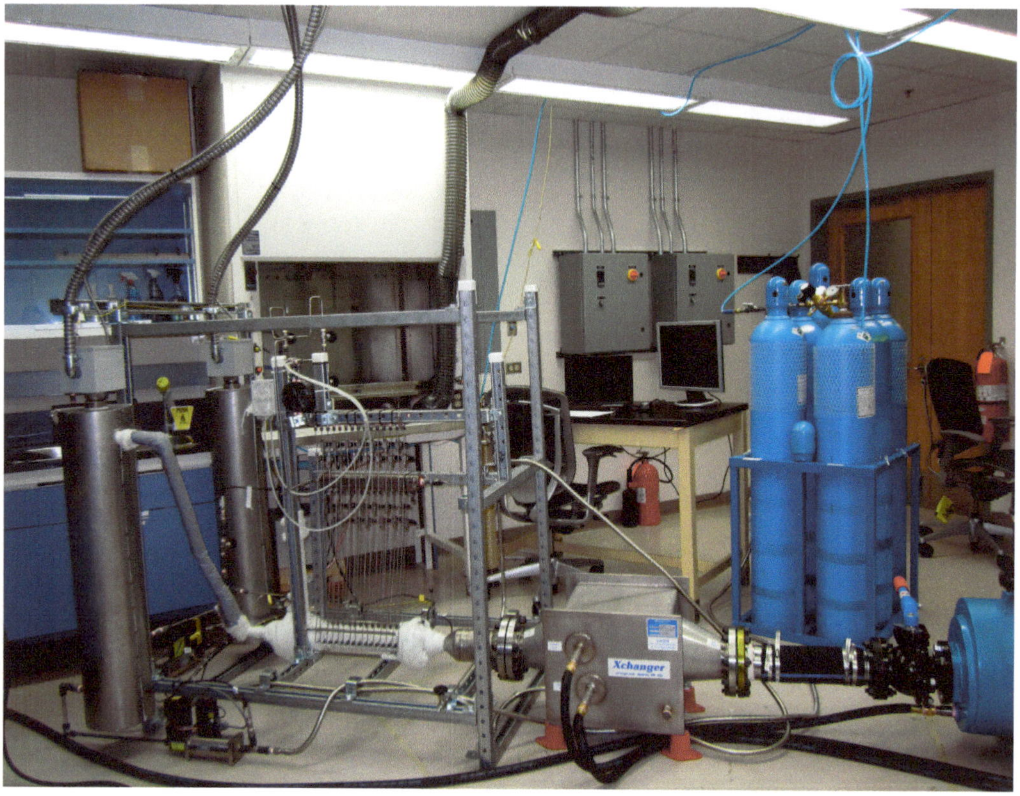

Figure 3. Experimental setup.

3. Experimental Procedures

In this study, two pairs of inert gases were tested: argon–helium and krypton–air. The primary and secondary flows were maintained at specific flow conditions, as follows: a temperature of 200 °C for the primary flow, a temperature of 50 °C for the secondary flow, primary flow pressure of 0.9 MPa, secondary flow pressure of 10 kPa and pressure at the ejector exit of 15 kPa. These pressures were regulated through a system of valves, pressure regulators and a vacuum pump. Initially, the vacuum pump was activated, and then the primary line was opened, with the primary pressure being adjusted using the pressure regulator and needle valve for fine tuning. Then, the secondary flow was regulated using a second pressure regulator and needle valve. While both lines were active, the nozzle pressure (PT2), secondary line pressure (PT6) and ejector exit pressure (PT4) were adjusted, until reaching given pressure operating conditions. To establish the temperature regime, two PID controllers were used in conjunction with thermocouples TC2 and TC6. The optimal parameters (proportional, integral and derivative) were defined for this set of experiments. The usual time of temperature stabilization was about 6 min. The temperature was monitored using LabVIEW software until stable temperatures were attained in both the primary and secondary lines. Throughout this process, slight adjustments were made to the pressure to account for temperature fluctuations and maintain the designated pressure regime described above. Subsequently, static pressure along the ejector's wall was recorded using a pressure manifold. Each port was opened sequentially, and pressure values were recorded after they had stabilized.

4. CFD Model

In this study, ANSYS FLUENT 18 was used to solve a two-dimensional axisymmetric problem. A mesh independence study was conducted, where the mesh was divided into regions and refined near walls, in the shear layer and around shock waves. The optimal number of cells for CFD analysis was determined to be between 373,000 and 434,000. The grid independence study was performed using the mass entrainment ratio for the krypton–air case (Figure 4).

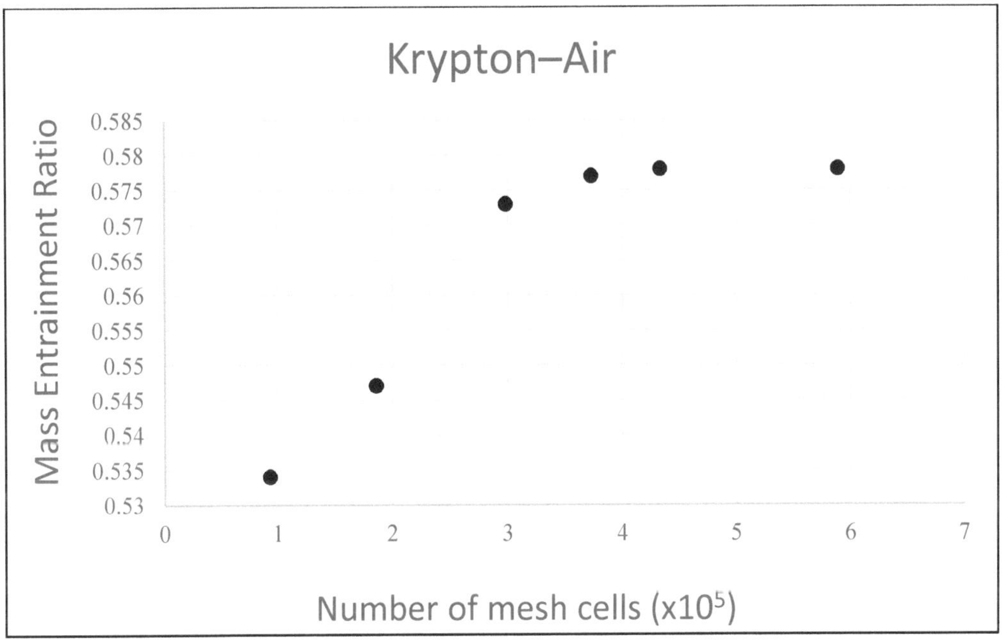

Figure 4. Mesh independence study.

The system of governing equations was solved using density-based and implicit methods, with a second-order upwind scheme employed for the spatial discretization of convective terms and the QUICK method used for the turbulent equations coupled with the energy equation. Various RANS turbulence models were studied, and it was found that the SST k-ω model provided the best agreement with experimental values for entrainment ratios across a wider range of operating conditions and geometries. The following equations of the turbulence model were used in this study [19–22]:

$$\frac{\partial(\rho k)}{\partial t} + \frac{\partial}{\partial x_i}(\rho k u_i) = \frac{\partial}{\partial x_j}\left(\Gamma_k \frac{\partial k}{\partial x_j}\right) + \widetilde{G_k} - Y_k + S_k$$

$$\frac{\partial(\rho \omega)}{\partial t} + \frac{\partial}{\partial x_i}(\rho \omega u_i) = \frac{\partial}{\partial x_j}\left(\Gamma_\omega \frac{\partial \omega}{\partial x_j}\right) + G_\omega - Y_\omega + D_\omega + S_\omega$$

In these equations, $\widetilde{G_k}$ represents the generation of turbulence kinetic energy due to mean velocity gradients, G_ω represents the generation of ω, and Γ_k and Γ_ω represent the effective diffusivity of k and ω, respectively. Y_k and Y_ω represent the dissipation of k and ω due to turbulence. D_ω represents the cross-diffusion term. S_k and S_ω are user-defined source terms.

Pressure inlet boundary conditions were set at the primary and secondary fluid entrances, with a pressure outlet at the diffuser exit. The temperature for the primary flow was 200 °C, and that of the secondary flow was 50 °C. Convergence was determined when the residual for each governing equation was less than 10^{-6}. The mesh structure and the boundary conditions are shown in Figure 5.

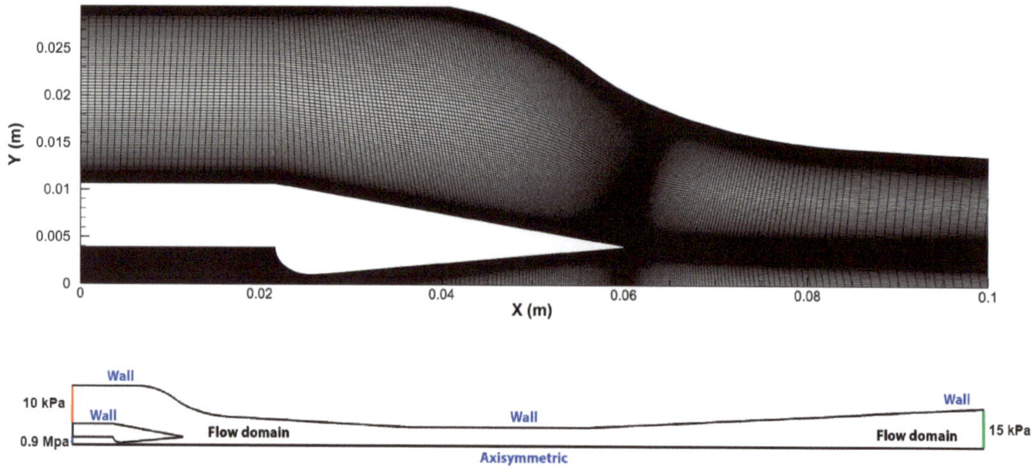

Figure 5. Mesh structure and boundary conditions.

5. Results and Discussion

Figure 6 and Table 1 illustrate the comparison between the experimental results and the CFD predictions. It can be seen that the SST k-ω model gives a good prediction of the static pressure distribution along the ejector wall and a good prediction of the exit temperature and entrainment ratio. It should be noted that the SST k-ω model provides a better prediction of the static pressure distribution for larger differences in molecular weight between the working fluids then in [5]. This could be related to the higher molecular diffusion rates that exist for fluid pairs with a small difference in molecular mass. Indeed, the SST k-ω model primarily focuses on capturing turbulent transport phenomena and may not adequately account for the influence of molecular diffusion.

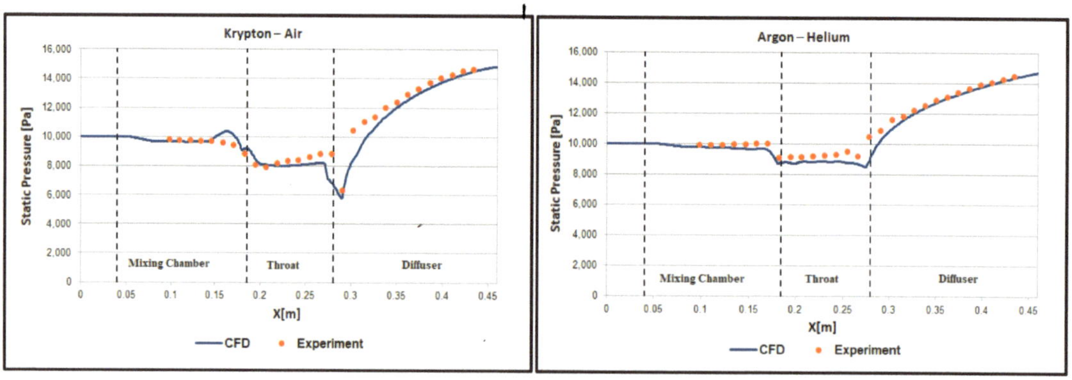

Figure 6. Static pressure profile along the wall. Experimental validation.

Table 1. Exit temperature and entrainment ratio experiment and CFD comparison.

	Krypton–Air		Argon–Helium	
	Diffuser Exit T [K]	**Mass (Molar) Entrainment Ratio, ER**	**Diffuser Exit T [K]**	**Mass (Molar) Entrainment Ratio, ER**
CFD	364	0.578 (1.67)	358	0.243 (2.43)
Experiment	352	0.534 (1.55)	351	0.226 (2.26)
CFDs Error (%)	3.41	8.24	1.99	7.52

CFDs Error (%) = 100 × (CFDs − Experiment)/Experiment; ER = (mass flow of secondary fluid)/(mass flow of primary fluid); molar entrainment ratio: $ER_{mol} = \frac{\dot{m}_2}{\dot{m}_1} \times \frac{M_1}{M_2}$, where $\dot{m}_2$ and $\dot{m}_1$ are the secondary and primary flow rates, respectively, and M_2 and M_1 are the molecular weights of the secondary and primary fluids, respectively.

Two cases were considered in this study for the pairs of inert gases with a significant difference in molar mass. In the first case, primary fluid molar mass was significantly increased by utilizing krypton (83.798 g/mol); in the second case, argon was used as the primary gas (molar mass: 39.948 g/mol) and the molar mass of the secondary gas was significantly decreased by utilizing helium (4.0026 g/mol). Table 1 shows that the mass entrainment ratio was significantly lower for the argon–helium case.

The molar entrainment ratio accounts for the molar mass of the secondary fluid being entrained, and it affects the density and specific volume of the fluid mixture, which in turn influence its compressibility and thermodynamic properties. By considering the molar entrainment ratio, the impact of different gases with varying molecular weights on the ejector's performance could be evaluated more accurately. In certain ejector applications, such as steam ejectors or gas ejectors, the entrainment of fluids with different molar masses might significantly affect the flow dynamics, pressure recovery and overall efficiency. The molar entrainment ratio could provide a deeper understanding of these effects and allow for a more comprehensive assessment of the ejector's performance. Both the mass entrainment ratio and the molar entrainment ratio are important for evaluating the performance of an ejector. The mass entrainment ratio reflects the quantity of entrained secondary fluid, while the molar entrainment ratio considers the molar mass of the entrained fluid, accounting for its thermodynamic properties and its influence on the overall system behavior. Furthermore, as was shown in [5], the COP is proportional to the mass entrainment ratio:

$$\mathrm{COP} = w \cdot f_h$$

where w is the mass entrainment ratio and f_h is the ratio of the evaporator over the boiler enthalpies (if pump energy is being neglected). Therefore, the outcomes of the present study can be useful in refrigerant selection for practical applications. Further analysis on the effect of other system energy exchanges on the system COP was discussed in [15,22].

It can be seen from Table 1 that the molar entrainment ratio is higher in the case of argon–helium compared to krypton–air, whereas the mass entrainment ratio is higher in the case of krypton–air. There could be several potential practical applications of this result, for example, in combustion or chemical reactors, where the composition of the gas mixture significantly affects the process outcomes. By utilizing higher molar entrainment with lower mass entrainment, it could be possible to precisely control the molar composition of the mixture, allowing for targeted gas mixing and achieving specific reaction conditions. Moreover, in gas separation processes, such as distillation or absorption by optimizing molar entrainment, it might be possible to selectively entrain gases with desired properties or components, enabling the efficient separation of the gas mixture. Higher molar entrainment can provide better selectivity in separating specific components based on their molar composition.

Figure 7 displays the Mach number distribution within the ejector for the two studied cases, where Mach number iso-values lower than one were omitted to emphasize the supersonic core's boundaries. The findings reveal a noticeable decrease in the supersonic core expansion within the ejector throat for the argon–helium case when compared to krypton–air. Moreover, a flow blockage is apparent in the krypton–air pair inside the ejector throat, which is also shown in the wall pressure profile (Figure 6) as a pressure drop at the throat exit. Additionally, oblique shockwaves can be observed in the diffuser for both cases.

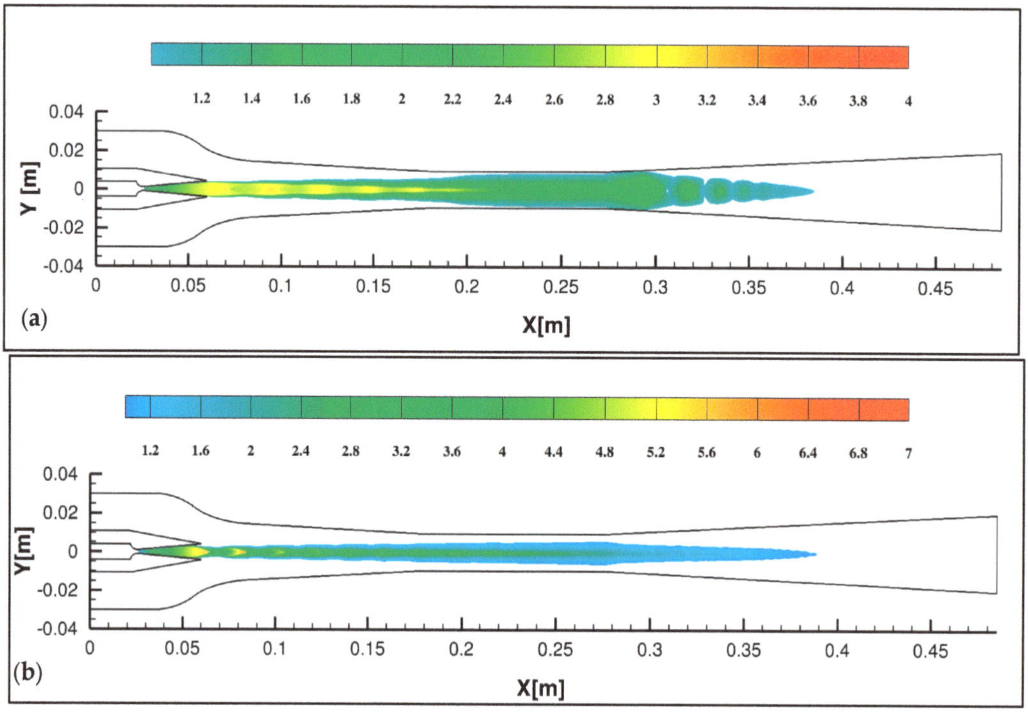

Figure 7. Mach number. Supersonic region (Ma $\geq$ 1). (**a**) krypton−air; (**b**) argon−helium.

Figure 8 shows that the wall shear stress (WSS) profile (obtained from the CFD simulations) exhibits two peaks in the krypton–air case, with one located at the throat inlet and the other located at the throat outlet. The initial peak corresponds to the contraction that occurs during the transition from the mixing chamber to the throat, where the fluids accelerate. The second peak is related to flow blockage and separation and correlates with the static pressure drop along the wall (Figure 6). In the case of argon–helium, only the first WSS peak is observed at the throat inlet since there is no flow separation or blockage, which allows the supersonic core to travel freely to the diffuser without any contact with the walls. This can also explain the absence of a sharp pressure drop that was observed for the krypton–air case (Figure 6). It should also be noted that the wall shear stress plateau presents a higher magnitude for the krypton–air case.

Figure 9 displays the Mach number profiles along the ejector centerline to demonstrate the quantitative variation in the Mach number distribution. It can be seen that the supersonic core length is the same in both cases; however, the argon–helium case exhibits considerably higher Mach numbers, which can be attributed to the notable increase in krypton's density. This leads to stronger shock waves and subsequently less mixing between the primary and secondary fluids, which can be seen from the entrainment ratio in Table 1.

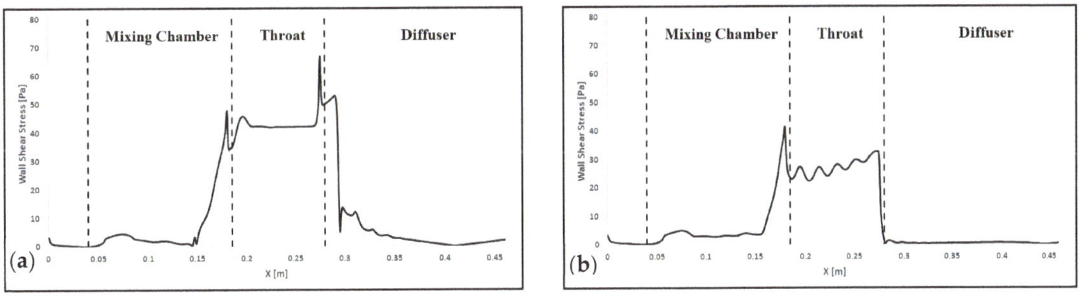

Figure 8. Wall shear stress. (**a**) krypton–air; (**b**) argon–helium.

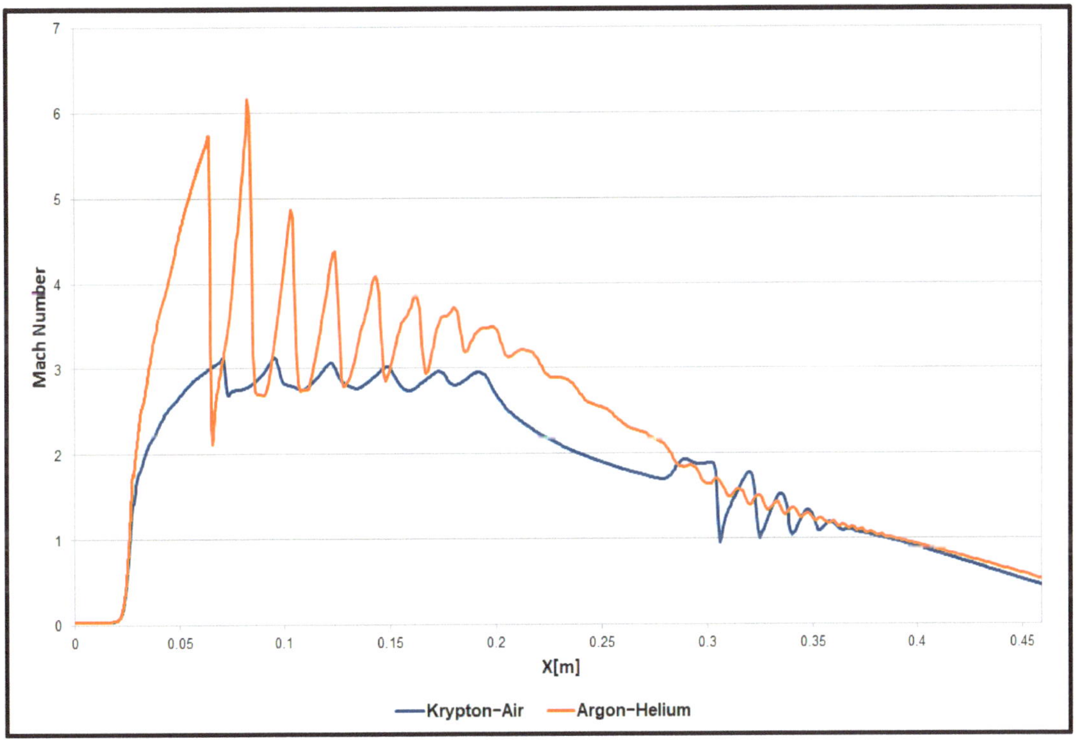

Figure 9. Mach number distribution along ejector centerline.

Figure 9 also shows that the argon–helium case shows significantly stronger fluctuations in Mach number distribution along the mixing chamber centerline, which indicate the presence of strong shockwaves in this region. In contrast, the krypton–air case displays a more intense sequence of oblique shockwaves inside the diffuser.

Further details about the flow dynamics around the diffuser inlet are shown in Figure 10. It is shown that small recirculation zones are observed along the diffuser wall for the krypton–air case. It is worth noting that the observed flow recirculation near the diffuser inlet has a negligible effect on the ejector's performance. However, substantial recirculation zones can significantly impact the ejector's performance, as reported in previous studies [23,24]; recirculation regions were not observed in the case of argon–helium.

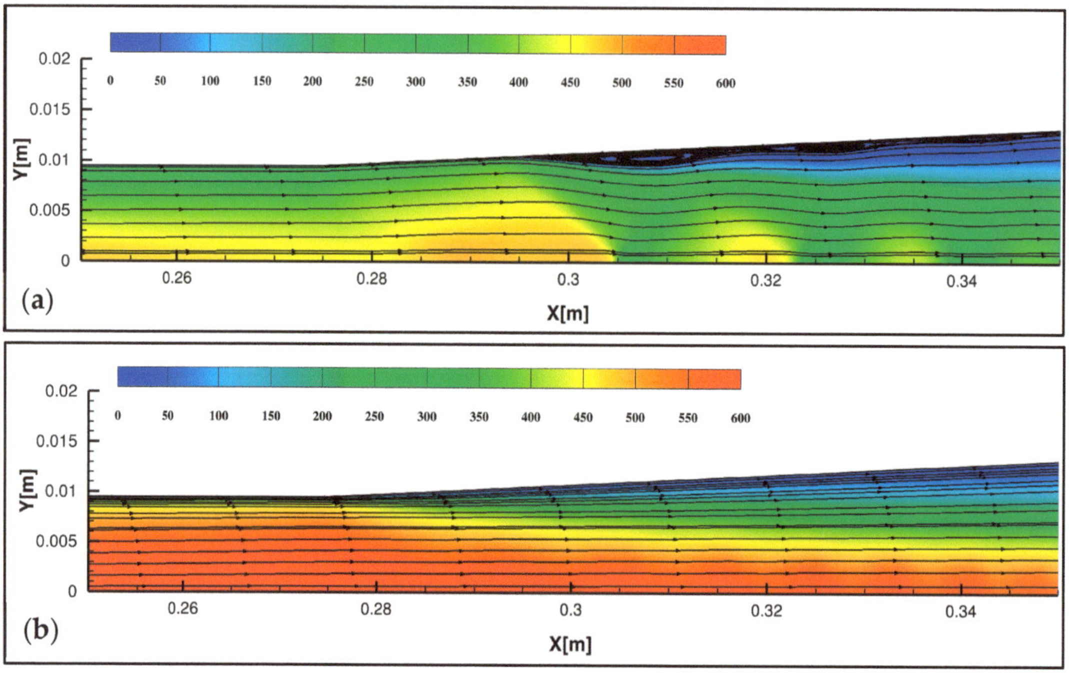

Figure 10. Velocity and streamlines. (**a**) krypton–air; (**b**) argon–helium.

6. Conclusions

CFD modeling was used in the present study to investigate the flow dynamics inside a supersonic gas–gas ejector with two different inert gas combinations. The experimental results were used to validate the numerical prediction of the entrainment ratio and static pressure. The main objective was to compare the ejector performance and flow structure between krypton–air and argon–helium combinations with a significant difference in molar mass.

The wall static pressure was well predicted along the ejector wall in both cases. It is interesting to note that the SST k-ω model made a better wall pressure prediction for the pairs of inert gases with a significantly higher difference in molar mass than in [5].

This study revealed that the molar entrainment ratio was notably greater in argon–helium than in krypton–air, with a 75% increase. Conversely, the mass entrainment ratio was substantially higher in the case of krypton–air. The potential applications of this outcome were discussed in the present paper.

The wall shear stress (WSS) profiles showed distinct characteristics in the krypton–air and argon–helium cases. This led to a better understanding of the flow dynamics inside the ejector. This was particularly the case for the krypton–air case, characterized by a flow blockage inside the ejector throat and flow separations near the diffuser inlet.

Finally, the Mach number distribution along the ejector centerline revealed that the centerline of the argon–helium case exhibited stronger shockwaves inside the mixing chamber, whereas the krypton–air case showed a more intense sequence of oblique shockwaves inside the diffuser.

Author Contributions: N.B.—experimental work, experimental procedure design, data postprocessing and participated in paper writing. M.E.H.—theoretical discussions, CFD, data postprocessing and participated in paper writing. All authors have read and agreed to the published version of the manuscript.

Funding: This research received no external funding.

Institutional Review Board Statement: Not applicable.

Informed Consent Statement: Not applicable.

Data Availability Statement: Not applicable.

Conflicts of Interest: The authors declare no conflict of interest.

References

1. El Hassan, M.; Gubanov, A.; May, W.; Martinuzzi, R.J. Numerical investigation of the flow dynamics of a supersonic fluid ejector. In Proceedings of the International Conference on Heat Transfer and Fluid Flow, Prague, Czech Republic, 11–12 August 2014. Paper No. 171.
2. Sun, F.; Fu, L.; Sun, J.; Zhang, S. A new waste heat district heating system with combined heat and power (CHP) based on ejector heat exchangers and absorption heat pumps. *Energy* **2014**, *69*, 516–524. [CrossRef]
3. Chunnanond, K.; Aphornratana, S. Ejectors: Applications in refrigeration technology. *Renew. Sustain. Energy Rev.* **2004**, *8*, 129–155. [CrossRef]
4. Fu, W.; Liu, Z.; Li, Y.; Wu, H.; Tang, Y. Numerical study for the influences of primary steam nozzle distance and mixing chamber throat diameter on steam ejector performance. *Int. J. Therm. Sci.* **2018**, *132*, 509–516. [CrossRef]
5. El Hassan, M.; Bukharin, N. Numerical investigation of the flow dynamics inside single fluid and binary fluid ejectors. *J. Phys. Conf. Ser.* **2019**, *1276*, 012012. [CrossRef]
6. Van Nguyen, V.; Varga, S.; Soares, J.; Dvorak, V.; Oliveira, A.C. Applying a variable geometry ejector in a solar ejector refrigeration system. *Int. J. Refrig.* **2020**, *113*, 187–195. [CrossRef]
7. Ruangtrakoon, N.; Thongtip, T.; Aphornratana, S.; Sriveerakul, T. CFD simulation on the effect of primary nozzle geometries for a steam ejector in refrigeration cycle. *Int. J. Therm. Sci.* **2013**, *63*, 133–145. [CrossRef]
8. El Zohbi, B.; Bukharin, N.; Assoum, H.H.; Abed-Meraim, K.; Sakout, A.; El Hassan, M. Investigation of the effects of the jet nozzle geometry and location on the performance of supersonic fluid ejectors. *Energy Rep.* **2022**, *8*, 228–233. [CrossRef]
9. Work, L.T.; Haedrich, V.W. Performance of Ejectors As a Function of the Molecular Weights of Vapors. *Ind. Eng. Chem.* **1939**, *31*, 464–477. [CrossRef]
10. Wang, P.; Ma, H.; Spitzenberger, J.; Abu-Heiba, A.; Nawaz, K. Thermal performance of an absorption-assisted two-stage ejector air-to-water heat pump. *Energy Convers. Manag.* **2021**, *230*, 113761. [CrossRef]
11. El Hassan, M.; Assoum, H.H.; Bukharin, N.; Abed-Meraim, K.; Sakout, A. Investigation of thermo-physical fluid properties effect on binary fluid ejector performance. *Energy Rep.* **2020**, *6*, 287–292. [CrossRef]
12. Rao, S.M.; Jagadeesh, G. Studies on the effects of varying secondary gas properties in a low entrainment ratio supersonic ejector. *Appl. Therm. Eng.* **2015**, *78*, 289–302. [CrossRef]
13. Van Vu, N.; Jan, K. CFD simulation of ejector: Is it worth to use real gas models? *EPJ Web Conf.* **2018**, *180*, 02075.
14. Kracik, J.; Dvorak, V. Preliminary study of the primary nozzle position of a supersonic air ejector with a constant-area mixing chamber. *EPJ Web Conf.* **2017**, *143*, 02056. [CrossRef]
15. El Hassan, M. System COP of Ejector-Based Ground-Source Heat Pumps. *Energies* **2022**, *15*, 8509. [CrossRef]
16. Sriveerakul, T.; Aphornratana, S.; Chunnanond, K. Performance prediction of steam ejector using computational fluid dynamics: Part 1. Validation of the CFD results. *Int. J. Therm. Sci.* **2007**, *46*, 812–822. [CrossRef]
17. Varga, S.; Oliveira, A.C.; Ma, X.; Omer, S.A.; Zhang, W.; Riffat, S.B. Comparison of CFD and experimental performance results of a variable area ratio steam ejector. *Int. J. Low-Carbon Technol.* **2011**, *6*, 119–124. [CrossRef]
18. Bartosiewicz, Y.; Aidoun, Z.; Desevaux, P.; Mercadier, Y. CFD-experiments integration in the evaluation of six turbulence models for supersonic ejector modeling. In Proceedings of the Integrating CFD and Experiments Conference, Glasgow, UK, 8–9 September 2003; pp. 14–24.
19. El Hassan, M. Numerical Investigation of the Flow Dynamics Inside Supersonic Fluid Ejector. *Arab. J. Sci. Eng.* **2019**, *45*, 909–919. [CrossRef]
20. Langtry, R.B.; Menter, F.R. Correlation-Based Transition Modeling for Unstructured Parallelized Computational Fluid Dynamics Codes. *AIAA J.* **2009**, *47*, 2894–2906. [CrossRef]
21. Menter, F.R.; Langtry, R.; Völker, S. Transition Modelling for General Purpose CFD Codes. *Flow Turbul. Combust.* **2006**, *77*, 277–303. [CrossRef]
22. El Hassan, M. Numerical Characterization of the Flow Dynamics and COP Estimation of a Binary Fluid Ejector Ground Source Heat Pump Cooling System. *Fluids* **2022**, *7*, 250. [CrossRef]

23. Hanafi, A.S.; Mostafa, G.M.; Waheeda, A.; Fathya, A. 1-D Mathematical Modeling and CFD Investigation on Supersonic Steam Ejector in MED-TVC. *Energy Procedia* **2015**, *75*, 3239–3252. [CrossRef]
24. Hemidi, A.; Henry, F.; Leclaire, S.; Seynhaeve, J.-M.; Bartosiewicz, Y. CFD analysis of a supersonic air ejector. Part II: Relation between global operation and local flow features. *Appl. Therm. Eng.* **2009**, *29*, 2990–2998. [CrossRef]

fluids MDPI

Article

Surrogate Models for Heat Transfer in Oscillating Flow with a Local Heat Source

Simon Knecht * Denislav Zdravkov and Albert Albers

IPEK—Institute of Product Engineering at Karlsruhe Institute of Technology (KIT), Kaiserstrasse 10, 76131 Karlsruhe, Germany
* Correspondence: simon.knecht@kit.edu; Tel.: +49-721-608-47032

Abstract: Simulative optimization methods often build on an iterative scheme, where a simulation model is solved in each iteration. To reduce the time needed for an optimization, finding the right balance between simulation model quality, and simulation time is essential. This is especially true for transient problems, such as fluid flow within a hydromechanical system. Therefore, we present an approach to building steady-state surrogate models for oscillating flow in a pipe with a local heat source. The main aspect is to model the fluid as a solid with an orthotropic heat transfer coefficient. The values of this coefficient are fitted to reproduce the temperature distribution of the transient case by parametric optimization. It is shown that the presented approach is feasible for different sets of parameters and creates suitable surrogate models for oscillating flow within a pipe with a local heat source. In future works, the presented approach will be transferred from the simplified geometry under investigation to industrial problems.

Keywords: oscillating flow; heat transfer; surrogate model; parametric optimization

Citation: Knecht, S.; Zdravkov, D.; Albers, A. Surrogate Models for Heat Transfer in Oscillating Flow with a Local Heat Source. *Fluids* **2023**, *8*, 80. https://doi.org/10.3390/fluids8030080

Academic Editors: Mehrdad Massoudi, Ernesto Benini and Francesco De Vanna

Received: 13 January 2023
Revised: 16 February 2023
Accepted: 20 February 2023
Published: 22 February 2023

1. Introduction

To comply with the task of the reduction in carbon emissions, modern aircraft development currently follows (among others) two trends: on the one hand, the so-called more electric aircraft, which focuses on replacing hydraulic and/or pneumatic power distribution systems within the aircraft with an electric power distribution [1], and on the other hand, high aspect ratio wings are used to reduce drag [2]. For a fixed wingspan, a high aspect ratio reduces the wing's thickness. Within the wing, complex systems to drive flaps, slats, and ailerons are positioned. The reduced space in the wings requires a highly compact design of these systems. Both trends result in the use of an electro-hydrostatic actuator (EHA) for primary flight control [3].

Since an EHA is a compact system with high power density, thermal aspects are more relevant [3,4]. An EHA comprises a closed hydraulic loop. Therefore, compared to a conventional hydraulic actuator, heat cannot be dissipated from the system through the hydraulic lines. However, this closed hydraulic loop distributes the heat within the system [5]. To develop an EHA for the application in primary flight control, it is therefore necessary to create a mechanical design that takes thermal aspects into account to ensure the full functionality of the EHA throughout all of the load cases [6]. To leverage the full potential during the design phase, computer-aided optimization methods can support the developer to create an optimized design.

Computer-aided optimization methods are widely used in research and industry [7]. There are different types of optimization methods, for example, parametric optimization [8], genetic algorithms [9], or shape and topology optimization [10]. A general optimization problem can be formulated as follows:

$$\begin{aligned} \text{minimize} \quad & f(x) \\ \text{subject to} \quad & g(x) \le 0 \\ & h(x) = 0 \end{aligned} \tag{1}$$

where f is the objective function, g is an optional inequality constraint, and h an optional equality constraint. x denotes the design variables that are subject to variation and are in most cases bounded by set values. [11] Many optimization algorithms for solving such problems work in iterations. This can lead to problems when each iteration requires a computationally expensive function evaluation [12]. The flow of the hydraulic fluid within an EHA is inherently transient in most load cases. When applying iterative optimization algorithms to create designs for an EHA, the computational cost is extremely high as each function evaluation requires a time-consuming transient conjugate heat transfer (CHT) simulation.

CHT-simulation models are the state of the art when dealing with simulation based analysis of heat transfer in both fluid and solid regions [13]. However, as already mentioned, they come at a high computational cost. Surrogate models can reduce the computational cost [14]. They have been successfully applied to various engineering optimization problems, including but not limited to aerospace design [15,16] and heat exchanger optimization [17–19], as well as coupled with neural networks for optimization problems [20,21].

The time scales of thermal conduction are much larger than those of convection [22]. This results in a quasi-steady thermal state when considering periodic flows, such as those found in an EHA with oscillating movements. Many publications investigated the thermal behavior of oscillating flow. One of the fundamental works in the area of oscillating flow is a publication of Womersley, who investigated the oscillating flow in blood vessels [23]. Named after him is the dimensionless Womersley number, defined as

$$\text{Wo} = \frac{\omega L}{\nu} \tag{2}$$

where ω is the oscillation's angular frequency, L is a characteristic length, and ν is the fluid's kinematic viscosity. The Womersley number gives insight into how the fluid behaves. For $\text{Wo} < 1$, the oscillation frequency is relatively low. In this case, a flow has time to develop in both phases of the oscillation. For $\text{Wo} > 10$, the frequency is relatively high, so that the fluid moves in a plug-like fashion within the pipe [23].

Many researchers investigated the application of oscillating flow and heat transfer to stirling engines [24–26]. There, the oscillation frequency was 100 Hz and higher. For our application, frequencies around 1 Hz and even lower are more common. Iwai et al. studied the heat transfer of a cylinder exposed to low-frequency oscillating flow [27]. More recently, Bouvier et al. conducted an experimental study of a heated cylinder, in which air creates an oscillating flow [28]. Choudari et al. conducted an extensive review of literature on oscillating flow heat transfer [29].

In all these publications, the heat was entered either through the whole wall or through one end of the pipe in order to investigate and derive heat transfer coefficients along the pipe. In the present article, we investigate the heat distribution within zero-mean velocity oscillating flow coming from a local heat source.

The aim of this article is to derive a surrogate model to replicate the resulting quasi-steady thermal state of oscillating flow with a local heat source in a steady-state simulation model. This surrogate model is derived on a simplified geometry consisting of a straight pipe with oscillating flow and a local heat source within the fluid. In the context of an EHA, this resembles a periodic movement of the piston, while the hydraulic fluid oscillates through a heat source as, e.g., the pump. Applying this surrogate model to industrial thermal problems, where oscillating flow transports heat away from a heat source enables a time-efficient iterative optimization. This can support the product developer in generating suitable thermo-mechanical designs.

The article is structured into three sections. Section 2 gives an overview of the proposed approach to derive a surrogate model for steady-state simulations. Section 3 explains the details of the numerical simulation setup used within this article for both the transient as well as the steady-state simulations as well as the parametric optimization. In Section 4, we present the results of the numerical simulations and evaluate the derived surrogate model.

2. Proposed Approach

The core idea of the proposed approach is to model the oscillating fluid's effect on the temperature distribution throughout the pipe with a steady-state simulation model. The fluid moving through the volume where heat enters the system ($V_{\dot{Q}}$) transports this heat by means of advection as well as diffusion along the pipe. The periodic movement is inherently quasi-stationary, and after a certain time t_{end}, the temperature distribution within the system is also quasi-stationary. The goal is to model the temperature distribution at the time t_{end} averaged over one oscillation period.

To achieve this goal, we assume the fluid in the surrogate model to be stationary since the average fluid movement over one period equals zero. Afterwards, the fluid's thermal conductivity and the length $l_{\dot{Q}}$ of the volume $V_{\dot{Q}}$ where heat enters the system are tuned through parametric optimization, so that the temperature distribution along the center line of the pipe equals the averaged values of the transient simulation. Results have shown that by using a single isotropic thermal conductivity in the fluid, the temperature distribution cannot be appropriately approximated. We therefore replaced the stationary fluid with a solid material with equal material properties (i.e., density, specific heat capacity, and thermal conductivity). This enables us to differentiate between a radial and an axial thermal conductivity λ_z and λ_ρ, respectively.

Figure 1 shows a graphical representation of the proposed approach, which will be explained in the following. A more detailed description of the individual steps taken for the practical application on the simplified pipe geometry is given in Section 3.

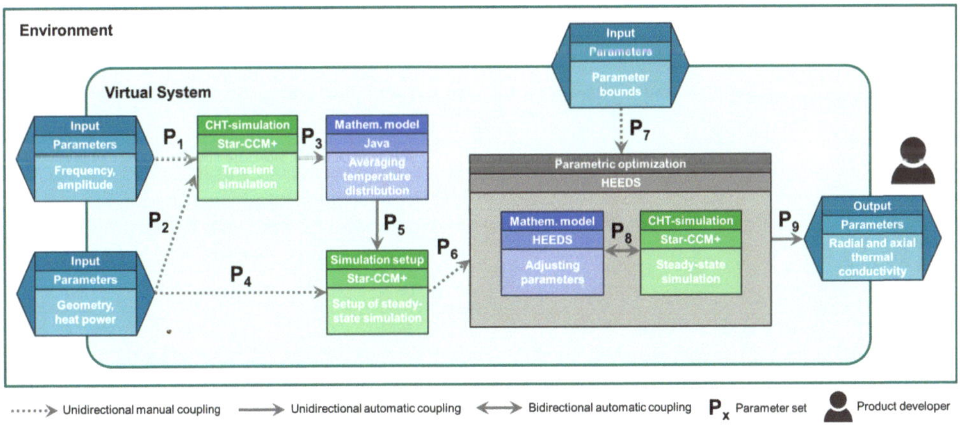

Figure 1. Graphical illustration of the proposed approach adapted from [30].

At first, a case is defined for which a surrogate model shall be derived. This definition consists of the geometric dimensions of the pipe, specifically the total length l_{total}, the inner (fluid) radius r_{inner}, the outer (solid) radius r_{outer}, and the width $l_{\dot{Q}}$ of the volume $V_{\dot{Q}}$, where heat enters the system. In addition, the heat $\dot{Q}$ and the frequency f and amplitude A of the fluid movement need to be defined. This input data will further be referred to as input variables.

All input variables are used to set up a transient simulation of the oscillating fluid flow. After running the transient simulation, we extract the time series of the temperature along the pipe's center line. This time series gets averaged over two oscillation periods to obtain a quasi steady-state temperature distribution from the transient simulation, which is written to a csv-file. The file is then imported into the basic simulation for the following parametric optimization as a reference for the optimization's objective function. Since Simcenter STAR-CCM+ provides a Java interface, the extraction, averaging, and exporting to a file is done in a Java-macro that is executed at the end of the transient simulation. Importing the file into the basic simulation and adjusting the objective function is also done in a Java-macro.

The basic simulation, which contains the geometry data and heat source but not the frequency and amplitude since it is steady-state, is then imported into HEEDS to perform the parametric optimization. In addition to the basic simulation, bounds for the parameters to be optimized are needed. This is the last step where interaction from the user is necessary. The parametric optimization then runs until a stopping criterion is met. Within the parametric optimization, HEEDS runs a number of steady-state simulations with different values for λ_z, λ_ρ, and $l_{\dot{Q}}$. For each simulation, the resulting value of the objective function is compared to previous results. The parameters are then adjusted according to a proprietary algorithm (see [31]) for a new number of steady-state simulations. The goal of the parametric optimization is to minimize the objective function given in Equation (3) as presented in Equation (1).

$$f(\mathbf{x}) = \sqrt{\sum_i (T_{\text{steady-state},i} - T_{\text{transient},i})^2} \tag{3}$$

In this equation, the design variables $\mathbf{x}$ consist of λ_z, λ_ρ, and $l_{\dot{Q}}$. $T_{\text{steady-state},i}$ and $T_{\text{transient},i}$ describe the temperature at points along the pipe's center line in the steady-state and transient simulation respectively. We investigated whether a normalization of the objective function is necessary. Results have shown that a normalization has no effect on the optimization result.

The result of using this approach is a set of parameters (λ_z, λ_ρ and $l_{\dot{Q}}$, in the following referred to as output parameters). The resulting values for the orthotropic thermal conductivity can be interpreted as an effective thermal conductivity resulting from the combination of heat transfer through diffusion within the fluid and convective heat transfer from the fluid's oscillating motion. To account for the limited extent of the volume where heat enters the system, the additional parameter $l_{\dot{Q}}$ needs to be adjusted. This set of parameters can then be applied by the product developer in a steady-state simulation to model the oscillating fluid's influence on the temperature distribution in a quasi-stationary state. The resulting set of parameters is valid for one set of input variables. The method can, however, be applied to different sets of input variables to generate sets of output parameters for each individual case.

3. Simulation Setup

In this section, we present the setup of both transient and steady-state simulations. This comprises the geometric setup, meshing, boundary conditions, and involved physics. We also give a short overview of the parametric optimization and the used hard- and software.

All of the numerical simulations were carried out in Simcenter STAR-CCM+ version 2021.2.1. The optimization was carried out in HEEDS [32]. Since our simulation models were relatively small, we were able to compute multiple simulations simultaneously on a 128-core 2 GHz server.

The points z_i at which the temperature is evaluated for the objective function are equidistant points along the pipes center line with a distance of 1 cm.

Table 1 shows the input variables that were varied. Starting from a so called basic set of input variables marked in bold, each parameter was varied while keeping the other two at their base value, to investigate whether our proposed approach works for different geometries. This resulted in 13 separate sets of input variables that were investigated in this article. In addition to these varied variables, the remaining necessary input variables were kept constant throughout all cases; their values are as follows: $l_{total} = 0.4\,\text{m}$, $r_{outer} = r_{inner} + 0.01\,\text{m}$, $l_{\dot{Q}} = 0.02\,\text{m}$ and $\dot{Q} = 5\,\text{W}$. The resulting Womersley number lies in the range of $1.089 < \text{Wo} < 27.22$. Note that $l_{\dot{Q}}$ is part of the input variables as well as the output variables.

Table 1. Overview of the varied input variables. Marked in bold are the values that are kept constant while varying the other variables.

Fluid Radius r_{inner} (m)	Frequency f (Hz)	Amplitude A (m)
0.002	0.1	0.01
0.005	**0.5**	**0.05**
0.01	1.0	0.1
0.02	1.5	0.15
0.05	2.0	0.2

An overview of all combinations of input variables that are investigated is given in Table 2. The resulting objective function values are discussed in Section 4.

Table 2. Combinations of input variables that were investigated.

Case Number	Fluid Radius r_{inner} (m)	Frequency f (Hz)	Amplitude A (m)	Objective Function Value
1	0.01	0.5	0.05	0.501
2	0.002	0.5	0.05	1.1169
3	0.005	0.5	0.05	0.303
4	0.02	0.5	0.05	0.206
5	0.05	0.5	0.05	0.037
6	0.01	0.1	0.05	0.4719
7	0.01	1.0	0.05	0.343
8	0.01	1.5	0.05	0.226
9	0.01	2.0	0.05	0.8477
10	0.01	0.5	0.01	20.06
11	0.01	0.5	0.1	0.0374
12	0.01	0.5	0.15	0.026
13	0.01	0.5	0.2	0.197

3.1. Geometry

Within this article, we derived surrogate models for oscillating flow within a straight pipe. We are evaluating our approach for different pipe geometries. The geometry, including the geometric parameters, is shown in Figure 2. The geometry consists of the pipe itself and the contained fluid. The pipe has a total length l_{total}, an inner radius r_{inner}, and an outer radius r_{outer}. The pipe is filled with the fluid. In the center of the pipe, the region where heat enters the system is located. This region is a cylindrical shape with a length of $l_{\dot{Q}}$. The fluid radius r_{inner} is part of the varied input variables. The corresponding values are listed in Table 1.

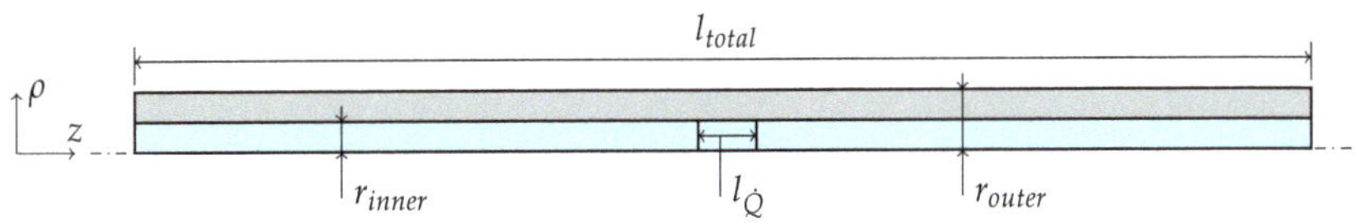

Figure 2. Cut view through the geometry under investigation. Half of the pipe and the center line of the rotational symmetry is shown. The fluid is shaded in blue and the pipe is shaded in gray. $l_{\dot{Q}}$ depicts the width of the volume where heat enters the fluid, situated at the middle of the pipe's length.

3.2. Transient Simulations

The transient simulations have shown that a large physical time needs to be simulated in order to obtain a quasi-steady thermal state. Paired with the low enough time-step needed for a stable simulation, this results in high computational costs. In order to save computation time, the transient simulations were carried out on a 30 degree wedge of the fluid and the pipe. This assumption is valid since the case under consideration is of transverse symmetry. Periodic boundary conditions on the wedge's sides ensure a proper representation of the pipe's rotational symmetry. The boundary condition at the inlet has been adapted to replicate the full pipe's mass flow.

The maximum Reynolds number was calculated for each set of input variables with the maximum velocity of the oscillation. With an overall maximum Reynolds number of $\mathrm{Re}_{\max} = 740.93$, we are well below the laminar-turbulent transition region at around $2300 < \mathrm{Re}_D$. Therefore, all simulations were set up with laminar flow.

The driving force of the oscillating flow is a mass flow boundary condition at one opening of the pipe. The second opening was set up as a pressure outlet, which enables a backflow into the pipe when a negative mass flow is prescribed at the first opening. The prescribed mass flow follows a sinusoidal function as defined in Equation (4), where ρ_f is the density of the fluid and t is the physical time.

$$\dot{m} = cos(2\pi f \cdot t) \cdot A \cdot 2\pi f \cdot \pi r_{\text{inner}}^2 \cdot \rho_f \tag{4}$$

This equation ensures that the oscillation of the fluid follows the given frequency and amplitude and oscillates around the pipes central region $V_{\dot{Q}}$.

On the outer surface of the pipe, we prescribed an arbitrary fixed temperature of 20 °C. Due to constant material properties, this temperature serves as as a reference point only. The left and right surfaces of the pipe allow for no heat transfer out of the system. The inflowing fluid on both ends of the pipe also has a prescribed temperature of 20 °C.

The pipe and fluid both were meshed using an extruded mesh. Starting from one end of the pipe, a polyhedral mesh was extruded along the pipe and the fluid to obtain the full mesh. At the fluid's boundary to the pipe, boundary layers were added. The cell size is approximately 1 mm perpendicular to the pipe's direction and 2 mm along the pipe's direction. The full mesh consists of around 38,500 cells for the transient basic simulation. For small values of r_{inner}, we reduced the cell size in order to properly discretize the region of the oscillating flow.

The time step was chosen to $\Delta t = \frac{1}{100f}$. This equals 100 time steps per oscillation period, which turned out to be sufficiently small for a stable simulation. Larger time steps resulted in non-physical simulation results. The stopping criterion was defined as the physical time of $t_{end} = \frac{500}{f}$. This equaled 500 oscillation periods, which resulted in a quasi steady-state throughout all variations of the input variables.

3.3. Steady-State Simulations

The steady-state simulations were set up similarly to the transient simulations. If not mentioned otherwise in the following, the setup is the same as in the transient simulations. Even though the fluid part is modeled as a solid with the fluid's altered material properties, we still refer to this region as the fluid region.

Compared to the transient simulations, the computational effort required to solve the steady-state simulations is relatively low (see Figure 3). We therefore modeled the geometry of the full pipe. Since the fluid is modeled as a solid, its left and right sides were changed to walls. On these sides, a constant temperature of 20 °C is prescribed. The volume where heat enters the system is parameterized with the value $l_{\dot{Q}}$ to allow for variation in the parametric optimization.

The pipe and the fluid were both meshed using a polyhedral mesh. The cell size was set to 2 mm, which results in an overall cell count of around 133,500 for the steady-state basic simulation.

The stopping criterion is a combination of the following three criteria. When all three criteria were met, the simulation was assumed to be converged:

- The residual needs to be equal to or less than $1 \cdot 10^{-6}$.
- The standard deviation of the average temperature in the solid volume within the last 10 iterations needs to be equal to or less than $1 \cdot 10^{-4}$ K.
- The standard deviation of the average temperature in the fluid volume within the last 10 iterations needs to be equal to or less than $1 \cdot 10^{-4}$ K.

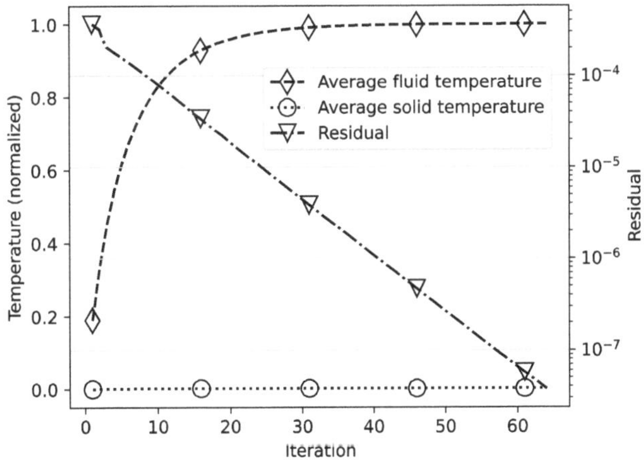

Figure 3. Residual and volume-averaged temperatures of fluid and solid for the steady-state simulation of the basic simulation. The simulation fulfilled all stopping criteria after 64 iterations.

Figure 3 shows the history of the residual and the volume-averaged temperatures of fluid and solid for the steady-state simulation of the basic simulation. The residual shows a logarithmic decrease, while the temperatures both converge. Due to the combination of a high thermal conductivity of the solid compared to the fluid and the fixed temperature boundary condition on the solid's outer surface, the heat within the solid can not contribute to a large rise in temperature but is dissipated from the system through the boundary. Therefore, the solid shows almost no increase in average temperature.

As already mentioned in Section 2, using an isotropic thermal conductivity for the surrogate models fluid did not yield adequate results. The oscillating fluid movement drastically increases the heat exchange throughout the fluid in the axial direction. Contrary to this, the radial heat exchange is barely affected by the fluid movement. Therefore, using an isotropic thermal conductivity results in either an overly large radial heat flux or an overly small axial heat flux with no value in between that satisfies our objective sufficiently. Since the fluid in the surrogate model is considered stationary, we can model this fluid as a solid part. This enables us to split the thermal conductivity into its axial and radial components. Both values can be tuned independently of one another to optimally parameterize the surrogate model.

3.4. Parametric Optimization

The optimization was carried out in HEEDS [32]. HEEDS is a software for design space exploration and optimization. It includes the SHERPA optimization algorithm [31]. The SHERPA algorithm is a proprietary algorithm that combines multiple optimization algorithms to find an optimal solution for a given optimization problem.

As mentioned in Section 2, our goal is to find optimal parameters (λ_z, λ_ρ, and $l_{\dot{Q}}$) for a surrogate model to reproduce the same temperature distribution as in the underlying transient simulation model. The objective function was given in Equation (3). Combining the objective function with the given parameter boundaries, the optimization formulation reads as follows:

$$\begin{aligned} \text{minimize} \quad & f(\lambda_z, \lambda_\rho, l_{\dot{Q}}) = \sqrt{\sum_i (T_{\text{steady-state},i} - T_{\text{transient},i})^2} \\ \text{subject to} \quad & 0.01\frac{\mathrm{W}}{\mathrm{mK}} \leq \lambda_z \leq 3000\frac{\mathrm{W}}{\mathrm{mK}} \\ & 0.01\frac{\mathrm{W}}{\mathrm{mK}} \leq \lambda_\rho \leq 1500\frac{\mathrm{W}}{\mathrm{mK}} \\ & 0.001\,\mathrm{m} \leq l_{\dot{Q}} \leq 0.19\,\mathrm{m} \end{aligned} \tag{5}$$

For each set of input variables, the optimization was instructed to create 250 different sets of parameters to find an optimal solution. This amount resulted in a sufficient convergence of the optimization and a low value for the objective function. During the optimization, 15 sets of input parameters were run simultaneously before the next 15 sets of input parameters were defined. After the 250 simulation models are computed, the set of parameters with the lowest objective function value is chosen as the solution to the optimization problem.

4. Results and Discussion

In this section, we will show the results of applying our proposed approach to different sets of input variables. We will further discuss the results.

Figure 4 shows the temperature distribution from both the steady-state simulation as well as the transient simulation for the basic set of input variables. A symmetric temperature distribution can be seen. This is to be expected since the fluid is oscillating around the pipe's center point at $z = 0\,\mathrm{m}$. The curve is bell-shaped due to the fact that heat enters the system in the volume around the center point. The fluids' oscillating movement as well as the heat conduction within the fluid then distributes this heat along the pipe. It is clearly visible that the temperature distribution resulting from the steady-state surrogate model almost perfectly replicates the time-averaged temperature distribution resulting from the transient simulation model. This shows that our proposed approach to construct a surrogate model for oscillating flow with a local heat source is feasible.

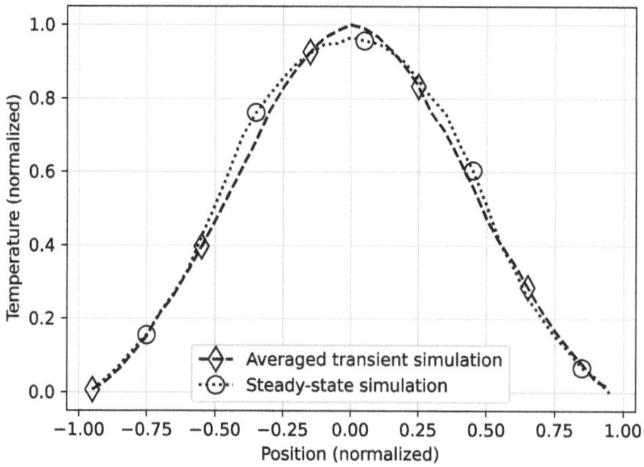

Figure 4. Results for the basic set of input variables. A comparison of the time-averaged temperature distribution of the transient simulation and the temperature distribution of the steady-state simulation is shown. Values on the x- and y-axis have been normalized to their respective minimum and maximum values.

Table 2 shows the resulting values of the objective function for the considered cases. As can be seen, the final objective function value is generally low, except for cases 2 and 10. Case 2 has a higher objective function value, which is due to the small fluid radius of $r_{\text{inner}} = 0.002\,\text{m}$. This small radius results in a Womersley number of Wo $= 1.089$, meaning that the oscillating fluid movement exhibits characteristics of a fully developed flow during each half of the periodic cycle. Case 10 has a comparably small amplitude that results in a thermal hotspot. The current formulation of the surrogate model and its parameters is not capable of fully replicating this temperature profile.

Since the objective function is not normalized, the objective function value is generally a comparable indicator for the temperature deviations. The mostly low values of the objective function correspond to a good agreement between the averaged temperature distribution resulting from the transient simulation and the temperature distribution resulting from the surrogate steady-state simulation model.

Figures 5 and 6 show the resulting values for λ_ρ over the frequency and the amplitude, respectively. Looking at the distribution of these discrete points, we suggest the assumption of a linear correlation of the radial thermal conductivity with the frequency and a quadratic correlation with the amplitude. These correlations are depicted by the dotted lines. Furthermore, it can be assumed that values of the radial thermal conductivity for frequencies and amplitudes that lie between the investigated points can be interpolated with the given linear and quadratic correlation. Care must be taken when interpolating near the extremes as, e.g., $f \to 0$, which correlates to Wo $\to 0$, where the y-intercept of the trend line strongly affects the results of the surrogate model. Additionally, for $A \to 0$, the quadratic correlation for the thermal conductivity goes below zero, which is physically not possible. This must be taken into account when using these correlations for interpolations.

In all result plots, deviations between the discrete points and the trend line can be seen. This is due to the optimization process, which generally does not find the single optimal solution but rather a near-optimal solution.

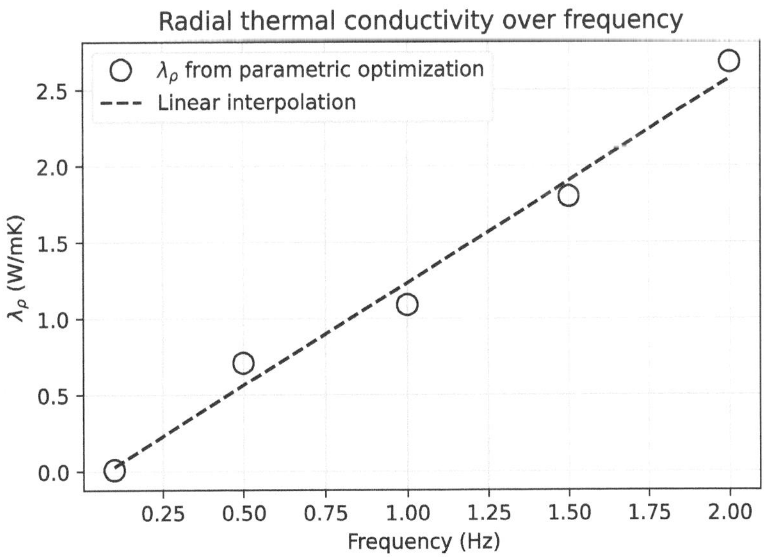

Figure 5. Resulting λ_ρ from the optimization over the frequency f. Shown also is a linear trend line.

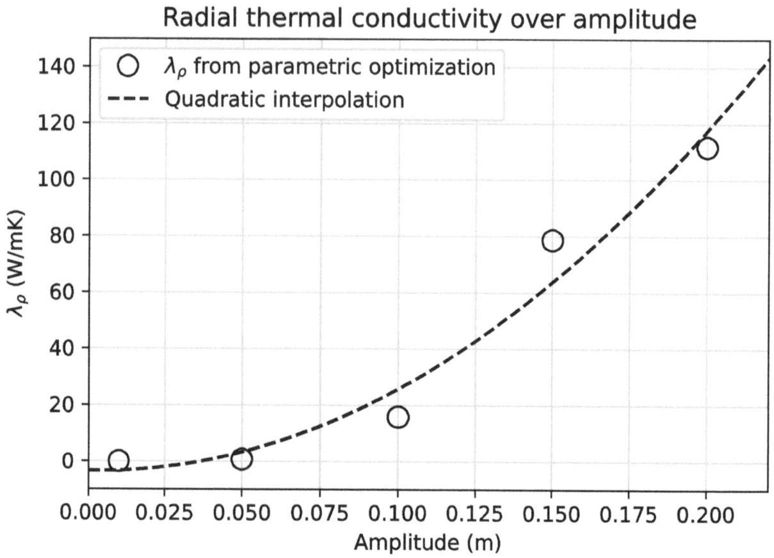

Figure 6. Resulting λ_ρ from the optimization over the amplitude A. Shown also is a quatratic trend line.

Figures 7 and 8 show the results for λ_z over the frequency and the amplitude, respectively. As in Figures 5 and 6, the axial thermal conductivity shows a correlation with the amplitude. This time we suggest the assumption of an exponential correlation. When plotting λ_z over the frequency, no clear correlation can be seen. The differences in axial thermal conductivity are quite large, with the maximum difference being 514.43 $\frac{\mathrm{W}}{\mathrm{mK}}$. A potential reason for this could be that the axial thermal conductivity is (nearly) independent of the frequency and has a small effect on the optimization's objective function. This would theoretically result in a constant axial thermal conductivity over the frequency. Due to its partly evolutionary nature, the optimization algorithm, however, alters all parameters , including the axial thermal conductivity, and small numerical errors lead to certain values yielding slightly better optimization results (in our case lower objective function values). This is supported by the fact that the objective function value is sufficiently low for all cases, which indicates a successful optimization. We also observed a large change in the resulting axial thermal conductivity when slightly changing the models input parameters. However, the objective function value changes marginally, indicating a negligible influence of the axial thermal conductivity on the objective function and therefore on the resulting temperature distribution.

The same largely differing axial thermal conductivites can be observed when plotting the resulting parameter over the fluid radius r_{inner}, as shown in Figure 9.

Figure 10 seems to involve some form of correlation, although no low-order (first or second order) polynomial fits the points. Higher-order polynomials could create a fitting trend line, however overfitting might occur.

Two different regions can be identified: for $r_{\text{inner}} > 0.01$ m, the radial thermal conductivity rises with an increasing fluid radius. For $r_{\text{inner}} < 0.01$ m, the radial thermal conductivity decreases with an increasing fluid radius. This is due to a low Womersley number of around 1, which indicates that the oscillatory effects on the fluid flow are replaced by regular pipe flow in both phases of the oscillation.

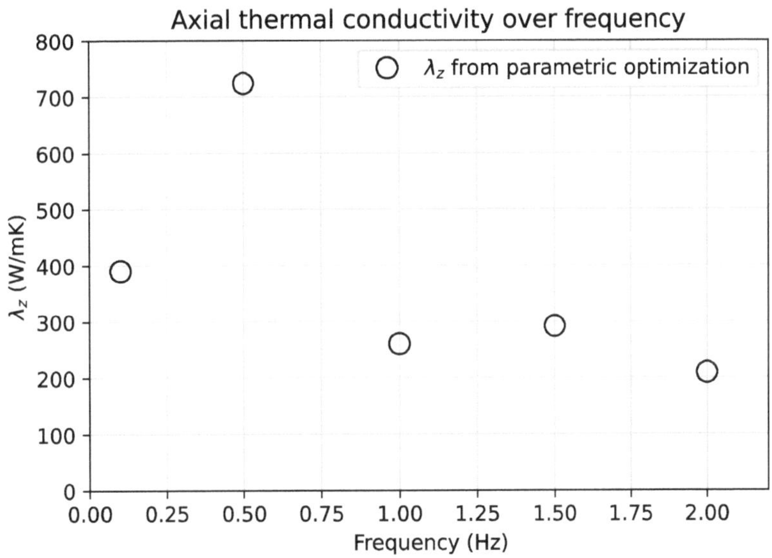

Figure 7. Resulting λ_z from the optimization over the frequency f.

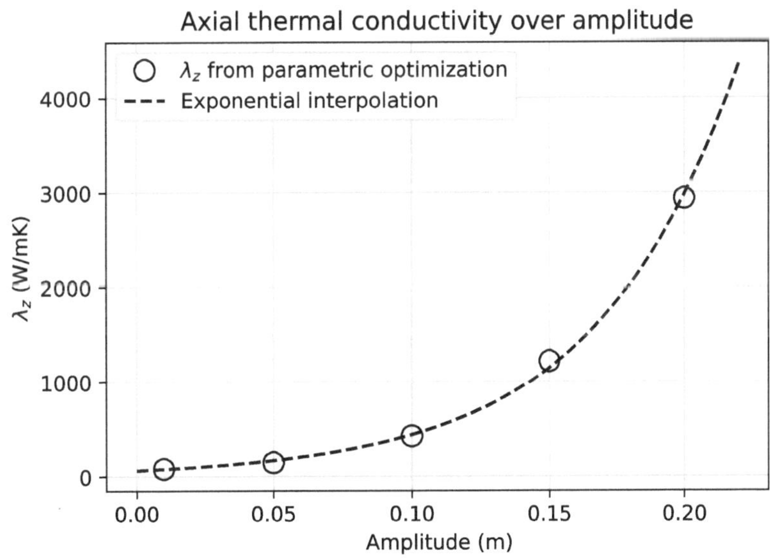

Figure 8. Resulting λ_z from the optimization over the amplitude A. Shown also is an exponential trend line.

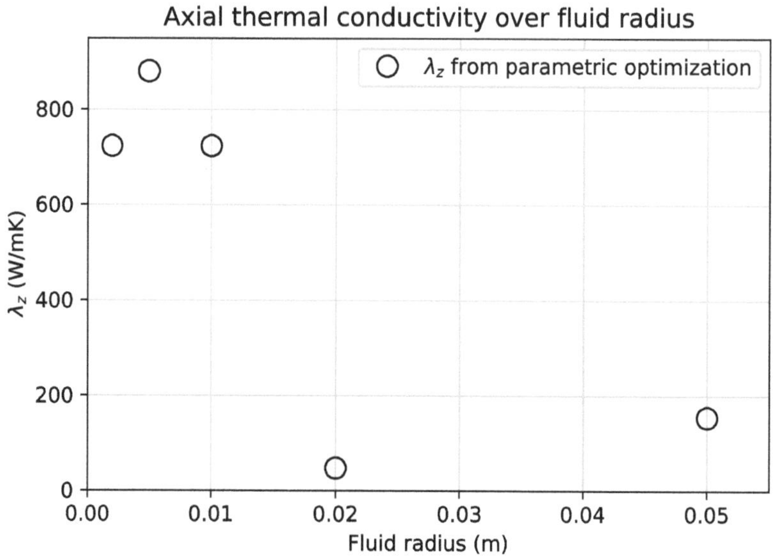

Figure 9. Resulting λ_z from the optimization over the fluid radius r_{inner}.

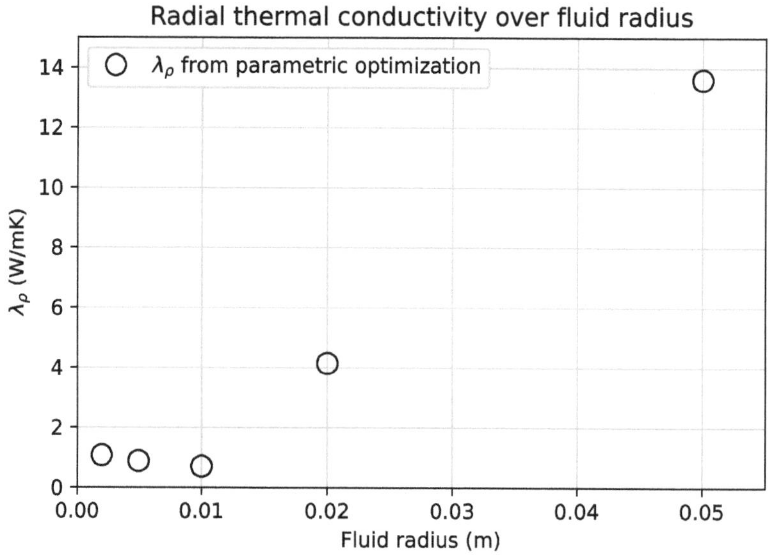

Figure 10. Resulting λ_ρ from the optimization over the fluid radius r_{inner}.

Figures 11–13 show the resulting $l_{\dot{Q}}$ over the frequency, the amplitude, and the fluid radius, respectively. For $l_{\dot{Q}}$ over f and r_{inner}, no clear correlation is recognizable. This is again due to the fact that the parameter $l_{\dot{Q}}$ is theoretically independent of the frequency and the fluid radius. Numerical deviations in the simulation and optimization then prefer certain values over others, without a noticeable change of the surrogate model. Even though certain parameters show large variations following from small changes of the input variables, the surrogate model still replicates the temperature profile for the given set of input parameters.

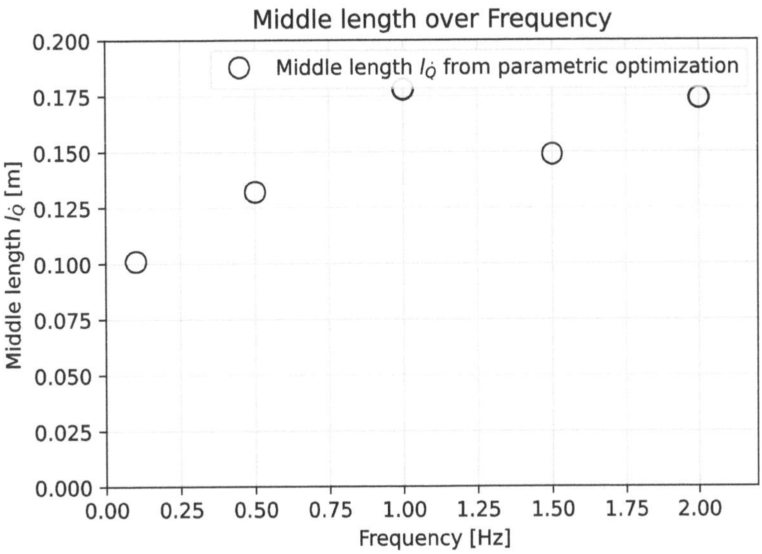

Figure 11. Resulting $l_{\dot{Q}}$ from the optimization over the frequency f.

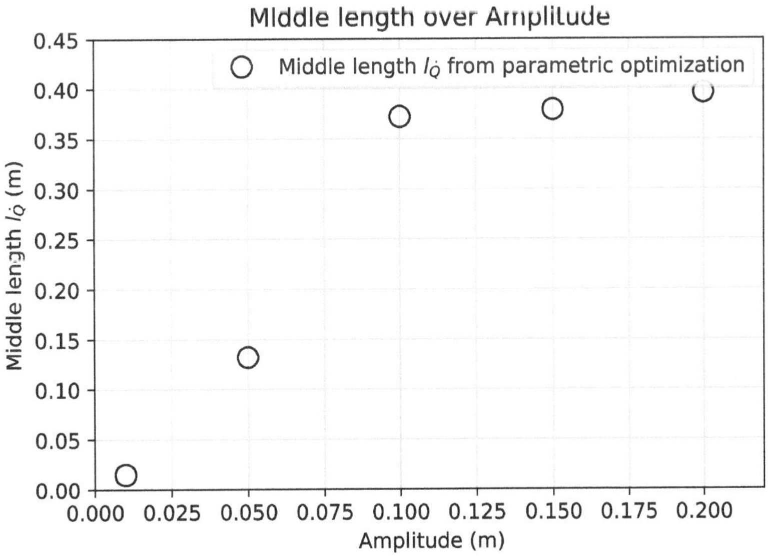

Figure 12. Resulting $l_{\dot{Q}}$ from the optimization over the amplitude A.

For $l_{\dot{Q}}$ over A, however, a clear dependence can be observed. With an increase in amplitude, the value for $l_{\dot{Q}}$ increases as well. For $A \geq 0.1$ m, $l_{\dot{Q}}$ stagnates slightly below 0.4 m. This is due to our total pipe length set to 0.4 m. Therefore, $l_{\dot{Q}}$ cannot climb above that value. At a first glance, this might seem like an erroneous simulation and/or optimization setup as the given input parameters and boundaries restrict the optimization algorithm. However, the original goal of the surrogate model was to replicate the temperature distribution within the given geometry. When applying this method to create a surrogate model for industrial applications, one might encounter similar restrictions due to geometric or procedural parameters. The created surrogate model then replicates the resulting quasi-steady thermal state within the given restrictions of the oscillating transient flow.

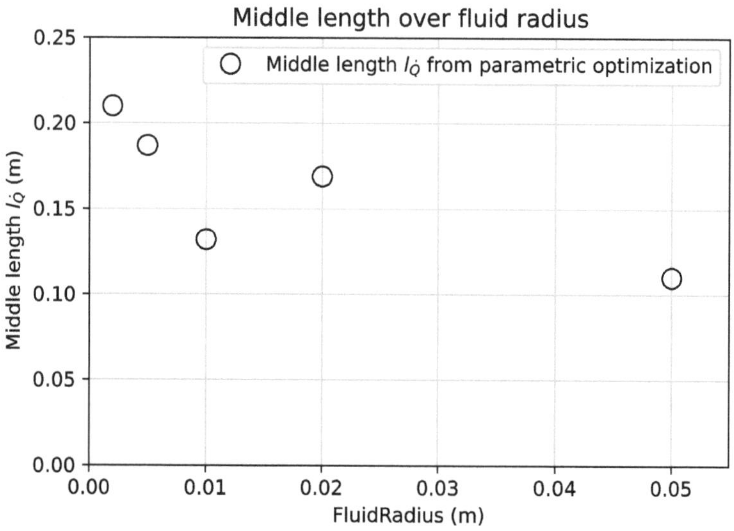

Figure 13. Resulting $l_{\dot{Q}}$ from the optimization over the fluid radius r_{inner}.

5. Conclusions and Outlook

In this article, we presented an approach to derive a surrogate model for oscillating flow with a local heat source. We applied the approach to the generic case of a straight pipe with an internally oscillating fluid. This surrogate model can then be applied to steady-state simulations, where, due to the periodic oscillation, a quasi-steady thermal state develops. With the surrogate model, the temperature distribution resulting from the transient oscillating flow is reproduced. By using a steady-state simulation, the computational cost for thermal investigations comprising oscillating flow and a local heat source can drastically be reduced. By reducing the computational cost, these surrogate models can serve as the foundation for the iterative optimization of industrial applications.

Certain dependencies of the surrogate model's parameters (axial and radial thermal conductivity and extent of the local heat source) on the input parameters (frequency, amplitude and fluid radius) could be observed:

- The radial thermal conductivity λ_ρ depends on the fluid radius, the frequency, and the amplitude.
- The axial thermal conductivity λ_z depends on the amplitude.
- The extent of the local heat source $l_{\dot{Q}}$ depends on the amplitude.

We suggest the assumption that λ_z and $l_{\dot{Q}}$ do not depend on the fluid radius and the frequency and have a small effect on the objective function. This is due to the seemingly random variations over the fluid radius and frequency. These variations come from the (partly) evolutionary nature of the optimization algorithm. In the optimization process, all variables are varied simultaneously. Thereby, when an optimum is found, values that have a very small effect on the objective function can exhibit an arbitrary behavior.

Except for λ_ρ over the fluid radius, all of the dependencies can be approximated by a first-order polynomial or an exponential curve. By formulating these dependencies, combinations of input parameters that differ slightly from those that were investigated can be derived. However, for small values of the fluid radius and the frequency, the deviations between the low-order approximations and the calculated parameters increased. This is due to the fact that the Womersley number decreases to 1 and below, which indicates a fully developed flow in both phases of the oscillation. Small amplitudes of the oscillation also posed problems in finding optimal parameters for the surrogate model. For input parameters that differ largely from those investigated, it is advised to set up a new parameter fitting as presented in this article since geometrical aspects and effects based on too low or too high Womersley numbers affect the behavior of the parameters.

In future works, we will apply this method to the optimization of an EHA. Due to the EHA's periodic movement, the internal hydraulic fluid creates an oscillating flow that distributes the heat from local heat sources, such as the motor or the pump, throughout the system. Difficulties might arise when applying the proposed approach to non-linear pipe geometries. Nevertheless, this should only increase the effort required to set up the simulation models. Based on the results from this article, it is reasonable to conclude that the proposed approach can successfully be applied to other geometries, such as the internal hydraulics of an EHA, as well. After deriving a surrogate model for the thermal behavior of the oscillating flow, optimization methods that require numerous function evaluations, such as the method of topology optimization or design space exploration, can then be applied with reasonable computational cost. This opens up further potential for the creation of mechanical designs that are thermally optimized.

Author Contributions: Conceptualization, S.K. and A.A.; methodology, S.K.; software, S.K. and D.Z.; investigation, S.K.; data curation, D.Z. and S.K.; writing—original draft preparation, S.K.; writing—review and editing, S.K. and D.Z.; visualization, S.K.; supervision, A.A.; and funding acquisition, A.A. All authors have read and agreed to the published version of the manuscript.

Funding: The project upon which this publication is based was funded by the Federal Ministry for Economic Affairs and Climate Action under the funding number 20Y1910F. The authors of this publication are responsible for its contents. The authors gratefully acknowledge the support.

Data Availability Statement: The data that support the findings of this manuscript are available from the corresponding author upon reasonable request.

Conflicts of Interest: The authors declare no conflict of interest.

Abbreviations

The following abbreviations are used in this manuscript:

MDPI	Multidisciplinary digital publishing institute
EHA	Electro-hydrostatic actuator
CHT	Conjugate heat transfer

References

1. Sarlioglu, B.; Morris, C.T. More Electric Aircraft: Review, Challenges, and Opportunities for Commercial Transport Aircraft. *IEEE Trans. Transp. Electrif.* **2015**, *1*, 54–64. [CrossRef]
2. Martins, J.R.R.A.; Kennedy, G.; Kenway, G.K. High Aspect Ratio Wing Design: Optimal Aerostructural Tradeoffs for the Next Generation of Materials. In Proceedings of the 52nd Aerospace Sciences Meeting, National Harbor, ML, USA, 13–17 January 2014; American Institute of Aeronautics and Astronautics: Reston, VA, USA, 2014. [CrossRef]
3. Alle, N.; Hiremath, S.S.; Makaram, S.; Subramaniam, K.; Talukdar, A. Review on electro hydrostatic actuator for flight control. *Int. J. Fluid Power* **2016**, *17*, 125–145. [CrossRef]
4. Li, K.; Wang, S. Analysis and modeling of the multidisciplinary coupling characteristics in EHA. In Proceedings of the 2011 International Conference on Fluid Power and Mechatronics, Beijing, China, 17–20 August 2011; pp. 602–608.
5. Li, K.; Lv, Z.; Lu, K.; Yu, P. Thermal-hydraulic Modeling and Simulation of the Hydraulic System based on the Electro-hydrostatic Actuator. *Procedia Eng.* **2014**, *80*, 272–281. [CrossRef]
6. Van Den Bossche, D. The A380 flight control electrohydrostatic actuators, achievements and lessons learnt. In Proceedings of the 25th International Congress of the Aeronautical Sciences, Hamburg, Germany, 3–8 September 2006; International Council of Aeronautical Sciences (ICAS): Hamburg, Germany, 2006; pp. 1–8.
7. Arora, J.S.; Elwakeil, O.A.; Chahande, A.I.; Hsieh, C.C. Global optimization methods for engineering applications: A review. *Struct. Optim.* **1995**, *9*, 137–159. [CrossRef]

8. Rao, R.V.; Saroj, A.; Ocłoń, P.; Taler, J. Design Optimization of Heat Exchangers with Advanced Optimization Techniques: A Review. *Arch. Comput. Methods Eng.* **2020**, *27*, 517–548. [CrossRef]
9. Lee, D.S.; Gonzalez, L.F.; Periaux, J.; Srinivas, K. Robust design optimisation using multi-objective evolutionary algorithms. *Comput. Fluids* **2008**, *37*, 565–583. [CrossRef]
10. Zhu, J.H.; Zhang, W.H.; Xia, L. Topology Optimization in Aircraft and Aerospace Structures Design. *Arch. Comput. Methods Eng.* **2016**, *23*, 595–622. [CrossRef]
11. Schumacher, A. *Optimierung Mechanischer Strukturen*; Springer: Berlin/Heidelberg, Germany, 2005.
12. Haftka, R.T.; Villanueva, D.; Chaudhuri, A. Parallel surrogate-assisted global optimization with expensive functions—A survey. *Struct. Optim.* **2016**, *54*, 3–13. [CrossRef]
13. John, B.; Senthilkumar, P.; Sadasivan, S. Applied and Theoretical Aspects of Conjugate Heat Transfer Analysis: A Review. *Arch. Comput. Methods Eng.* **2019**, *26*, 475–489. [CrossRef]
14. Alizadeh, R.; Allen, J.K.; Mistree, F. Managing computational complexity using surrogate models: A critical review. *Res. Eng. Des.* **2020**, *31*, 275–298. [CrossRef]
15. Mack, Y.; Goel, T.; Shyy, W.; Haftka, R. Surrogate Model-Based Optimization Framework: A Case Study in Aerospace Design. In *Evolutionary Computation in Dynamic and Uncertain Environments*; Studies in Computational Intelligence; Kacprzyk, J., Yang, S., Ong, Y.S., Jin, Y., Eds.; Springer: Berlin/Heidelberg, Germany, 2007; Volume 51, pp. 323–342. [CrossRef]
16. Yondo, R.; Bobrowski, K.; Andrés, E.; Valero, E. A Review of Surrogate Modeling Techniques for Aerodynamic Analysis and Optimization: Current Limitations and Future Challenges in Industry. In *Advances in Evolutionary and Deterministic Methods for Design, Optimization and Control in Engineering and Sciences*; Computational Methods in Applied Sciences; Minisci, E., Vasile, M., Periaux, J., Gauger, N.R., Giannakoglou, K.C., Quagliarella, D., Eds.; Springer International Publishing: Cham, Switzerland, 2019; Volume 48, pp. 19–33. [CrossRef]
17. Ghosh, S.; Mondal, S.; Fernandez, E.; Kapat, J.S.; Roy, A. Parametric Shape Optimization of Pin-Fin Arrays Using a Surrogate Model-Based Bayesian Method. *J. Thermophys. Heat Transf.* **2021**, *35*, 245–255. [CrossRef]
18. Li, C.; Li, Y.; Gao, L.; Garg, A.; Li, W. Surrogate model–based heat dissipation optimization of air–cooling battery packs involving herringbone fins. *Int. J. Energy Res.* **2021**, *45*, 8508–8523. [CrossRef]
19. Lebaal, N.; SettaR, A.; Roth, S.; Gomes, S. Conjugate heat transfer analysis within in lattice-filled heat exchanger for additive manufacturing. *Mech. Adv. Mater. Struct.* **2022**, *29*, 1361–1369. [CrossRef]
20. White, D.A.; Arrighi, W.J.; Kudo, J.; Watts, S.E. Multiscale topology optimization using neural network surrogate models. *Comput. Methods Appl. Mech. Eng.* **2019**, *346*, 1118–1135. [CrossRef]
21. Sun, G.; Wang, S. A review of the artificial neural network surrogate modeling in aerodynamic design. *Proc. Inst. Mech. Eng. Part G: J. Aerosp. Eng.* **2019**, *233*, 5863–5872. [CrossRef]
22. Errera, M.P.; Turpin, G. Temporal multiscale strategies for conjugate heat transfer problems. *J. Coupled Syst. Multiscale Dyn.* **2013**, *1*, 89–98. [CrossRef]
23. Womersley, J.R. Method for the calculation of velocity, rate of flow and viscous drag in arteries when the pressure gradient is known. *J. Physiol.* **1955**, *127*, 553. [CrossRef]
24. Kuosa, M.; Saari, K.; Kankkunen, A.; Tveit, T.M. Oscillating flow in a stirling engine heat exchanger. *Appl. Therm. Eng.* **2012**, *45-46*, 15–23. [CrossRef]
25. Ni, M.; Shi, B.; Xiao, G.; Luo, Z.; Cen, K. Heat transfer characteristics of the oscillating flows of different working gases in U-shaped tubes of a Stirling engine. *Appl. Therm. Eng.* **2015**, *89*, 569–577. [CrossRef]
26. Zhao, Y.; Yu, G.; Tan, J.; Mao, X.; Li, J.; Zha, R.; Li, N.; Dang, H. CFD modeling and experimental verification of oscillating flow and heat transfer processes in the micro coaxial Stirling-type pulse tube cryocooler operating at 90–170 Hz. *Cryogenics* **2018**, *90*, 30–40. [CrossRef]
27. Iwai, H.; Mambo, T.; Yamamoto, N.; Suzuki, K. Laminar convective heat transfer from a circular cylinder exposed to a low frequency zero-mean velocity oscillating flow. *Int. J. Heat Mass Transf.* **2004**, *47*, 4659–4672. [CrossRef]
28. Bouvier, P.; Stouffs, P.; Bardon, J.P. Experimental study of heat transfer in oscillating flow. *Int. J. Heat Mass Transf.* **2005**, *48*, 2473–2482. [CrossRef]
29. Choudhari, M.; Gawali, B.S.; Patil, J.D. Oscillating flow heat transfer: A comprehensive review. *Energy Sources Part A Recover. Util. Environ. Eff.* **2022**, *44*, 7598–7619. [CrossRef]
30. Albers, A.; Haberkern, P.; Holoch, J.; Joerger, A.; Knecht, S.; Renz, R.; Revfi, S.; Schulz, M.; Spadinger, M. Strategische Planung des Entwicklungsrisikos gekoppelter CAE-Methoden. *Konstruktion* **2022**, *74*, 72–77. . [CrossRef]
31. Red Cedar Tech. SHERPA—An Efficient and Robust Optimization/Search Algorithm. Available online: http://www.redcedartech.com/pdfs/SHERPA.pdf (accessed on 21 February 2023).
32. Siemens. Simcenter HEEDS | Siemens Software. Available online: https://plm.sw.siemens.com/de-DE/simcenter/integration-solutions/heeds/ (accessed on 21 February 2023).

fluids

Article

Coupled Simulation of Fluid Flow and Vibro-Acoustic Processes in the Channel with a Circular Cylinder

Konstantin Volkov

Institute of Hydrodynamics and Control Processes, St. Petersburg State Marine Technical University, 190121 St. Petersburg, Russia; k.volkov@kingston.ac.uk

Abstract: Vibro-acoustic processes are an interacting set of pulsations of the working fluid and vibrations of mechanical structural elements. The simulation of vibro-acoustic processes in a long pipe with an elastic round cylinder is considered. The mathematical model is developed in a coupled formulation, when not only pressure pulsations cause pipe vibrations, but also vibrations of the mechanical subsystem affect sound wave propagation in the working fluid. The influence of vortex formation processes in the channel on the system dynamics is taken into account. The fluid flow is found using delayed detached eddy simulation. The flow regimes around a single round cylinder corresponding to various Reynolds numbers are investigated to validate the computational algorithm. The distributions of the flow quantities and vibro-acoustic behavior of the system are discussed.

Keywords: numerical simulation; coupled problem; channel; cylinder; turbulence

Citation: Volkov, K. Coupled Simulation of Fluid Flow and Vibro-Acoustic Processes in the Channel with a Circular Cylinder. *Fluids* **2022**, *7*, 382. https://doi.org/10.3390/fluids7120382

Academic Editors: Ernesto Benini and Mehrdad Massoudi

Received: 8 November 2022
Accepted: 7 December 2022
Published: 11 December 2022

1. Introduction

Turbulent pressure fluctuations on a streamlined surface play an important role in acoustic measurements in moving media. In particular, in flexible extended towed antennas, which are used in seismic exploration to search for minerals under the seabed, sensitive receiving elements are enclosed in an acoustically transparent cylindrical shell and are located close enough to the streamlined surface. Sources of hydrodynamic noise and vibrations are the phenomena associated with the flow around various structural elements [1] (boundary layer formed on a streamlined surface, separation regions near bluff bodies, and jet streams). An important role in excitation of vibrations is played by pressure fluctuations in the boundary layer on the walls and acoustic oscillations excited by the separation of vortices. Pseudo-sonic pulsations of near-wall pressure on the streamlined surface of carriers of hydro-acoustic equipment lead to the appearance of hydrodynamic noise.

The generation of coherent vortices and their interaction with each other and with the streamlined surface lead to the emission of powerful pressure pulsations into the environment. The variety of scales of vortex structures in a turbulent boundary layer and their transfer rates lead to a broadband spectrum of velocity and pressure fluctuations. The appearance of dominant vortex formations, generated in one way or another, determines the presence of discrete or tonal peaks in the spectral dependencies of pressure fluctuations measured both above the streamlined surface and on its surface.

The hydrodynamic noise of the turbulent boundary layer, the sources of which are coherent vortex structures interacting with each other and with the streamlined surface, has sonic and pseudo-sonic components [2]. Unlike sound, which has a wave nature and propagates into the environment at the speed of sound, pseudo-sonic pressure pulsations are transferred at a speed close to the flow speed. Pseudo-sound generated by the nonlinear interaction of the vortex structures of the boundary layer does not obey the superposition principle. The pseudo-sound pressure does not depend on the average ambient pressure and decreases inversely with the square of the distance from the source. When gas flows around elongated bodies (strings, cables, long rods, high pipes), acoustic vibrations arise due to the separation of vortices (Aeolian tones). The appearance of Aeolian tones is associated with fluctuations in the lift force acting on a streamlined body and its drag [3].

Pseudo-sonic pressure fluctuations are localized within the turbulent flow, while sonic pressure fluctuations propagate over large distances outside the region occupied by the turbulent flow. The study of the physics of turbulent pressure fluctuations is important not only because they are sources of hydrodynamic noise. They affect the surface in a flow, generating vibrational oscillations, but they are also present in the correlation dependencies in the Reynolds stress transfer equations and the dissipation tensor energy.

One of the test problems that make it possible to determine the characteristic features of hydrodynamic acoustic processes is the problem of flow past a round cylinder. The fluid flow around a circular cylinder is considered in a large number of experimental and computational studies [4–7]. A round cylinder in a transverse flow of fluid is a typical element of many structures and is the object of intensive theoretical, numerical, and experimental research. The formation and shedding of vortices behind the cylinder and other bluff bodies leads to unwanted vibrations and destruction of the structure. In this regard, it is necessary to be able to control the process of vortex shedding, reduce the drag force of the streamlined body and the amplitude of fluctuations of the forces applied to it [8,9]. The results of calculations of the loads experienced by an obstacle during its flow are used to optimize the methods of attaching the cylinder to the channel walls.

Modeling and simulation of the flowfield around cylinders have been tackled by many researchers at various Reynolds and Mach numbers [10–23]. The results obtained form a basis for validation of new models and computational algorithms.

Studies of the flow around a cylinder near the wall include data on the effect of the gap between the cylinder and the wall on the frequency of vortex formation (von Karman road), as well as information on the interaction of shear layers on the surface of the cylinder and on the wall depending on the gap and the Reynolds number [5,24–26]. A system of large-scale horseshoe-shaped vortices is formed at the junction of the shedder body and the walls parallel to the flow. Such vortex formations have a significant impact on the flow, determining local shear, acoustic noise, drag and lift coefficients [27,28].

Pseudo-sonic pulsations of near-wall pressure behind an annular obstacle on a longitudinally streamlined flexible extended cylinder are studied in [29]. The integral and spectral statistical characteristics of the pressure fluctuation field behind the obstacle are obtained, and its influence on the structure of the turbulent boundary layer is studied. An increase in the obstacle diameter and flow velocity leads to an increase in the low-frequency spectral components of pressure fluctuations and a weakening of high-frequency pressure fluctuations in comparison with the boundary layer on a hydraulically smooth cylinder.

The results of theoretical, numerical, and experimental studies of the characteristics of pseudo-sonic pressure oscillations caused by the interaction of the flow and the sound field inside a three-dimensional spherical dimple are presented in [30]. Symmetric and asymmetric large-scale vortex systems inside the dimple are found, the existence of which depends on the flow regime, their location and periodicity of ejection to the outside are indicated.

An important task in assessing the vibration strength of technical systems is to determine their vibration characteristics. Many of the developed approaches are empirical in nature and are not universal. The available methods for calculating the vibration parameters of a system under its force loading by a pulsating flow of the working fluid are characterized by significant idealization. Mathematical modeling of vibro-acoustic processes in technical systems is an important task, the solution of which will make it possible to predict vibro-acoustic loads at the design stage and ensure the operability of the system by rationally choosing its parameters, locations, and characteristics of supports, devices for correcting dynamic characteristics (vibration dampers, dynamic vibration dampers, vibration dampers of the working liquids).

In [31], simulation of vibro-acoustic interaction processes is implemented for a rectangular channel with one compliant and three rigid bounding surfaces. The form of vibration of the compliant plate, excited by the pulsations of the pressure of the working medium, is a superposition of at least two natural modes of vibration. The possibilities of eddy-resolving approaches to modeling turbulent flow near a system of ordered circular cylinders and predicting their flow characteristics, including pressure fluctuations, are shown in [32]. One of the main requirements for the methods of solving vibro-acoustics problems is the high resolution of the difference scheme and its low dissipation in the calculation of convective flows, which is important for the correct transfer of acoustic disturbances with a small amplitude.

The solution of coupled problems is characterized by large velocity and pressure gradients. In such problems, there is also a wide range of flow quantity variations, which creates challenges for numerical methods. Many previous studies, for example, Refs. [29–32] are limited to the assumption that it is possible to describe acoustic processes in a working medium by a one-dimensional wave equation. In addition, in many previous works, an assumption is made about the harmonic law of changing the parameters of the working environment. Insufficient attention is also paid to the issues of determining variable stresses from the impact of vibro-acoustic loads when calculating the strength of systems.

In this study, large vortices of vibro-acoustic processes are simulated in a pipe with a transversely located round cylinder. The generalized mathematical model is considered in a coupled formulation, when not only pressure pulsations cause pipe vibration, but also vibrations of the mechanical subsystem affect wave processes in the working fluid. Based on the results of numerical simulation, the vibro-acoustic behavior of the system is discussed.

2. Coupled Problems

Vibro-acoustic problems (Acoustic–Structure Interaction, ASI) imply joint modeling of elastic waves in solids, acoustic pressure waves in a liquid, and taking into account the interaction between them. Acoustic waves in different types of materials are described by different equations.

The loads on the surfaces found from the CFD problem are applied to the boundaries of the computational domain. These loads define boundary conditions for the solution of the stress–strain state. With a single transfer of data on the impact of a flow on a certain structural element from a CFD package to an FEA package, the interaction is unidirectional and may be used when the reverse effect of the deformation process on the flow is weak. In the general case, a bidirectional data exchange is required (the CFD package transmits loads, but receives the displacement values of the nodes of the interface boundary).

With a single data exchange during a time step, an explicit scheme for pairing software systems is implemented. Multiple data exchange at the time step of solving the adjoint problem leads to an implicit conjugation scheme. The need for multiple data exchange is associated with the problem of stability of the calculation process, which is explained by the high inertia of the liquid due to its density and leads to large reaction forces acting from the flow on the moving interface. The iterative procedure [33,34] is used to obtain a convergent solution of the adjoint problem at a time step.

The need to modify the grid structure is due to the displacement of the boundaries that form the object and are the boundaries of the computational domain of the hydrodynamic problem. Modification of the grid consists in its local rebuilding or deformation (stretching/compression) of the set of grid lines while maintaining the relationships of nodes, edges, and faces of control volumes.

For situations that do not express a high degree of interaction nonlinearity, the explicit method seems to be more efficient and flexible, allowing flow calculations and finite element analysis to be performed independently and then iteratively linked. Pairing on each time step is implemented by a recursive procedure based on iterations between different calculated modules until the specified level of convergence is reached.

3. Governing Equations

Modeling of vibro-acoustic processes consists of description of the non-stationary movement of the working medium, the interaction of the working medium with moving or stationary surrounding or surrounding elements of the mechanical subsystem, their interaction with the environment, and other elements that, for some reason, were not included in this system.

3.1. Sound Propagation in Fluid

Flows are considered in whichever period of change of flow parameters is much greater than the frequency of sound propagating in the flowfield. The flow quantities are presented as sum of average and fluctuating components ($u = \overline{u} + u'$). The prime corresponds to small fluctuations of flow quantities.

The governing equations describing small fluctuations of the flow parameters have the form (the line above the average values is omitted)

$$\begin{aligned} &\frac{\partial \mathbf{v}'}{\partial t} + [\mathrm{rot}\,\mathbf{v}, \mathbf{v}'] + [\mathrm{rot}\,\mathbf{v}', \mathbf{v}] + \nabla(\mathbf{v}, \mathbf{v}') = -\frac{1}{\rho}\nabla p' + \frac{1}{\rho^2}\nabla(p \cdot \rho'); \\ &\frac{\partial \rho'}{\partial t} + (\mathbf{v}, \nabla\rho') + (\mathbf{v}', \nabla\rho) + \rho\,\mathrm{div}\,\mathbf{v}' + \rho'\mathrm{div}\,\mathbf{v} = 0; \\ &\frac{\partial s'}{\partial t} + (\mathbf{v}, \nabla s') + (\mathbf{v}', \nabla s) = 0. \end{aligned} \tag{1}$$

The equation of state is written as

$$p' = c^2\rho' + hs', \quad c^2 = \left(\frac{\partial p}{\partial \rho}\right)_s, \quad h = \left(\frac{\partial p}{\partial s}\right)_\rho.$$

The system of Equation (1) is valid for a flow with non-uniform entropy ($\nabla s \neq 0$) and in the presence of vorticity ($\mathrm{rot}\,\mathbf{v} \neq 0$). Approximations consist only in the fact that only linear terms are taken into account, and irreversible processes in the sound wave are not taken into account; therefore, the propagation of sound waves is considered as an adiabatic process $d(s + s')/dt = 0$. Auxiliary relations, which are necessary for the transition from entropy to temperature, have the form

$$\nabla s = \frac{c_p}{T}\nabla T - \frac{\beta_p}{\rho}\nabla p, \quad \left(\frac{\partial p}{\partial s}\right)_\rho = \frac{\rho c^2}{c_p}\beta_p T,$$

where c_p is the specific heat capacity, $\beta_p = (1/V)(\partial V/\partial T)_p$ is the volume expansion coefficient.

Assuming that the flow is isentropic ($\nabla s = 0$) and irrotational ($\mathrm{rot}\,\mathbf{v} = 0$), for a small adiabatic compression or expansion of the gas, the equation of state takes the form

$$p' = c^2\rho'.$$

The pressure potential, Π, and the velocity potential of sound vibrations, φ, are introduced using the relations

$$\Pi = \frac{p'}{\rho}, \quad \mathbf{v}' = -\nabla\varphi.$$

The equation for the velocity potential is

$$\frac{d^2\varphi}{dt^2} = c^2\Delta\varphi + (\nabla\Pi_0, \nabla\varphi) + \frac{d\varphi}{dt}(\mathbf{v}, \nabla \ln c^2). \tag{2}$$

The potential of the initial flow is

$$\Pi_0 = \int \frac{dp}{\rho}.$$

The convective derivative is defined as

$$\frac{d\varphi}{dt} = \frac{\partial\varphi}{\partial t} + (\mathbf{v}, \nabla\varphi).$$

It follows from Equation (2) that in the absence of vorticity and entropy gradient, no sound is generated in the medium.

In a medium at rest $\mathbf{v} = 0$ and $\operatorname{rot} z, \mathbf{v} = 0$, the wave equation for the velocity potential (2) takes the form

$$\frac{\partial^2\varphi}{\partial t^2} - c^2\Delta\varphi = 0,$$

where $c^2 = \gamma p/\rho$ is the speed of sound.

The boundary condition on the surface of an elastic body has the following form

$$\frac{\partial\varphi}{\partial n} = -v_{0n},$$

where v_{0n} is the normal velocity of the body surface. For an absolute rigid body,

$$\frac{\partial\varphi}{\partial n} = 0.$$

For the correct statement of the boundary value problem, it is necessary to set the initial conditions for φ and $\partial\varphi/\partial t$.

3.2. Oscillations of an Elastic Body

The equation of motion for an elastic body is written in the following form

$$\rho\mathbf{a} = \rho\mathbf{F} + \operatorname{div}\sigma,$$

where $\mathbf{a}$ is acceleration, $\mathbf{F}$ is external force, σ is stress tensor. Acceleration is represented in terms of derivatives of speed

$$\mathbf{a} - \frac{d\mathbf{V}}{dt} - \frac{\partial\mathbf{V}}{\partial t} + V_i\frac{\partial\mathbf{V}}{\partial x_i}.$$

Slow deformations are considered (velocities V_i are small) and it is assumed that $\mathbf{a} \approx \partial\mathbf{V}/\partial t$. The equations in displacements $\mathbf{u}$ for an elastic body take the form

$$\rho\frac{\partial^2\mathbf{u}}{\partial t^2} = \rho\mathbf{F} + \operatorname{div}\sigma.$$

The stress tensor is determined by Hooke's law for an isotropic elastic medium in the Lame form

$$\sigma = \lambda I\,\mathrm{trace}(\varepsilon) + 2\mu\varepsilon.$$

The function $\mathrm{trace}(\varepsilon)$ denotes the trace of the strain tensor. The strain tensor has the following form

$$\varepsilon = \frac{1}{2}\left[\operatorname{grad}\mathbf{u} + (\operatorname{grad}\mathbf{u})'\right].$$

The stress tensor satisfies the relation

$$\sigma = \lambda I \operatorname{div}\mathbf{u} + \mu\left[\operatorname{grad}\mathbf{u} + (\operatorname{grad}\mathbf{u})'\right].$$

The Lame coefficients are determined by the relations

$$\mu = \frac{E}{2(1+\nu)}, \quad \lambda = \frac{\nu E}{(1+\nu)(1-2\nu)},$$

where E is Young's modulus, ν is Poisson's ratio.

Substituting the stress tensor from Hooke's law into the equations of motion, we obtain the equation in small displacements

$$\rho \frac{\partial^2 \mathbf{u}}{\partial t^2} = \rho \mathbf{F} + (\mu + \lambda) \mathrm{grad\,div}\, \mathbf{u} + \mu \Delta \mathbf{u}. \tag{3}$$

Equation (3) is used to describe small oscillations in an elastic body.

4. Coupling Scheme

The pressure fluctuations obtained from CFD calculations under the assumption of stationary pipe walls are not sound, since the signal drifts with the flow velocity. The fluid acts on the pipe walls and barriers, which are made of elastic material and begin to deform and oscillate. In a solid material, not only body waves propagate, as in a fluid, but also shear waves. Interaction of wall vibrations with the fluid leads to the generation of acoustic waves and their propagation. Acoustic waves also propagate in the surrounding air. Simulation of vibro-acoustic characteristics is a coupled problem, in which waves in a solid material and acoustic waves in the fluid interact. To simplify the model, it is assumed that acoustic waves have a small effect on the flow. In this case, the vibro-acoustic problem is considered as three independent problems: wave propagation in a solid material, acoustic wave propagation in water and ambient air.

At stage 1, a water flow in a channel with rigid walls and unsteady pressure and friction forces on the pipe walls are found. At stage 2, the problem of the propagation of vibro-acoustic waves is solved under the influence of non-stationary forces on the wall from the inside of the pipe.

The internal diameter of the channel is $d = 50$ mm. Wall thickness is $h = 5$ mm, and length of the channel is $L = 500$ mm. The density of water is $\rho = 998.5$ kg/m^3. Kinematic viscosity of water is $\nu = 1.003 \cdot 10^{-3}$ m/s^2. The pipe ends are fixed (Figure 1). The diameter of cylinder is $d_c = 8$ mm. The cylinder is located at $l_1 = 0.1$ m from the inlet. The inlet water velocity is $V = 5$ m/s. Relative pressure at the outlet is zero. In the solution of the CFD problem, the pipe walls are stationary. No-slip and no-penetration boundary conditions are applied to the walls. Air pressure and temperature are $p = 10^5$ Pa and $T = 300$ K. Material of the pipe is steel. When calculating the vibro-acoustic problem on the inner walls of the pipe, the pressure fluctuations (pressure amplitudes for each frequency after the Fourier transform) are interpolated from the hydrodynamic problem.

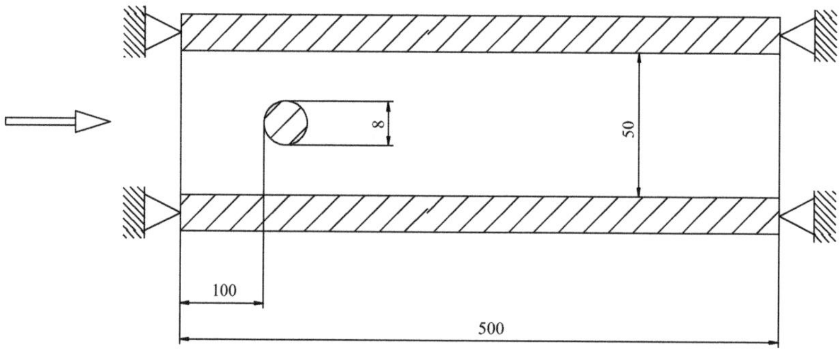

Figure 1. Scheme of the computational domain in the plane of symmetry (dimensions are given in millimeters).

5. Estimations of Flow Quantities

The flow regime is defined by the Reynolds number $\mathrm{Re}_h = \rho V d_h / \mu$, where ρ is density, V is flow velocity, d_h is hydraulic diameter, μ is dynamic viscosity. If the flow velocity is $V = 5$ m/s, the Reynolds number equals $\mathrm{Re}_h = 248,804$ (turbulent regime of flow).

In fluid flows past a cylinder, the vortex shedding frequency depends on the Reynolds number [10–12]. The Strouhal number is found as $\mathrm{St} = fd_c/V$, where f is characteristic frequency, d_c is cylinder diameter, V is flow velocity. The Reynolds number calculated from the cylinder diameter is $\mathrm{Re}_c = 39,809$ (turbulent regime of flow). For Reynolds numbers larger than about 1000, the Strouhal number remains almost constant ($\mathrm{St} = 0.21$). The vortex street during the flow in a pipe with a cylinder is the dominant non-stationary process; therefore, the fundamental frequency is approximately equal to $f_m = \mathrm{St}V/d_c = 131$ Hz, and the oscillation period is $T_m = 7.62 \times 10^{-3}$ s.

To describe a periodic process, it is necessary to measure at least two points per period (Nyquist's theorem). The calculations use approximately 100 points per period to resolve the dominant frequency. The time step is chosen equal to $\Delta t = T_m/100 = 7.62 \times 10^{-5}$ s. In this case, processes with frequency $f_{\max} = 1/(10\Delta t) = 1312$ Hz are calculated with 10 points per period.

To perform a Fourier transform on the fundamental tone of a signal, it is necessary to have data for a certain number of periods of this tone. A total of 100 periods of the pitch signal is stored in the calculations. After the oscillations reach a stable mode, the data are saved for $t_f = 100T_m = 0.76$ s. In this case, it is necessary to perform $N = t_f/\Delta t = 10^4$ steps in time. During the time t_f, there are 10 periods with frequency $f_{\min} = 10/t_f = 12.5$ Hz. Proper resolution is observed in the frequency range from $f_{\min}$ to $f_{\max}$ (greater errors occur above or below these frequencies).

6. Scheme of Simulation

The delayed detached-eddy simulation (DDES) method with a shear stress transport (SST) k–ω turbulence model is used to resolve vortex structures. DDES is an alternative to Reynolds-averaged Navier Stokes (RANS) equations and large-eddy simulation (LES). RANS models are not capable of providing acceptable accuracy for calculating flows with large separation zones [7], and LES imposes high requirements on computational resources when modelling separated flows, which makes their practical use difficult [19].

The following approach is used to simulate the unsteady flowfield. A preliminary stationary calculation of the flow in the pipe is performed using the RANS equations with the SST-model. The semi-implicit method for pressure-linked equations-consistent (SIMPLEC) method is used to integrate the equations. Then, unsteady calculations of the flow in the pipe are performed with DDES and the SST-model. The calculations are carried out until a steady state flow is realized. To integrate the equations of motion, a modified pressure implicit with a splitting of operator (PISO) method is used. The pressure correction step is performed more times by the SIMPLEC method. The time step is found from the stability conditions. An unsteady flowfield in a pipe with a fixed time step is found. The pressure values are preserved in each cell bordering the pipe and cylinder walls at every time step.

A block-structured mesh is used. Nodes are clustered near the cylinder surface, wake region, and pipe walls to resolve the boundary layer (Figure 2). In the pipe, DDES operates in the LES mode, and the mesh is uniform. In the DDES approach, inviscid fluxes are discretized with a hybrid scheme and the Roe method to maintain accuracy in the LES region and stability of the scheme in the RANS region with a non-uniform and coarse mesh. The distributions of flow quantities found from the RANS solution with the SST-model are used as initial conditions.

The mesh nodes are distributed uniformly along the circumferential coordinate. In the inner region ($0.5 \leqslant r \leqslant 5$), which is a circle, an O-type mesh is used. In the near-wall region ($0.5 \leqslant r \leqslant 1$), the minimum step along the radial coordinate is $r_{\min} = 0.003$, the growth factor is $q = 1.032$ (about 16 nodes along the radial coordinate are used to resolve the boundary layer). In the outer region, the minimum step in radial direction is $r_{\min} = 0.2$, and the growth factor is $q = 1.04$. In the buffer region, which is located at a considerable distance from the cylinder and allows avoiding the reflection of sound waves from the boundaries of the computational domain, the maximum step in radial direction is $r_{\min} = 20$, and the growth factor is $q = 1.09$.

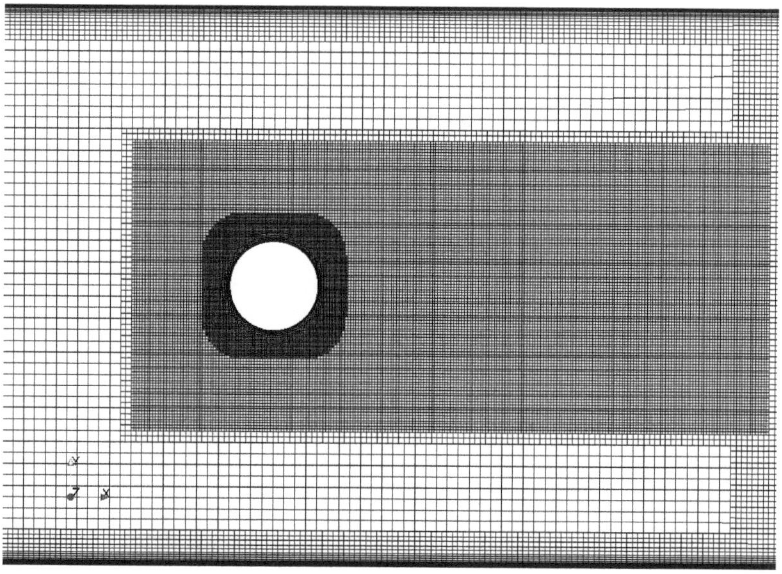

Figure 2. A block-structured mesh.

The total number of mesh nodes is about 1.5 million. In this case, the condition $y^+ < 1.2$ is satisfied in the near-wall region. The dimensionless time step is 0.1056. The calculations are carried out on a time interval equal to 162.

Calculations are based on a fully compressible solver [34]. Convergence to a steady state is accelerated by the use of multigrid techniques, and by the application of block-Jacobi preconditioning for high-speed flows, with a separate low Mach number preconditioning method for use with low-speed flows.

With a known flowfield, the calculations of pressure fluctuations that occur on the channel walls due to the influence of an unsteady fluid flow are performed. The boundary condition is set in the form of a known amplitude of pressure fluctuations for each frequency.

The fast Fourier transformation (FFT) allows to obtain dependencies of pressure on time over the pipe walls and cylinder. The data obtained are input for solving the vibro-acoustic problem. The solution of the fluid problem, which describes the distribution of amplitudes of pressure fluctuations along the walls of the pipe and cylinder, contains data with a fine frequency step. For vibro-acoustic calculation, such a set of frequencies is redundant, and therefore, the data are averaged over a coarser frequency grid ($1/n$ octave spectrum filters). The most common are 1/3-octave and 1/8-octave filters.

7. Flow around Cylinder

A numerical simulation of an unsteady flow of a viscous compressible fluid around a cylinder is considered. The calculation results are given for two Reynolds numbers equal to Re $= 150$ (regime 1) and Re $= 3900$ (regime 2), which are widely used for testing numerical methods in the literature. In both cases, the Mach number is set equal to M $= 0.2$. The Reynolds numbers correspond to the flow regime near the cylinder with flow separation from its surface. Using the fast Fourier transform, the values of the Strouhal numbers, the amplitudes of oscillations of the aerodynamic coefficients are determined, and the phases of the formation of vortices and separation of the flow are found from the values of the maximum amplitudes.

7.1. Computational Domain and Boundary Conditions

A transverse flow around a circular cylinder located in a viscous compressible gas, which is at rest for $t \leqslant 0$, is simulated. For $t > 0$, the gas is set in motion with a constant speed U_0. When calculating the Reynolds number, the diameter of the cylinder and the velocity of the undisturbed flow are used as characteristic parameters of the problem. The equations are solved in the region external to the cylinder boundary.

The center of the cylinder with diameter D is located at the origin. In the (x, y) plane, the computational domain has the shape of a rectangle ($-20 < x < 80$, $-20 < y < 20$). The extent of the computational domain in the direction of the z axis is $4D$.

The potential flow field is used as the initial conditions, which is modified in the boundary layer region in such a way as to satisfy the no-slip conditions on the cylinder surface. At the inlet boundary, the velocity of the undisturbed flow is set, which is found from the given Reynolds number, and the temperature is equal to 290 K. At the outlet boundary of the computational domain, non-reflecting boundary conditions are used. A static pressure equal to 10^5 Pa is fixed at the upper and lower boundaries of the computational domain. The surface of the cylinder is assumed to be thermally insulated. In the direction of the z axis, periodic boundary conditions are applied.

7.2. Regime 1

The distribution of pressure over time at various points on the surface of the cylinder is shown in Figure 3. The lines correspond to the numerical simulation data, and the symbols correspond to the calculated data of [10]. There is a fairly good agreement between the calculation results and the data of [10]. The amplitude of pressure fluctuations at $\theta = -90°$ (symbols △, solid line) and $\theta = +90°$ (symbols □, dotted line) is several times greater than the amplitude pressure fluctuations observed at $\theta = 0°$ (symbols ◇, dash-dotted line).

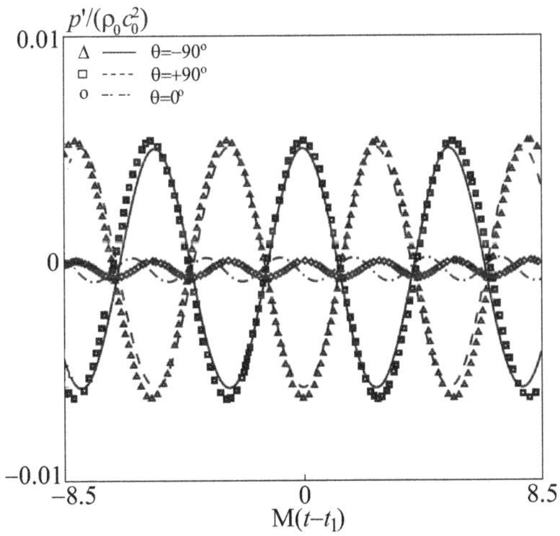

Figure 3. Change in pressure over time for Re = 150.

The distributions of the lift coefficient, drag coefficient, and pressure coefficient are shown in Figure 4. The lines correspond to the numerical simulation data, the symbols ∘ correspond to the data of [10]. It achieves good agreement with the available data for all characteristics.

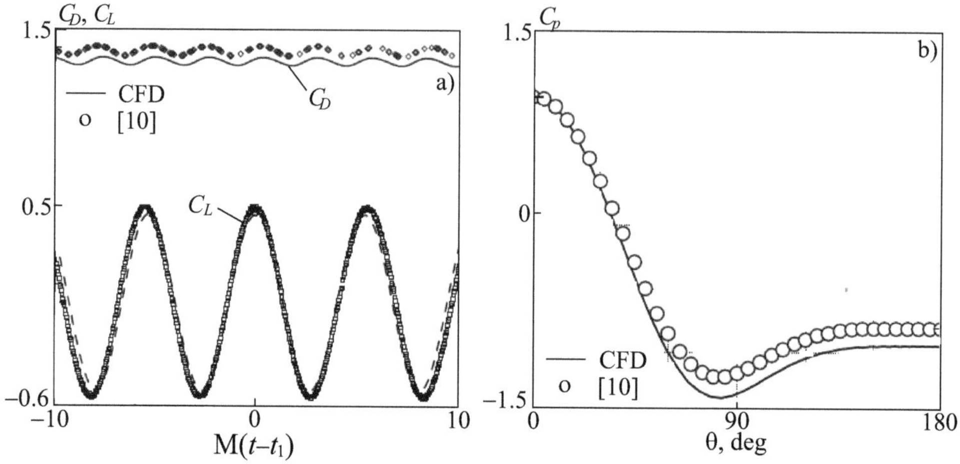

Figure 4. Changes in the lift coefficient (dashed line) and drag coefficient (solid line) over time (**a**), and the distribution of the average pressure coefficient over the cylinder surface (**b**) for Re = 150.

The characteristics of the flow in regime 1 are in good agreement with the available numerical and experimental data (Table 1, where C_{pb}, C_L, and C_D are the mean pressure coefficient, the mean square lift coefficient, and the mean drag coefficient). In particular, in calculations the Strouhal number is 0.186, while the data of a physical experiment give a value of 0.184 [8] and 0.183 [12]. The angular coordinate of the flow separation point is $\theta_s = 108.5°$. Averaging is carried out over a time interval equal to 150. To find the Strouhal number, the correlation relation $\mathrm{St} = 0.2684 - 1.0356\mathrm{Re}^{-1/2}$ from [15] is used. The Strouhal number equals St = 0.1838 for Re = 150, which is in good agreement with the obtained value. The results of calculations of the Strouhal number in comparison with experimental data are shown in Figure 5.

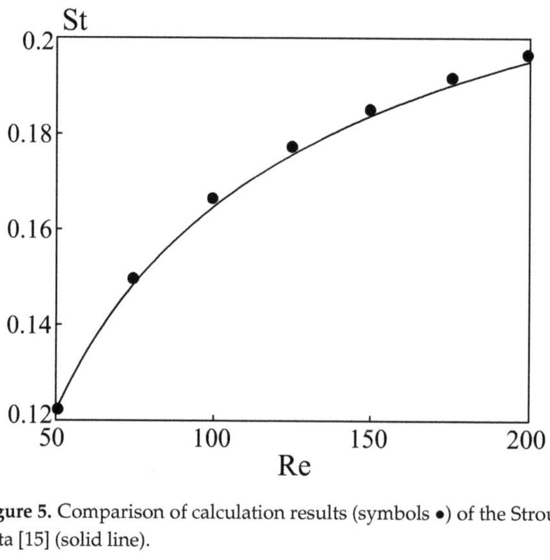

Figure 5. Comparison of calculation results (symbols •) of the Strouhal number with experimental data [15] (solid line).

Table 1. Comparison of calculated data with available numerical and experimental results (for Re = 150).

Source	St	$-\overline{C}_{pb}$	C_L	C_D	L_z/D
CFD	0.186	0.850	0.342	1.224	20
[18]	0.184	0.848	—	1.313	140
[13]	0.185	0.885	—	1.333	56
[16]	0.181	—	0.338	1.268	16.46
[14]	0.191	—	0.403	1.418	12
[12]	0.183	—	0.356	—	—
[8]	0.184	0.856	—	—	—
[10]	0.184	—	—	1.334	—
[20]	0.184	0.846	0.355	1.305	—

The instantaneous pressure distribution near a circular cylinder is shown in Figure 6, which clearly shows oscillations of the wake behind the cylinder.

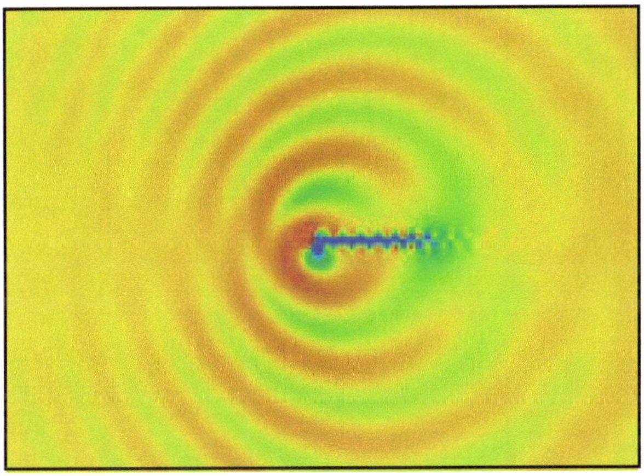

Figure 6. Instantaneous pressure distribution around a cylinder for Re = 150.

7.3. Regime 2

The characteristics of the flow in mode 2 are in good agreement with the available numerical and experimental data (Table 2). C_{pb}, C_L, C_D, $\overline{C}_D$, θ_s are understood as mean pressure coefficient, mean square lift coefficient, mean square drag coefficient, mean drag coefficient, mean angular coordinate of the lift-off point, average length of the recirculation region. In particular, in the calculations, the Strouhal number is 0.186, while the data of a physical experiment give values of 0.184 [8] and 0.183 [12]. The angular coordinate of the flow separation point is $\theta_s = 108.5°$. Averaging is carried out over a time interval equal to 150.

Table 2. Comparison of calculated data with available numerical and experimental results (for Re = 3900.

Source	St	$-\overline{C}_{pb}$	C_L	C_D	$\overline{C}_D$	$\overline{\theta}_s$	L_r/D	L_z/D
CFD	0.209	0.860	0.074	0.029	0.963	88.3	1.57	20
[17]	0.219	0.840	—	—	—	—	1.59	20
[19]	0.220	0.920	—	—	1.03	85.7	1.30	18

The cause of noise generation is the deterministic tone sound with discrete frequencies and turbulent fluctuations in the boundary layer and wake, which are stochastic in nature and have a continuous frequency spectrum. The sound intensity is found from the relation $I = \overline{p'^2}/(\rho_0 c_0)$. The RMS pressure is determined by the relation

$$\left\langle p'^2(\mathbf{x}) \right\rangle = \lim_{T\to\infty} \frac{1}{T} \int_0^T p'^2(\mathbf{x},t)dt,$$

where $T = t_n - t_0$ is the averaging period. Under t_0 and t_n, we mean the moments of time corresponding to the beginning and end of saving the samples.

RMS pressure distributions obtained using various approaches are shown in Figure 7. The control surface is located at $r_0/R = 1.2$, and the observation point—on a circle with radius $r_c/D = 250$. Symbols □ correspond to simulation results, dash-dotted line—experimental data [12] at Re = 3000, double dash-dotted line corresponds to experimental data [12] at Re = 5100. Good agreement between the calculated and experimental data takes place at $30 < L_c/D < 150$.

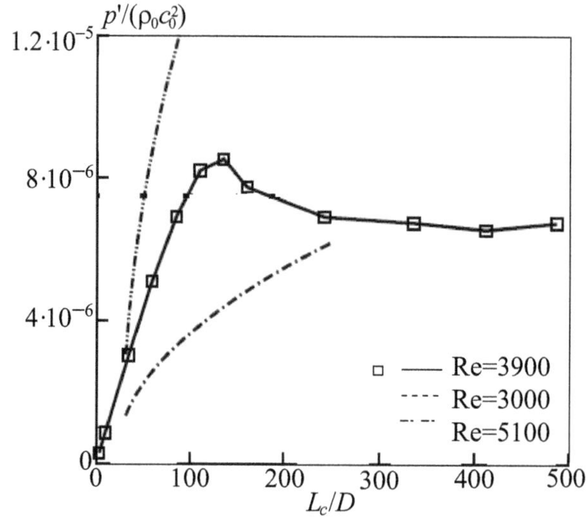

Figure 7. RMS pressure distributions depending on the length of the cylinder at $\theta = 90°$ for Re = 3900.

8. Results and Discussion

The structure of the flow behind a transversely streamlined cylinder in a stationary external flow has been well studied. In the case of a transverse flow around a cylinder, a boundary layer is formed on its surface, the thickness of which gradually increases upstream. The main determining parameters of this layer are the Reynolds number and the turbulence of the oncoming flow. The separation of the boundary layer and the formation of vortices in the wake of cylindrical bodies are periodic processes. The frequency of separation is characterized by the Strouhal number, which depends on the Reynolds number, the turbulence of the external flow, the degree of obstruction of the channel, and other factors.

The flow that is formed when the boundary cylinder is separated from the surface loses its stability and interacts with the pipe walls. As a result, self-sustaining pressure oscillations arise, which lead to the presence of intense peaks in the spectrum of the generated noise. The field of the vorticity modulus and the z-component of the vorticity vector in the symmetry plane are shown in Figure 8. The vortices form an extensive wake, grow downstream, and interact with the pipe walls. The level lines of the z-component of the vorticity vector give an idea of the direction of rotation of the vortices.

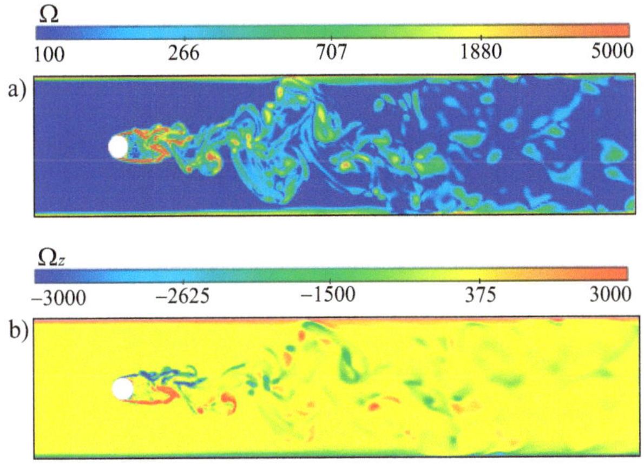

Figure 8. Contours of the modulus of the vorticity vector (**a**) and the z-component of the vorticity vector (**b**) in the symmetry plane for $\mathrm{Re}_c = 39,809$.

The distribution of overpressure on the channel walls is shown in Figure 9. The pressure is distributed in a significantly non-uniform way, and therefore, one should expect the occurrence of oscillations of the elastic walls of the pipe.

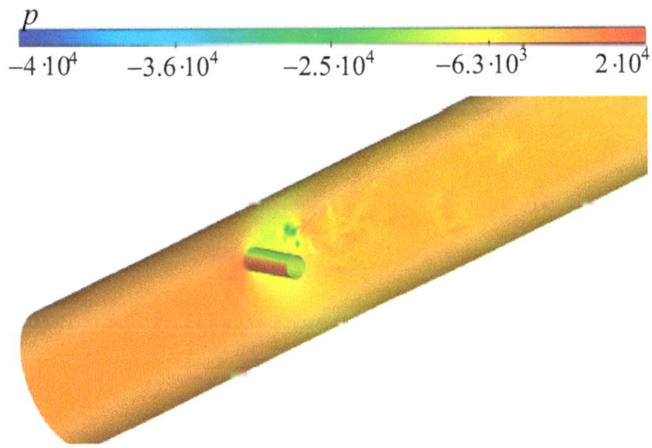

Figure 9. Overpressure field on pipe walls for $\mathrm{Re}_c = 39,809$.

One of the criteria used to identify eddy flows is the Q-criterion

$$Q = \frac{1}{2}\left(|\Omega|^2 - |S|^2\right),$$

where $|S| = [\mathrm{trace}(SS')]^{1/2}$, $|\Omega| = [\mathrm{trace}(\Omega\Omega')]^{1/2}$. The tensor invariant is

$$Q = -\frac{1}{2}(\lambda_1 + \lambda_2 + \lambda_3),$$

where λ_i are the eigenvalues of the velocity gradient tensor ($i = 1, 2, 3$). A vortex is defined as a flow region in which the inequality $Q > 0$ is satisfied (the flow region in which the norm of the vorticity tensor exceeds the norm of the strain rate tensor). The connection between the pressure and the Q criterion is established using the Poisson equation for the pressure $\nabla^2 p = 2\rho Q$. The condition $Q > 0$ does not necessarily correspond to the pressure maximum (for $Q > 0$, the pressure maximum is observed at the boundary).

To identify eddies, the λ_2-criterion is used, based on the decomposition of the velocity gradient tensor into symmetric (strain rate tensor) and antisymmetric (rotation tensor) tensors [35]. The $S^2 + \Omega^2$ tensor is symmetric and has real eigenvalues ($\lambda_1 \geqslant \lambda_2 \geqslant \lambda_3$), two of which are negative. The vortex flow region is defined as the region in which

$$\lambda_2\left(S^2 + \Omega^2\right) < 0,$$

where $\lambda_2(A)$ is an eigenvalue of the symmetric tensor A. Under adiabatic conditions, this criterion guarantees an instantaneous pressure minimum in a two-dimensional flow [36].

The level surfaces of the Q and λ_2 criteria, which are different combinations of the velocity gradient tensor invariants [37], colored according to the value of the velocity modulus, make it possible to visualize the vortex structure of the flow (Figure 10). The figure clearly shows both the three-dimensional flow structure and the resulting large-scale vortex component of the Karman street.

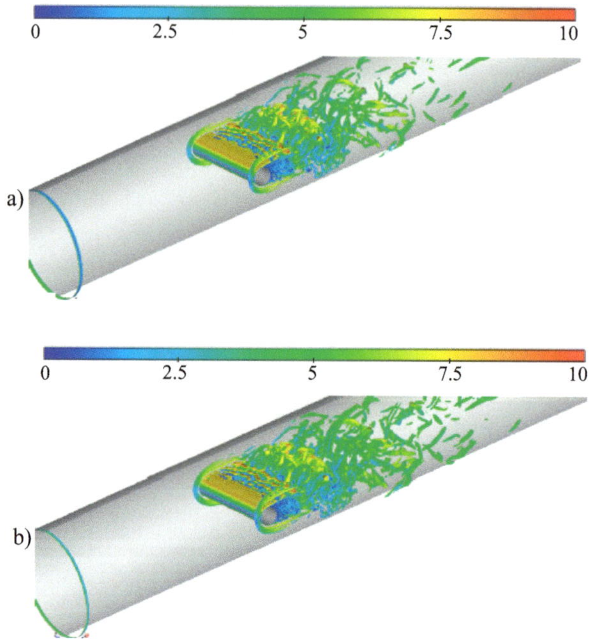

Figure 10. Contours of Q-criterion (**a**) and λ_2-criterion (**b**) for $\mathrm{Re}_c = 39,809$.

The presented fields of physical quantities illustrate the general picture of the separated flow near the cylinder. As quantitative estimates of the accuracy of solving this problem, the most interesting are the spectra of pressure fluctuations on the cylinder surface and acoustic loads on the pipe walls. At each point on the wall, as well as at some selected points, the Fourier transform of the data is performed, expressing the dependence of pressure on time. The time dependence of overpressure on the pipe wall is shown in Figure 11 at $x = 0.1$ m. Dotted lines show pressure levels corresponding to ± 200 Pa.

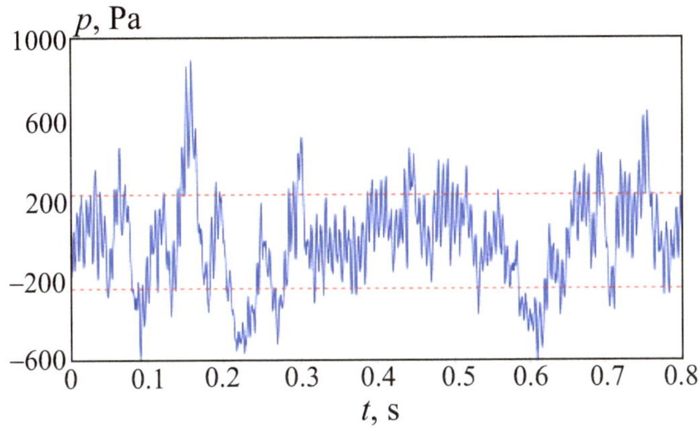

Figure 11. Time dependence of pressure on the wall at point $(0.1, 0, 0)$.

The Fourier transform of the pressure signal at the same point is shown in Figure 12. Line 1 shows the results of calculations, line 2—smoothed spectrum, and columns—results of calculations using the $1/12$ filter. A peak with a frequency corresponding to the fundamental tone of the pulsations stands out well.

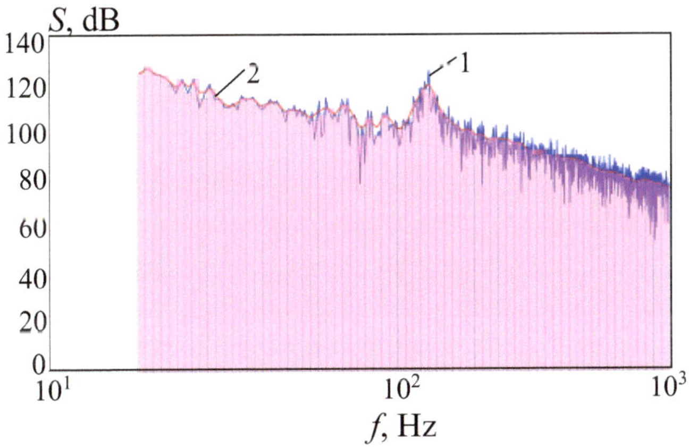

Figure 12. Spectrum of pressure fluctuations on the wall at point $(0.1, 0, 0)$.

The distribution of the spectrum over all points on the channel wall for a certain frequency makes it possible to judge the places of the most intense impact from the flow on the walls at a given frequency (Figure 13).

The power spectral densities of near-wall pressure fluctuations on the streamlined surface and on the channel walls have clear discrete peaks that correspond to the nature of the vortex motion over the surface under study. Discrete peaks of the spectral levels of near-wall pressure fluctuations, which have the most energy-intensive main frequency harmonic, have higher-order harmonics, as well as sub-harmonics, due to the nonlinear interaction of vortex structures (sources of pseudo-sonic fluctuations) with each other and with the surface in a stream. This occurs due to the merging of the vortex structures dominating here with each other (generation of sub-harmonics) and their destruction in accordance with the mechanism of the cascade process of vortex transformation.

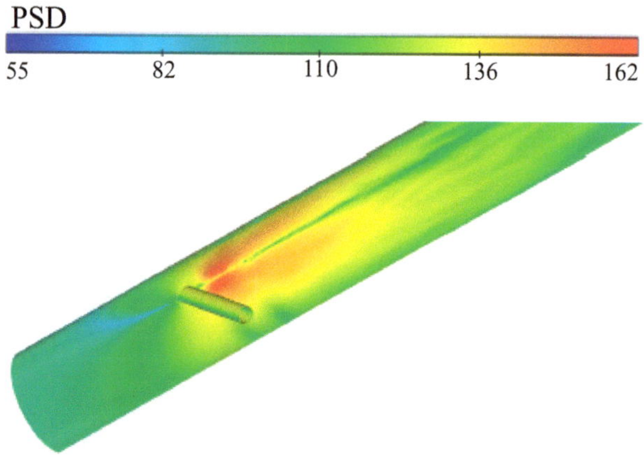

Figure 13. Spectrum of pressure fluctuations on the walls at $f = 167.5$ Hz.

As a result of the calculations, the displacement amplitudes were obtained depending on the frequency. The dependence of the displacement amplitude at the point $(0.5, 0, 0)$ is shown in Figure 14. There is a clear peak corresponding to the frequency of the fundamental tone.

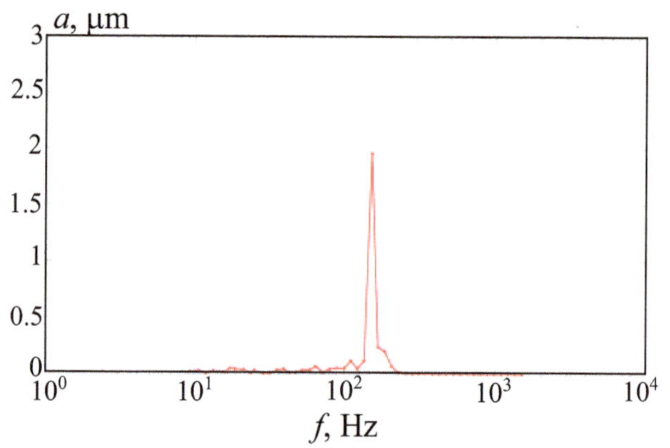

Figure 14. The dependence of the displacement amplitude on the frequency at the point $(0.5, 0, 0)$.

9. Conclusions

A mathematical model and computational algorithm to perform coupled simulations of unsteady processes in the circular channel with a round cylinder were developed and validated. At the first stage, the velocity and pressure fields are calculated with delayed detached eddy simulation. At the second stage, the problem of calculating the stress–strain state of the pipe is solved using the finite element method.

Stand-alone CFD simulation of unsteady viscous flowfield around a circular cylinder were performed, and results of CFD calculations were validated against experimental and computational data from the literature. The values of Strouhal numbers and amplitudes of oscillations of the aerodynamic coefficients are determined with the fast Fourier transformation technique.

When fluid flow separates from the cylinder and flow loses its stability, self-sustaining pressure oscillations arise, which lead to the presence of intense peaks in the spectrum of the generated noise. The spectrum of pressure oscillations on the cylinder surface and wall of the pipe have clear discrete peaks. The behavior of spectral dependencies is generally preserved if the flow velocity (Reynolds number) increases.

Funding: The research is partially funded by the Ministry of Science and Higher Education of the Russian Federation as part of World-class Research Center program: Advanced Digital Technologies (contract No. 075-15-2020-903 dated 16 November 2020).

Institutional Review Board Statement: Not applicable.

Informed Consent Statement: Not applicable.

Data Availability Statement: Not applicable.

Conflicts of Interest: The author declares no conflict of interest.

References

1. Kudashev, E.B. Experimental studies of flow noise on a pop-up device. *Acoust. J.* **2005**, *51*, 488–499.
2. Snarski, S.R.; Lueptow, R.M. Wall pressure and coherent structures in a turbulent boundary layer on a cylinder in a axial flow. *J. Fluid Mech.* **1995**, *286*, 137–171. [CrossRef]
3. Khasanov, H.A. Acoustic oscillations of a gas near nested thin-walled cylindrical obstacles in a channel. *J. Appl. Mech. Tech. Phys.* **2011**, *4*, 91–99. [CrossRef]
4. Mazo, A.B.; Morenko, I.V. Numerical simulation of viscous separated flow around a rotating circular cylinder. *J. Eng. Phys. Thermophys.* **2006**, *79*, 75–81. [CrossRef]
5. Okhotnikov, D.I.; Molochnikov, V.M.; Mazo, A.B.; Malyukov, A.V.; Goltsman, A.E.; Saushin, I.I. Viscous flow in the wake of a circular cylinder near the wall at moderate Reynolds numbers. *Thermophys. Aeromech.* **2017**, *24*, 897–907. [CrossRef]
6. Isaev, S.A.; Sudakov, A.G.; Zhukova, Y.V.; Usachov, A.E. Simulation of drag reduction and removal of alternating load on a circular cylinder due to the throttling effect. *J. Eng. Phys. Thermophys.* **2014**, *87*, 904–907. [CrossRef]
7. Isaev, S.A.; Zhukova, Y.V.; Popov, I.A.; Sudakov, A.G. Vortex intensification of heat transfer in laminar flow around a circular and elliptical cylinders with air and M-20 oil. *J. Eng. Phys. Thermophys.* **2018**, *91*, 664–672.
8. Williamson, C.H.K. Oblique and parallel modes of vortex shedding in the wake of a circular cylinder at low Reynolds numbers. *J. Fluid Mech.* **1989**, *206*, 579–627. [CrossRef]
9. Williamson, C.H.K. Vortex dynamics in the cylinder wake. *Annu. Rev. Fluid Mech.* **1996**, *28*, 477–539. [CrossRef]
10. Inoue, O.; Hatakeyama, N. Sound generation by a two-dimensional circular cylinder in a uniform flow. *J. Fluid Mech.* **2002**, *471*, 285–314. [CrossRef]
11. Etkin, B.; Korbacher, G.K.; Keefe, R.T. Acoustic radiation from a stationary cylinder in a fluid stream (Aeolian tones). *J. Acoust. Soc. Am.* **1957**, *29*, 30–36. [CrossRef]
12. Norberg, C. An experimental investigation of the flow around a circular cylinder: Influence of aspect ratio. *J. Fluid Mech.* **1994**, *258*, 287–316. [CrossRef]
13. Henderson, R.D. Details of the drag curve near the onset of vortex shedding. *Phys. Fluids* **1995**, *7*, 2102–2104. [CrossRef]
14. Zhang, H.-Q.; Fey, U.; Noack, B.R.; Konig, M.; Eckelmann, H. On the transition of the cylinder wake. *Phys. Fluids* **1995**, *7*, 779–794. [CrossRef]
15. Fey, U.; Kőnig, M.; Eckelmann, H. A new Strouhal–Reynolds-number relationship for the circular cylinder in the range $47 < Re < 2 \times 10^5$. *Phys. Fluids* **1998**, *10*, 1547–1549.
16. Persillon, A.; Braza, M. Physical analysis of the transition to turbulence in the wake of a circular cylinder by three-dimensional Navier–Stokes simulation. *J. Fluid Mech.* **1998**, *365*, 23–88. [CrossRef]
17. Ma, X.; Karamanos, G. Dynamics and low-dimensionality of a turbulent wake. *J. Fluid Mech.* **2000**, *410*, 29–65. [CrossRef]
18. Posdziech, O.; Grundmann, R. Numerical simulation of the flow around an infinitely long circular cylinder in the transition regime. *Theor. Comput. Fluid Dyn.* **2001**, *15*, 121–141.
19. Tremblay, F.; Manhart, M.; Friedrich, R. DNS of flow around a circular cylinder at a subcritical Reynolds number with cartesian grids. In Proceedings of the 8th European Turbulence Conference, Barcelona, Spain, 27–30 June 2002; pp. 659–662.
20. Qu, L.; Norberg, C.; Davidson, L.; Peng, S.-H.; Wang, F. Quantitative numerical analysis of flow past a circular cylinder at Reynolds number between 50 and 200. *J. Fluids Struct.* **2013**, *39*, 347–370. [CrossRef]
21. Choi, W.-S.; Choi, Y.; Hong, S.-Y.; Song, J.-S.; Kwon, H.-W.; Jung, C.-M. Turbulence-induced noise of a submerged cylinder using a permeable FW–H method. *Int. J. Nav. Archit. Ocean. Eng.* **2016**, *8*, 235–242. [CrossRef]
22. Koval, K.A.; Sukhorukov, A.L.; Chernyshev, I.A. Verification results of the numerical method for calculating the hydrodynamic and hydroacoustic characteristics of a fin propulsor. *Fundam. Appl. Hydrophys.* **2016**, *9*, 60–72.
23. Liu, X.; Thompson, D.J.; Hu, Z. Numerical investigation of aerodynamic noise generated by circular cylinders in cross-flow at Reynolds numbers in the upper subcritical and critical regimes. *Int. J. Aeroacoust.* **2019**, *18*, 470–495. [CrossRef]
24. Kiselev, M.G.; Nifagin, V.A.; Drozdov, A.V. On one model of interaction of a vibroacoustic system with a rigid base. *J. Eng. Phys. Thermophys.* **1958**, *77*, 20–27.
25. Valles, N.G.; Kaplunov, R.M. Mathematical modeling of non-stationary separated flow past a lattice of tubes of heat exchangers. *J. Eng. Phys. Thermophys.* **1987**, *53*, 57–64.
26. Khabbouchi, I.; Guellouz, M.S.; Nasrallah, S.B. A study of the effect of the jet-like flow on the near wake behind a circular cylinder close to a plane wall. *Exp. Therm. Fluid Sci.* **2013**, *44*, 285–300. [CrossRef]

27. Armellini, A.; Casarsa, L.; Giannattasio, P. Separated flow structures around a cylindrical obstacle in a narrow channel. *Exp. Therm. Fluid Sci.* **2009**, *33*, 604–619. [CrossRef]
28. Palkin, E.V.; Mulyadzhanov, R.I. Low-frequency pulsations in the problem of fluid flow around a cylinder in a narrow gap at Re = 3750. *Sib. Phys. J.* **2017**, *12*, 43–49.
29. Voskoboynik, V.A.; Grinchenko, V.V.; Makarenkov, A.P. Pseudosound behind an obstacle on a longitudinally streamlined cylinder. *Acoust. Bull.* **2002**, *5*, 22–36.
30. Voropaev, G.A.; Voskoboinik, A.V.; Voskoboinik, V.A.; Grinchenko, V.V.; Isaev, S.A.; Rozumnyuk, N.V. Sources of pseudosonic pressure fluctuations in the flow around a spherical dimple. *Acoust. Bull.* **2008**, *11*, 27–49.
31. Martin, V.; Cummings, A.; Gronier, C. Discrimination of coupled structural/acoustic duct modes by active control: Principles and experimental results. *J. Sound Vib.* **2004**, *274*, 583–603. [CrossRef]
32. Shinde, V.; Marcel, T.; Hoarau, Y.; Deloze, T.; Harran, G.; Baj, F.; Cardolaccia, J.; Magnaud, J.-P.; Longatte, E.; Braza, M. Numerical simulation of the fluid—Structure interaction in a tube array under cross flow at moderate and high Reynolds number. *J. Fluids Struct.* **2014**, *47*, 99–113. [CrossRef]
33. Volkov, K.N.; Emelaynov, V.N. *Gas Flows Induced by Fluid Injection in Energy Systems*; Publishing House of Physical and Mathematical Literature: Moscow, Russia, 2011.
34. Volkov, K.N. Simulation of coupled heat transfer on unstructured meshes. *Comput. Math. Math.* **2013**, *53*, 396–416.
35. Nikushchenko, D.; Pavlovsky, V. Fluid motion equations in tensor form. In *Advances on Tensor Analysis and their Applications*; IntechOpen: London, UK, 2020.
36. Haller, G. An objective definition of a vortex. *J. Fluid Mech.* **2005**, *525*, 1–26. [CrossRef]
37. Volkov, K.N.; Zapryagaev, V.I.; Emelyanov, V.N.; Gubanov, D.A.; Kavun, I.N.; Kiselev, N.P.; Teterina, I.V.; Yakovchuk, M.S. *Visualization of Data of Physical and Mathematical Simulation in Gas Dynamics*; Publishing House of Physical and Mathematical Literature: Moscow, Russia, 2018.

fluids

Article

CFD Modeling of Wind Turbine Blades with Eroded Leading Edge

Michael Carraro [1], Francesco De Vanna [1,*], Feras Zweiri [2], Ernesto Benini [1], Ali Heidari [2] and Homayoun Hadavinia [2]

1 Department of Industrial Engineering, University of Padova, Via Venezia 1, 35100 Padova, Italy
2 Department of Mechanical Engineering, Kingston University, London SW15 3DW, UK
* Correspondence: francesco.devanna@unipd.it

Abstract: The present work compares 2D and 3D CFD modeling of wind turbine blades to define reduced-order models of eroded leading edge arrangements. In particular, following an extensive validation campaign of the adopted numerical models, an initially qualitative comparison is carried out on the 2D and 3D flow fields by looking at turbulent kinetic energy color maps. Promising similarities push the analysis to consequent quantitative comparisons. Thus, the differences and shared points between pressure, friction coefficients, and polar diagrams of the 3D blade and the simplified eroded 2D setup are highlighted. The analysis revealed that the inviscid characteristics of the system (i.e., pressure field and lift coefficients) are precisely described by the reduced-order 2D setup. On the other hand, discrepancies in the wall friction and the drag coefficients are systematically observed with the 2D model consistently underestimating the drag contribution by around 17% and triggering flow separation over different streamwise locations. Nevertheless, the proposed 2D model is very accurate in dealing with the more significant aerodynamics performance of the blade and 30 times faster than the 3D assessment in providing the same information. Therefore the proposed 2D CFD setup is of fundamental importance for use in a digital twin of any physical wind turbine with the aim of carefully and accurately planning maintenance, also accounting for leading edge erosion.

Keywords: leading edge erosion; wind turbine; CFD modeling; reduced-order modeling; digital twin

Citation: Carraro, M.; De Vanna, F.; Zweiri, F.; Benini, E.; Heidari, A.; Hadavinia, H. CFD Modeling of Wind Turbine Blades with Eroded Leading Edge. *Fluids* **2022**, *7*, 302. https://doi.org/10.3390/fluids7090302

Academic Editor: Mehrdad Massoudi

Received: 12 August 2022
Accepted: 10 September 2022
Published: 14 September 2022

1. Introduction

Wind energy is playing a significant role in producing clean and eco-friendly electricity while mitigating climate change. In 2021, the EU installed 17.4 GW of new wind power facilities, raising its total capacity to 236 GW. Over the next five years, it is predicted that the Old EU will install 116 GW of new wind power farms for an average of 23.1 GW per year [1]. In 2021, the USA installed 14 GW of new wind capacity, expanding its wind energy amount to 132.7 GW. Globally, installed wind energy in 2021 was estimated to be 93 GW, thus increasing the cumulative world capacity to around 825 GW [2]. The need to find alternative energy sources to fossil fuels and the importance of wind energy in this transition is also highlighted at the scientific level. Starting from experimental activities, in the early 1990s, Butterfield et al. [3] and Somers and Tangler [4] made the first experiments of a wind turbine aerofoil and since then many tests have been conducted [5–12]. On the other hand, looking at numerical approaches and thanks to the increasing availability of computing power, Computational Fluid Dynamics (CFD) have gained popularity for simulating wind turbine aerodynamics [13–24]. A clear survey about experimental and numerical works related to wind energy is provided by Vermeer et al. [25]. Generally speaking, the increasing trend in wind energy has created a demand for greater energy capture per utility-scale wind turbine, and as the energy production by wind turbine is proportional to swept area, a significant increase in rotor diameter has occurred over the past decades [26]. As the diameter increases, the blade tip of the wind turbine achieves higher linear speed ranges, thus presenting several challenges for wind turbine operations and maintenance. One major issue encountered as the blade speed increases is Leading Edge Erosion (LEE). The problem consists of progressive damage to the wind turbine blades due to climatic and meteorological factors. Most wind farms, in fact, are sited in tricky operating environments, e.g., hilly terrain where heavy precipitation is common or offshore locations where the turbines are exposed to droplets of salted water. As a result, wind turbines are subjected to various environmental factors throughout their operational lives: extreme wind/gusts, frequent rain showers, hailstone showers, snow, icing, extreme temperatures, and ultraviolet light exposure [27]. Thus, high blade tip velocities and harsh climates increase the blade's erosion susceptibility.

Unfortunately, LEE begins early in the blade life. Rempel [28] reported that blades as young as three years old would start to show signs of wear and, if repairs are not done early, the damage to the underlying laminate will be present as early as year five. Dalili et al. [29] investigated a wide range of surface engineering issues concerning the performance of wind turbine blades, with a particular emphasis on the problems presented by icing in Nordic climates. Keegan et al. [27] examined the potential degradation posed by the various environmental variables, focusing on the impact of rain droplets and hailstones on the blade's leading edge. To better understand how roughness affects performance, Ehrmann et al. [30] made field measurements of turbine-blade roughness and reproduced these measurements on an airfoil section tested in a wind tunnel. These data are used to validate and calibrate a one equation roughness amplification model that works in conjunction with the Langtry–Menter transition model. According to wind tunnel experiments by Sareen et al. [31], LEE on a wind turbine airfoil can result in significant aerodynamic performance degradation. In their study, DU 96-W-180 airfoils with varying degrees and types of LEE were tested to assess the effects of erosion on performance. It was shown that LEE leads to a significant increase in airfoil drag and an earlier onset of stall at lower angles of attack. The study's findings revealed a 6–500% increase in drag due to varying levels of LEE (light-to-heavy). According to further research, the predicted losses in Annual Energy Production (AEP) are between 5% and 25%, depending on the LEE severity. Gaudern [32] conducted a thorough experimental campaign on 18%-thick commercial wind turbine airfoils with various erosion stages, measuring full lift and drag polars in a wind tunnel. Later on, Langel et al. [33] examined the effects of airfoil leading edge roughness on lift and drag, performing experimental and computational studies on a NACA $63_3 - 418$ airfoil. Results show that LEE produces airfoil performance degradation by increasing the drag coupled with a significant loss in lift near the upper corner of the drag polar. Maniaci et al. [34] used a profilometer to measure the roughness of operational wind turbine blades, and measurements were statistically replicated in the wind tunnel. Simultaneously, a computational model was developed and calibrated to capture the effect of roughness and erosion on flow transition and performance. Han et al. [35] conducted a quantitative analysis of the impact of contamination and erosion at the leading edge of wind turbine blades on AEP. The lift and drag coefficients of blades subjected to LEE decreased and increased by up to 53% and 314%, respectively. Elhadi Ibrahim and Medraj [36] developed an analytical surface fatigue model to predict the initiation of LEE on wind turbine blade coatings due to rainfall, then combined this LEE forecast model with an efficiency reduction model to predict AEP loss over time on different sites. Dashtkar et al. [37] provided an extensive review of the liquid erosion mechanism, water erosion testing procedures, and the contributing factors to the LEE. Techniques for improving the erosion resistance of the leading edge are also discussed. Later on, Shankar Verma et al. [38] showed that a significantly higher erosion damage rate is found for blades exposed to offshore rainfall conditions than for blades under onshore rainfall conditions. Law and Koutsos [39] investigated the energy losses associated with LEE on an operational wind farm, discovering that the average AEP is decreasing by 1.8% due to medium levels of erosion, with the worst affected turbine experiencing losses of 4.9%. Castorrini et al. [40] analyzed blades with LEE consisting of pits and gouges using a 3D Reynolds-Averaged Navier–Stokes (RANS) approach. An AEP loss between 1 and

2.5% is encountered for the considered LE damages. Later, Wang et al. [41] undertook CFD simulations of wind turbine blades with LEE, showing that the degree of erosion significantly impacts flow separation, aerodynamic coefficients, and wind turbine power output. Finally, Ortolani et al. [42] presented RANS analyses of the NACA $63_3 - 618$ airfoil with LEE. The erosion pattern studied are extracted from real offshore wind turbine after about six years of operation. For an overview of the latest research in LEE, based on meteorology, aerodynamics, materials science, and computational mechanics, refer to the recent survey by Mishnaevsky Jr et al. [43].

In this work we report just one of the many real examples of LEE. In particular, Figure 1 depicts the potential damage caused by LEE on a wind turbine blade by showing four levels of erosion damage. According to Sareen et al. [31], LEE starts with the formation of small pits near the blade leading edge. As the density of these pits grows over time, they join together to form larger and deeper gouges. The final stage of the erosion process is delamination, which begins at the leading edge and grows in chordwise extent over time. The result of this process poses a complex geometrical configuration of the leading edge, and a three-dimensional flow behavior over the airfoil is expected.

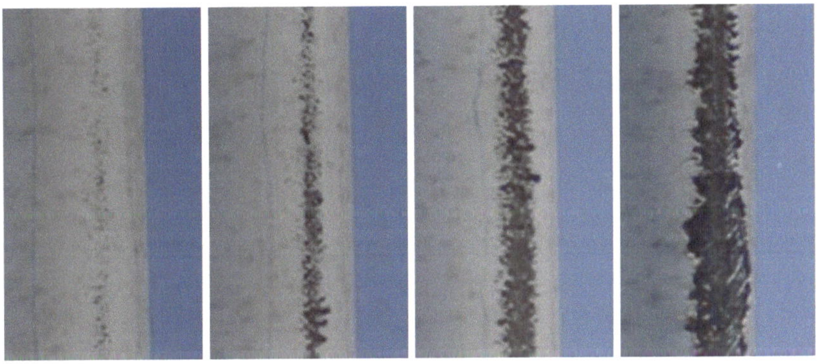

(**a**) Level 1 (**b**) Level 2 (**c**) Level 3 (**d**) Level 4

Figure 1. Various levels of erosion from lighter (**left**) to heavier (**right**), adapted from [44].

Despite the numerous contributions in the literature relating to LEE, and generally to wind turbine applications, no explicit references are available concerning whether the LEE modeling should be approached with CFD methods. In particular, no guidelines are given in respect of how to model the problem efficiently; whether adopting a fully 3D approach (which at first sight would seem the most reasonable given the numerous geometric complexities posed by an eroded airfoil) or if simplified 2D models can describe the blade aerodynamics with sufficient degree of accuracy. This issue is of fundamental importance for predicting the wind turbine operating life in view of planning routine maintenance interventions. Thus, the present research aims to fill this research gap by comparing 2D and 3D CFD models of damaged wind turbine blades. It will be shown that a simplified and properly tuned 2D model can reproduce wind turbine blades' complex 3D aerodynamics behavior with severe LEE for a wide range of angles of attack. In particular, the analyzed LEE is compatible with the Type C/Stage 5 case as defined by Sareen et al. [31], finding a condition observed in wind turbine blades that had been operating for 10+ years (see [27]). Thus, in the present work, a two-dimensional flow configuration is compared to a realistic three-dimensional model configuration. The analysis has been done on NACA 0012 airfoils. Results highlight that the reduced-order model well represents the complex 3D physics of the damaged blade but at a much lower computational cost. Therefore, to forecast wind turbine life using predictive digital twin models, the present study results allow us to state that LEE can be fruitfully approached even with 2D CFD models.

The structure of the current paper is as follows: Section 2 details the computational setup and numerical methodology. Section 3 explains how the models are tuned and validated. Section 4 discusses the results, while Section 5 states the conclusions.

2. Computational Setup and Numerical Methodology

The current work employs the commercial suite Ansys Fluent® 19.2 [45] for the simulations' pre-processing, meshing, solution, and post-processing phases. MATLAB® R2019b [46] scripts are also used for results post-processing.

2.1. Governing Equations

The flow dynamics of the systems addressed in this work are simulated according to the uncompressible RANS system of equations in the hypothesis of steady flows. Thus, with $\phi = \bar{\phi} + \phi'$ being the Reynolds decomposition associated with the generic flow quantity, ϕ, the models dynamics are governed by the following equations:

$$\frac{\partial \bar{u}_i}{\partial x_i} = 0 \tag{1a}$$

$$\frac{\partial \bar{u}_i \bar{u}_j}{\partial x_j} = -\frac{1}{\rho}\frac{\partial \bar{p}\delta_{ij}}{\partial x_j} + \frac{\partial}{\partial x_j}\left(\nu\left(\frac{\partial \bar{u}_i}{\partial x_j} + \frac{\partial \bar{u}_j}{\partial x_i}\right) - \overline{u_i' u_j'}\right) \tag{1b}$$

Here, $\bar{u}_i$ is the ensemble average velocity component in the i-th direction, $\bar{p}$ is the ensemble average mechanical pressure, while $\nu = \mu/\rho$ denotes the molecular kinematic viscosity with μ the molecular flow viscosity. Reynolds stress components, $\rho\overline{u_i' u_j'}$, are modeled via the canonical Boussinesq's hypothesis

$$\rho\overline{u_i' u_j'} = \rho\frac{2}{3}k\delta_{ij} - \rho\bar{\nu}_t\left(\frac{\partial \bar{u}_i}{\partial x_j} + \frac{\partial \bar{u}_j}{\partial x_i}\right) \tag{2}$$

with $k = 1/2u_i'u_i'$ being the turbulent kinetic energy and $\bar{\nu}_t$ the turbulent kinematic viscosity, respectively. The latter is modeled according to three increasingly complex turbulence models and, in particular, we use the one-equation Spalart–Allmaras (SA) model by Spalart and Allmaras [47], the two-equation $k-\omega$ Shear Stress Transport ($k-\omega$ SST) model by Menter [48] and the four-equation Transition SST (TSST) model by Menter et al. [49]. Such models are among the best choices for external aerodynamics applications. Here, a brief overview of these models is reported, while for a more in-depth description, the reader is addressed to the Ansys Fluent theory guide [50]. For all the models, the default setups are kept as suggested by the Ansys Fluent manual.

2.1.1. Spalart–Allmaras Model

The one-equation SA model is widely used for external aerodynamic flows and provides improved performance relative to $k-\epsilon$ models for flows with adverse pressure gradients and separation. Overall, the accuracy of predicting separation is lower than for optimal two-equation models such as SST models. According to the model, the turbulent kinematic viscosity, $\bar{\nu}_t$, is defined by the following transport equation:

$$\frac{\partial}{\partial t}(\rho\bar{\nu}_t) + \frac{\partial}{\partial x_i}(\rho\bar{\nu}_t\bar{u}_i) = G_\nu + \frac{1}{\sigma_\nu}\left[\frac{\partial}{\partial x_j}\left\{(\mu + \rho\bar{\nu}_t)\frac{\partial \bar{\nu}_t}{\partial x_j} + C_{b2}\rho\left(\frac{\partial \bar{\nu}_t}{\partial x_j}\right)^2\right\}\right] - Y_\nu + S_\nu \tag{3}$$

Here, G_ν is the production of turbulent viscosity, Y_ν is the destruction of turbulent viscosity that occurs in the near-wall region due to the wall blocking and viscous damping, σ_ν and C_{b2} are the constants while S_ν is a user-defined source term.

2.1.2. $k-\omega$ Shear Stress Transport Model

The $k-\omega$ SST model uses a blending function to gradually transition from the standard $k-\omega$ model near the wall to a high Reynolds number version of the $k-\epsilon$ model in the outer portion of the boundary layer. The model, in particular, accounts for the transport of turbulent shear stress and gives highly accurate predictions of the onset and the amount of flow separation under adverse pressure gradients. The drawbacks are dependency on wall distance makes this less suitable for free shear flows compared to standard $k-\omega$. In addition, the resolution of the boundary layer in the wall-normal direction is required. The model is based on two transport equations: one for the turbulence kinetic energy, k, and one for the specific dissipation rate, $\omega = \epsilon/k$. Here, ϵ denotes the turbulent dissipation, $\epsilon = \nu\overline{\partial u_i'/\partial x_j \partial u_i'/\partial x_j}$. Thus, the turbulent kinetic viscosity, $\bar{\upsilon}_t$, is computed by combining k and ω as follows:

$$\bar{\upsilon}_t = \alpha^* \frac{k}{\omega} \tag{4}$$

where α^* is a damping coefficient to account for wall locations. The following equations hold:

$$\frac{\partial}{\partial t}(\rho k) + \frac{\partial}{\partial x_i}(\rho k \bar{u}_i) = \frac{\partial}{\partial x_j}\left[\rho\left(\nu + \frac{\bar{\upsilon}_t}{\sigma_k}\right)\frac{\partial k}{\partial x_j}\right] + G_k - Y_k + S_k \tag{5a}$$

$$\frac{\partial}{\partial t}(\rho \omega) + \frac{\partial}{\partial x_i}(\rho \omega \bar{u}_i) = \frac{\partial}{\partial x_j}\left[\rho\left(\nu + \frac{\bar{\upsilon}_t}{\sigma_\omega}\right)\frac{\partial \omega}{\partial x_j}\right] + G_\omega - Y_\omega + D_\omega + S_\omega \tag{5b}$$

Here, G_k and G_ω represent the generation of k and ω, respectively, while Y_k and Y_ω represent their dissipation. D_ω is the cross-diffusion term. σ_k and σ_ω are turbulent Prandtl numbers for k and ω, respectively, while S_k and S_ω are user-defined source terms. All of the above terms are discussed in detail in [50].

2.1.3. Transition Shear Stress Transport Model

The TSST model, also known as the $\gamma - Re_\theta$ model, is based on the coupling of the $k-\omega$ SST equations with two other transport equations, one for the intermittency, γ, and one for the transition onset criteria, in terms of transported momentum–thickness Reynolds number, $Re_{\theta t}$. The role of the two additional parameters is to account for the transition from a laminar to a turbulent flow, while in the $k-\omega$ SST model, the flow is in a turbulent state as it exerts the wall regions in the problem. The following equations hold for γ and $Re_{\theta t}$:

$$\frac{\partial}{\partial t}(\rho \gamma) + \frac{\partial}{\partial x_i}(\rho \gamma \bar{u}_i) = \frac{\partial}{\partial x_j}\left[\rho\left(\nu + \frac{\bar{\upsilon}_t}{\sigma_\gamma}\right)\frac{\partial \gamma}{\partial x_j}\right] + P_{\gamma 1} - E_{\gamma 1} + P_{\gamma 2} - E_{\gamma 2} \tag{6a}$$

$$\frac{\partial}{\partial t}(\rho Re_{\theta t}) + \frac{\partial}{\partial x_i}(\rho Re_{\theta t} \bar{u}_i) = \frac{\partial}{\partial x_j}\left[\sigma_\theta \rho(\nu + \bar{\upsilon}_t)\frac{\partial Re_{\theta t}}{\partial x_j}\right] + P_\theta \tag{6b}$$

Here, $P_{\gamma 1}$ and $E_{\gamma 1}$ are transition sources while $P_{\gamma 2}$ and $E_{\gamma 2}$ are relaminarization sources. P_θ is the source term of the $Re_{\theta t}$ transport equation. For an in-depth analysis of these terms, the reader is addressed to [50]. These two equations interacts with the $k-\omega$ SST turbulence model through modification of the transport equation for k (Equation (5a)), which is recast as follows:

$$\frac{\partial}{\partial t}(\rho k) + \frac{\partial}{\partial x_i}(\rho k \bar{u}_i) = \frac{\partial}{\partial x_j}\left[\rho\left(\nu + \frac{\bar{\upsilon}_t}{\sigma_k}\right)\frac{\partial k}{\partial x_j}\right] + \gamma_{eff} G_k - Y_k^* + S_k \tag{7}$$

where

$$Y_k^* = \min(\max(\gamma_{eff}, 0.1), 1.0) Y_k \tag{8}$$

and $\gamma_{eff} = \gamma_{eff}(\gamma, Re_{\theta t})$ modulates the flow separation/transition. The rationale behind the above model formulation is given in detail by Menter [48].

2.2. Three-Dimensional Setup

At first, we present the three-dimensional model. The latter is obtained by extruding the clean NACA 0012 airfoil and generating a damage pattern compatible with severe leading edge delamination. This stage of LEE is compatible with the damage observed in real blades after 10+ years of operating life [27,31]. Figure 2 reports the damaged blade model's representation. With c being the chord length, this geometry has a span, b, of $0.65c$, an average delamination length of $0.1c$, and a delamination depth of $0.014c$. Since this model represents a blade element, no twist or tip effects have been considered. This approach is commonly used for this type of analysis (see, e.g., [34,51,52]). The model is simulated using a C-mesh with a radius of $12c$ and a wake region length of $12c$ in the streamwise direction. The fluid domain is obtained using Boolean subtraction between the C region and the blade.

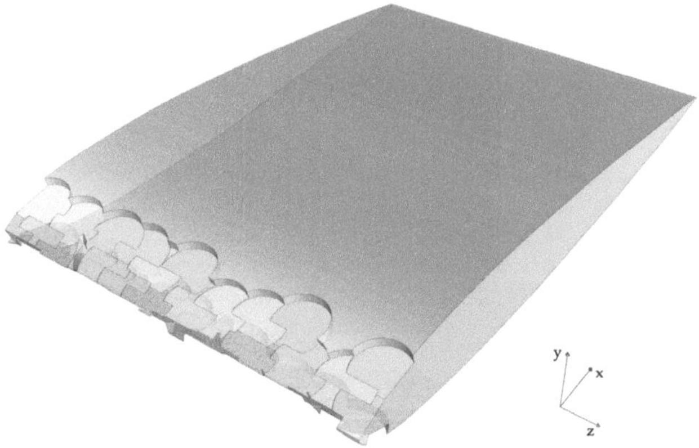

Figure 2. Geometrical representation of the three-dimensional blade with LEE.

A total of three hybrid increasingly refined grids are produced. These meshes are referred to as *Coarse*, *Medium*, and *Fine* with sizes of roughly 1.25, 2.50, and 5 million elements, respectively. The meshing process is carried out using the meshing software of the Ansys suite. Each grid quality is tested, resulting in a mean skewness of 0.34 and a mean orthogonal quality of 0.65. As previously stated, the meshing technique employs a hybrid structured–unstructured approach. An unstructured mesh is required to capture the blade's complex three-dimensional features. Structured cells are used in the flow areas near the walls to improve resolution in the boundary layer and ensure that proper wall y-plus values, $y_w^+ = \rho_w u_\tau y_w / \mu_w$, are kept. Here, ρ_w and μ_w denote the wall fluid density and viscosity, respectively, $u_\tau = \sqrt{\tau_w / \rho_w}$ is the wall friction velocity, y_w is the first-off-the-wall cell distance and $\tau_w = \mu_w \partial u_p / \partial y$ is the local wall friction, with u_p being the local wall parallel speed. In particular, first guess of the first-off-the-wall cell distance, y_w^*, is obtained by enforcing a y-plus value of $y_w^+ = 0.5$ to flat-plate empirical estimations according to the following equation:

$$y_w^* = 5.87 y_w^+ \frac{\mu_\infty}{\rho_\infty u_\infty} Re^{1/10} \tag{9}$$

Here, u_∞, ρ_∞ and μ_∞ denote the free stream velocity, flow density and viscosity, respectively, while Re is the Reynolds number of the flow. The near-wall resolution, in terms of y_w^+ distribution, is verified a posteriori as reported in Section 3.2. Figure 3 reports a leading edge overview of the *Fine* grid, as well as an enlargement of the damaged zone.

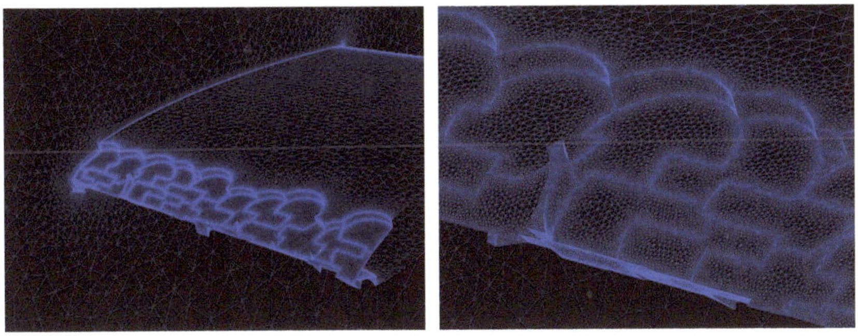

(a) (b)

Figure 3. Mesh overview (**a**) and detail (**b**) of the three-dimensional geometry. The figures refer to the fine grid of 5 million elements.

The simulations are carried out using the *Coupled* scheme implemented in Ansys Fluent. The spatial discretization is treated with a 2nd-order upwind scheme, while the Least Squares Cell-Based (Ansys) approach is employed for gradients' reconstruction. A precomputed tolerance of 1×10^{-6} is set as a stopping point for the iterative Navier–Stokes residuals dropping.

In terms of boundary conditions, the following are specified: *velocity-inlet* is enforced on the domain C edge while a *pressure-outlet* condition is set on the outlet section by selecting a reference value for the static pressure. The *symmetry* of the flow is enforced on the lateral boundaries. Figure 4 depicts an overview of the entire grid as well as boundary conditions. The inflow setup corresponds to a free stream Reynolds number of $Re = \rho_\infty u_\infty c/\mu_\infty = 3 \times 10^5$.

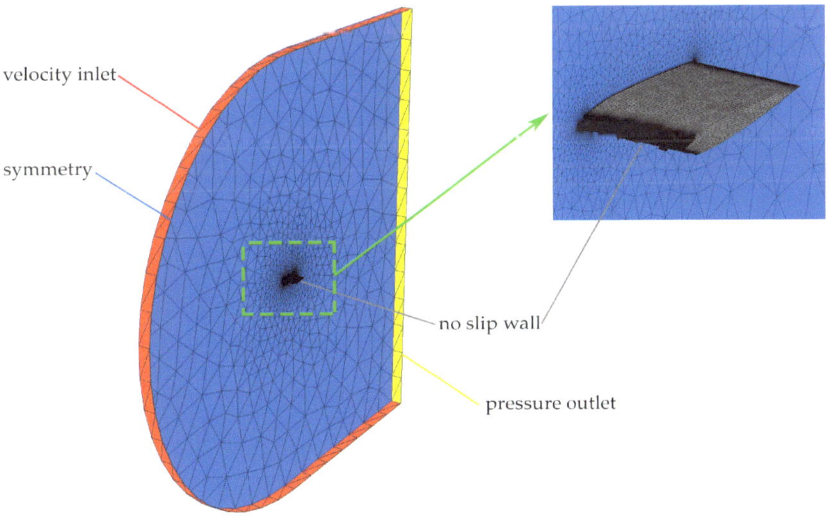

Figure 4. Boundary conditions overview: the red surface is the velocity-inlet while the pressure-outlet is highlighted in yellow. To enhance clarity only one side boundary, with symmetry condition imposed, is shown in light blue.

Finally, initial conditions assume the flow at rest in the whole domain. From such a state, the Ansys Fluent *hybrid initialization* is used together with 10 iterations of the *fmg-initialization* procedure aiming to speed up the residuals convergence.

2.3. Two-Dimensional Setup

Concerning the two-dimensional model, Figure 5 reports the damaged airfoil's geometrical properties. In particular, LEE is accounted for in a severely damaged NACA 0012. Thus, the 2D clean geometry is eroded accordingly, as seen in the sketch reported in Figure 5. With c being the chord of the airfoil, a delamination length of $x = 0.1c$ and a delamination depth of $h = 0.014c$ are considered, respectively. Such values fit the 3D arrangement at the average level. The model is simulated using a C-mesh with a radius of $12c$ and a wake region length of $12c$ in the streamwise direction. The fluid domain is obtained using Boolean subtraction between the C region and the airfoil, then it is divided into eight blocks with rectangular topology to produce a mapped grid.

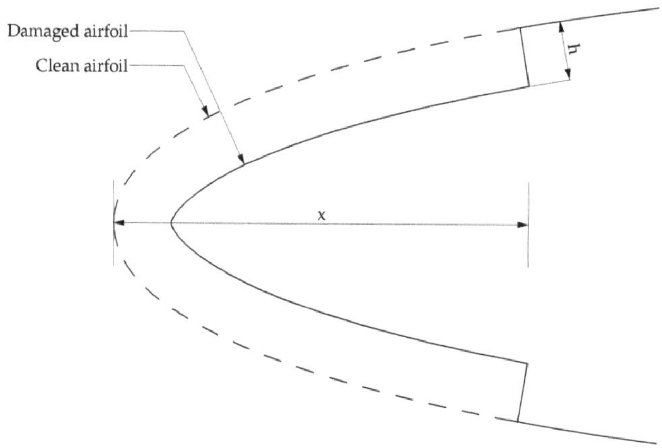

Figure 5. Geometrical representation of the LEE configuration. The chordwise delamination extent is denoted with the letter x and the delamination depth is denoted with the letter h.

Three mapped and increasingly refined meshes are used. The sizes are roughly 125 k, 250 k and 500 k for the *Coarse*, *Medium* and *Fine* grid, respectively. The meshing process is carried out using the mesh software of the Ansys suite. The quality of each grid is tested, resulting in a mean skewness of 0.34 and a mean orthogonal quality of 0.96. Each structured mesh is generated with a growth rate of 1.03 in the wall-normal direction to improve the boundary layer's resolution and assure a proper wall y-plus distribution. The $y_w^+ < 1$ threshold is pursued to provide a high-quality resolution of the near-wall flow. Similar procedures reported for the 3D setups are conducted to estimate the first guess of the first-off-the-wall cell distance. A posteriori verification of the y-plus distribution is reported in Section 3.1. Figure 6 reports an overview of the fine grid, as well as some enlargements of the leading edge and the delamination zone (Figure 6b,c).

Simulations are still carried out using the *Coupled* scheme implemented in Ansys Fluent, and the spatial discretization is still treated with a second-order upwind scheme. The same convergence tolerance, i.e., 1×10^{-6}, is used for monitoring the residual drop.

The boundary conditions are also kept equivalent to the 3D case. Thus, *velocity-inlet* is enforced on the domain C edge by specifying the inflowing speed such that $Re = 3 \times 10^5$. *Pressure-outlet* condition is set on the outlet section by specifying a reference value for the static pressure. Flow initializations follow the same procedure addressed for the 3D case.

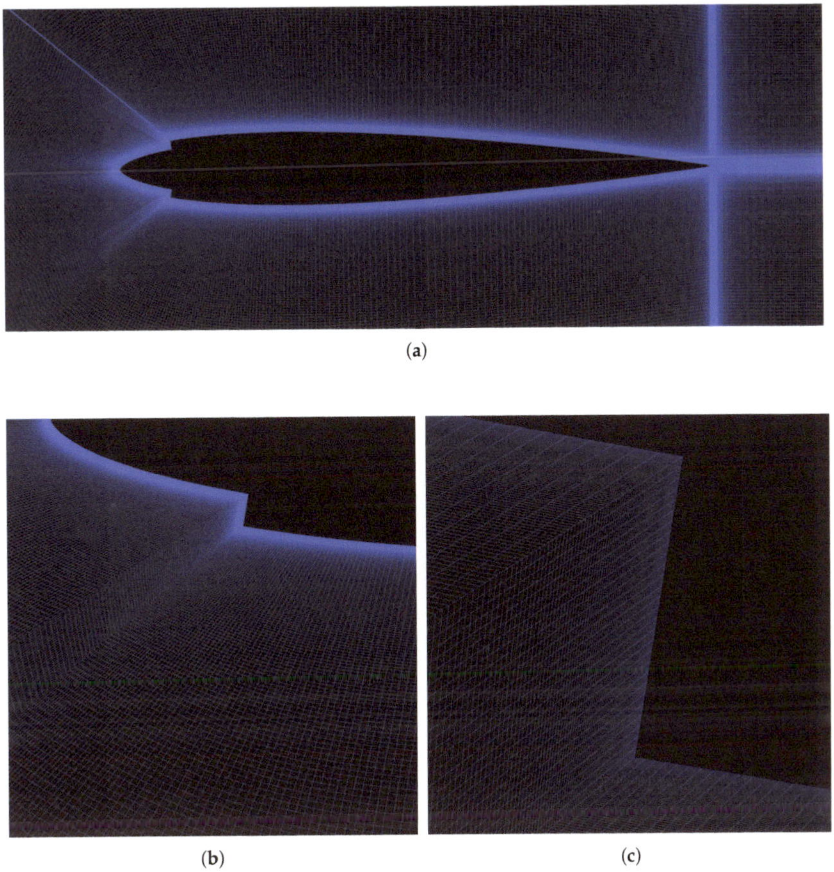

Figure 6. Mesh overview (**a**) and details (**b**,**c**). The figures refer to the fine grid of 500 k elements.

3. Numerical Models Validation

This section aims at tuning and validating the two- and three-dimensional numerical models according the best CFD practice [53–55]. The analyses aim to determine the most appropriate mesh refinement level and establish the influence of the turbulence model on system behavior. For clarity, results related to the two-dimensional model are firstly presented.

3.1. Two-Dimensional CFD Model Validation

As initial step, we compare the global drag coefficient values of the clean airfoil at zero incidence as a function of the grid refinement and parametrically to the turbulence model. Thus, the drag coefficient per unit length, $C_d = D/q_\infty c$, is compared to the experimental value C_{d-Exp}, the latter acquired in a in-house wind tunnel facility setting $Re = 3 \times 10^5$. Here, D is the drag force while $q_\infty = 1/2\rho_\infty u_\infty^2$ is the free stream dynamic pressure. Results are reported in Figure 7a. According to the rescaling procedure, C_d/C_{d-Exp} value close to 1 equals the experimental prediction. Due to the low-speed aerodynamics, the TSST model fits the best prediction. This result is consistent since the TSST model is tuned to describe the laminar-to-turbulent transition process, a key point for correctly predicting the wind turbine airfoils' aerodynamic performance in low-Reynolds conditions [14,15]. Figure 7b also depicts the pressure coefficient, $C_p = (p_w - p_\infty)/q_\infty$, trend obtained with the finest grid parametrically to the turbulence model. Since the airfoil is symmetrical and data are acquired at zero incidence ($\alpha = 0°$), the pressure distribution is symmetric, and a single curve is reported. Here, p_w is the wall pressure while p_∞ denotes the free stream static pressure. The increasingly darker blue lines represent the data acquired with SA, $k-\omega$ SST, and TSST model, respectively. The black dashed line denotes the XFOIL prediction [56], whereas the red triangles denote the experimentally acquired data by Gregory and O'Reilly [57]. Results are confined to a relatively tight range around experiments, except in a region near the trailing edge, where the TSST model overestimates the C_p. In this concern, it should be recalled that pressure distribution by Gregory and O'Reilly [57] is acquired at a different Reynolds number. Simulations, in fact, are carried on at $Re = 3 \times 10^5$, while experiments refer to $Re = 2.88 \times 10^6$, leading to different behaviors in the boundary layer mechanics. Nevertheless, XFOIL viscous predictions at $Re = 3.00 \times 10^5$ confirm the CFD pressure coefficient distribution. After these considerations, compliance with the experimental data assures that the pressure field is well captured by the CFD model, proving the quality of the numerical model via both global and local quantities.

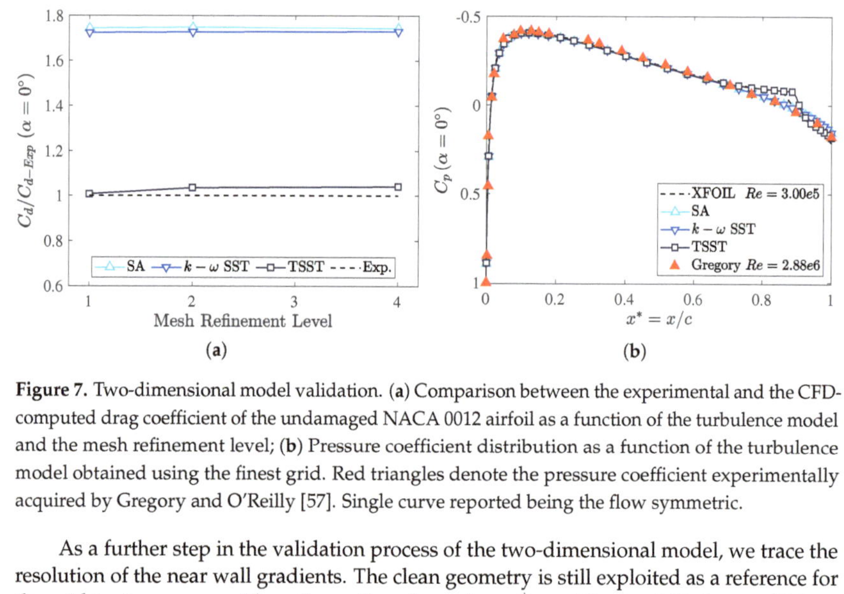

Figure 7. Two-dimensional model validation. (**a**) Comparison between the experimental and the CFD-computed drag coefficient of the undamaged NACA 0012 airfoil as a function of the turbulence model and the mesh refinement level; (**b**) Pressure coefficient distribution as a function of the turbulence model obtained using the finest grid. Red triangles denote the pressure coefficient experimentally acquired by Gregory and O'Reilly [57]. Single curve reported being the flow symmetric.

As a further step in the validation process of the two-dimensional model, we trace the resolution of the near wall gradients. The clean geometry is still exploited as a reference for the grid tuning process. Thus, the wall y-plus value, y_w^+, and the local friction coefficient, $C_f = \tau_w/q_\infty$, are investigated as a function of the turbulence model. Figure 8a reports the y_w^+ distribution along the clean airfoil obtained with the finest grid. As the reader can observe, each of the three adopted models provides values around the recommended threshold for wall-resolved RANS simulations (i.e., $y_w^+ < 1$), which allows us to state that, within the boundary layer, there is a sufficient number of points to resolve the velocity gradients near the walls accurately. The fine grid provides an average y_w^+ value of 0.18 using the TSST model. This value is well below the recommended threshold for wall-turbulence estimation, allowing a large margin also for the simulation of the damaged geometry. In that case, in fact, different wall dynamics are expected, especially in the delamination zone, and y_w^+ values closer to 1 are expected. Figure 8b reports the friction coefficient distribution. It can be observed that the SA and the $k-\omega$ SST models' predictions along the airfoil are very similar, while the TSST model provides for a recirculation zone on the final section of the airfoil. This fact is more consistent with the airfoil wall dynamics, making the TSST model outperform the other models concerning the experimental measurements. Such behavior is due to the low-Reynolds conditions of the flow. Thus, the ability of the TSST model to account for transition phenomena of the boundary layer is a fundamental aspect of predicting the flow physics correctly.

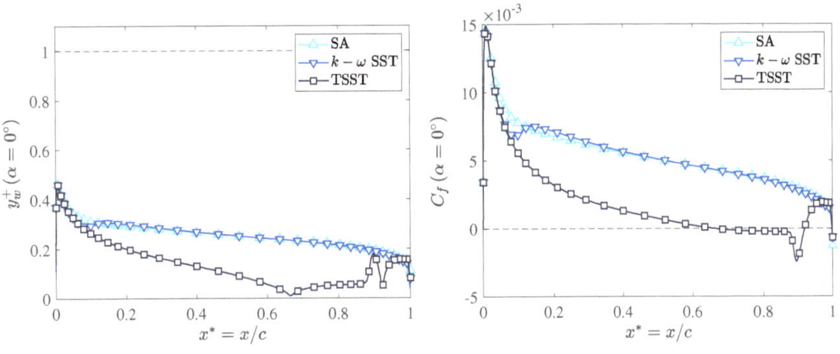

Figure 8. Wall y^+ and friction coefficient trends as a function of the turbulence model obtained with the finest grid. (**a**) y_w^+ distribution on the airfoil surface; (**b**) C_f distribution on the airfoil surface.

Finally, we conclude the two-dimensional model tuning by tracing its behavior in off-design conditions. Thus, the lift, C_l, and drag, C_d, coefficients distributions are obtained as a function of the angle of attack and compared with experimental data. Figure 9 reports the results of the analysis. Numerical data are still obtained with the most refined grid combined with the TSST model. High compatibility with experiments in a wide range of angles of attack is observed. In particular, the linear portion of the lift curve is very well captured by the steady-state RANS-TSST model (Figure 9a). While approaching stall conditions, RANS modeling struggles to capture the wake turbulent mechanics. This is why the CFD drag results do not perfectly fit the experimental data at high angles of attack (Figure 9b).

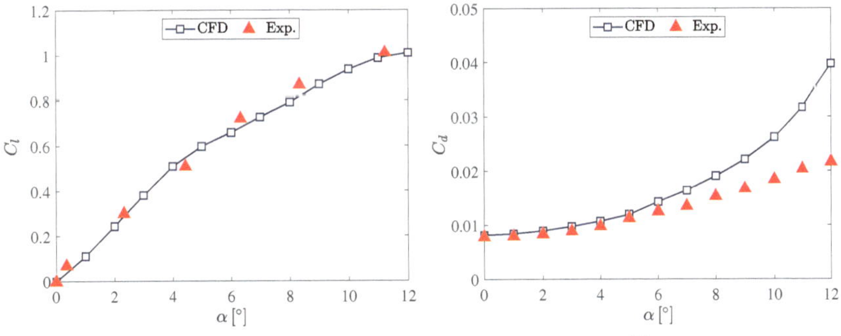

Figure 9. Comparison between lift and drag polar diagrams obtained with the finest grid (using the TSST turbulence model) and the experimental data for NACA 0012 airfoil. (**a**) C_l as a function of the incidence; (**b**) C_d as a function of the incidence.

Based on these findings, the most refined grid combined with the TSST turbulence model is selected for the subsequent analyses. The model provides excellent agreement with the experimental data. In addition, offering an average y_w^+ value of around 0.18, the model is promising to trace the flow physics even in off-design conditions.

3.2. Three-Dimensional CFD Model Validation

Following the two-dimensional model tuning, here we report the validation process concerning the three-dimensional arrangement. Since no experimental data are available to compare CFD outcomes, a mesh sensitivity study is carried out to establish the grid refinement level and the turbulence model in light of making results independent from the numerics. Thus, the global drag coefficient of the blade is gathered from three turbulence models (i.e., the SA, the $k-\omega$ SST and the TSST models) and three increasingly refined meshes, fitting approximatively 1.25, 2.5, and 5 million elements, respectively. The drag coefficient values are scaled with the corresponding outcome obtained with the most refined setup combined with the TSST turbulence model, $C_{d-Fine,TSST}$. Such a rescaling choice is motivated by the fact that TSST model has shown the best results in the two-dimensional setup compared to the available experimental data.

Figure 10a summarize the outcome of the analysis. The reader can observe that very high compatibility is recovered among whole methodologies. The scaled C_d values, in fact, are clustered to 1 as the mesh refinement level increases, regardless of how turbulence is accounted for. We observe that compared to the 2D case, in three-dimensional conditions, the TSST model does not outperform the others. The reason has to be found in the geometry damage, the latter triggering the separation of the boundary layer almost independently of the turbulence model, and even the SA one-equation model fits results independent of grid resolution. Thus, the fine mesh combined with the TSST is used to head the three-dimensional system mechanics.

Finally, we verify the mesh quality by tracing the y-plus wall values. Figure 10b provides the contour plot of the y_w^+. Again, the model confirms its quality since the near-wall resolution is consistently higher than the recommended thresholds for wall-resolved RANS simulations.

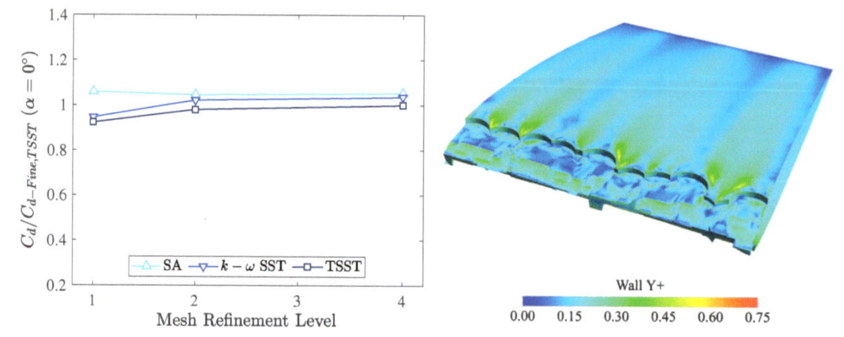

Figure 10. Three-dimensional model validation. (**a**) CFD-computed drag coefficient of the damaged blade as a function of the turbulence model and the mesh refinement level; (**b**) Wall y^+ contour obtained with the finest grid and the TSST model at 0° incidence.

4. Results and Discussion

This section compares the 2D and the 3D damaged blades' aerodynamic behaviour to highlight the differences and compatibilities between the two modeling approaches.

4.1. Flow Assessments Qualitative Comparison between 2D and 3D Models

Firstly, we remark on the flow configuration in a 3D path. Figure 11 shows the streamlines for the damaged blade with 4° incidence. The damaged blade's flow mechanics is very complex since pathlines reveal several flow detachments and vortical structures, the latter induced by the erosion and the rough surface of the leading edge. In particular, geometrical features such as spikes and steps avoid flow streamlining, making the system's macroscopic behavior strongly three-dimensional.

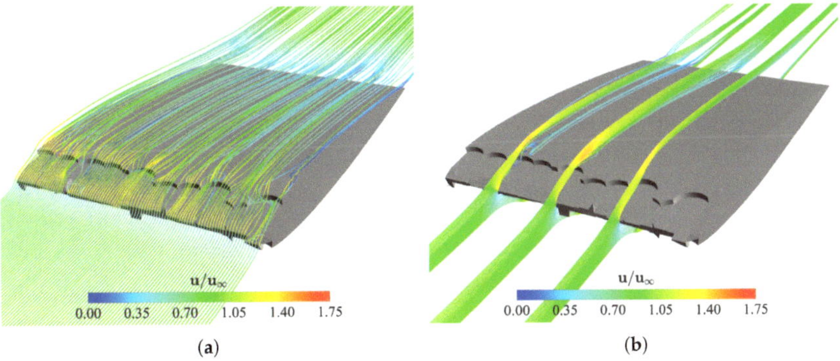

Figure 11. Streamlines around the damaged blade at 4° incidence. (**a**) Streamlines on the suction side of the blade; (**b**) Streamlines in three xy planes passing through 1/4, 1/2, and 3/4 of the span.

The friction coefficient distribution reveals further insights into the three-dimensional flow field. In particular, Figure 12 provides the comparison between the 0° and the 4° incidence configurations. Friction values abruptly increase on the various delaminated edges, the latter acting as an intense source of instability for the local boundary layer. The C_f map suggests some clear patterns of the flow. In fact, vortices generated by the eroded leading edge are leaving a trace on the blade surface. The higher a vortex is, the tighter the friction coefficient trace becomes. Thus, since at $\alpha = 4°$ the boundary layer is substantially thicker than the zero incidence case, vortical structures are driven further from the surface, leading to narrower friction coherent traces.

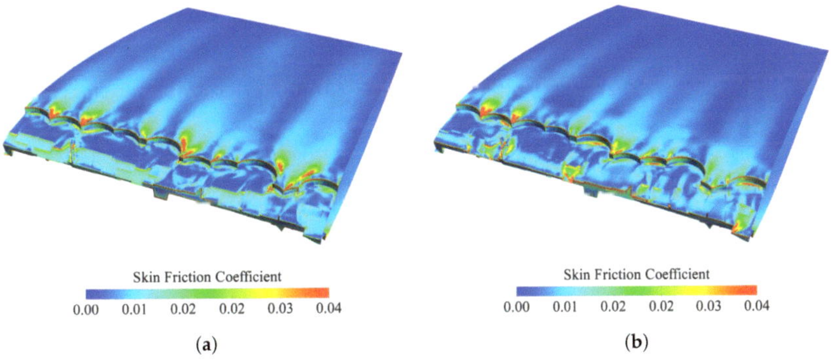

Figure 12. Comparison between contour plots of the friction coefficient on the damaged blade at 0° incidence and 4° incidence. (**a**) C_f contour at $\alpha = 0°$; (**b**) C_f contour at $\alpha = 4°$.

Now a question arises. Can such a complex flow be modeled with a two-dimensional approach? Having a sufficiently accurate two-dimensional model is an undeniable advantage. The 3D setup, in fact, consists of about 5 million elements against the 500 k of the most resolved 2D model. Thus, working with a reliable 2D model would allow easily analyze any concern relating to the safe operation and expected life of a damaged wind turbines with an affordable computational cost. To answer this question first, we can look at the turbulent kinetic energy contours by comparing the 2D setup and several 3D damaged isosurfaces along the blade span. Figure 13 compares the contours of the dimensionless turbulent kinetic energy, $k^* = k/u_\infty^2$, as a function of the blade span for 2D and 3D models. The results refer to the 4° incidence configuration. Looking at the k^* field, the two-dimensional model reasonably reproduces the three-dimensional assessment of the flow. Fluctuation patterns, in fact, look very similar regardless of the geometrical model on both the pressure and the suction sides, and the magnitude of the dimensionless turbulent kinetic energy are quantitatively comparable. The three-dimensional model exhibits slightly higher turbulent kinetic energy values immediately after the leading edge, and generally, the fluctuation zone is broader. This is due to the roughness effect induced by the leading edge's sharp edges, which introduces more turbulence in the flow field. Conversely, having a rounded surface, the two-dimensional model produces lower turbulent energy. Thus, regardless of the strongly three-dimensionality of the flow field, the 2D and the 3D models show similar aerodynamics behaviors, at least at the velocity fluctuation level.

The following sections aim to trace the characteristics of the 2D damaged airfoil and compare its aerodynamics behavior with the clean reference geometry and the 3D arrangement. Pressure and friction coefficients, together with polar diagrams, will be examined.

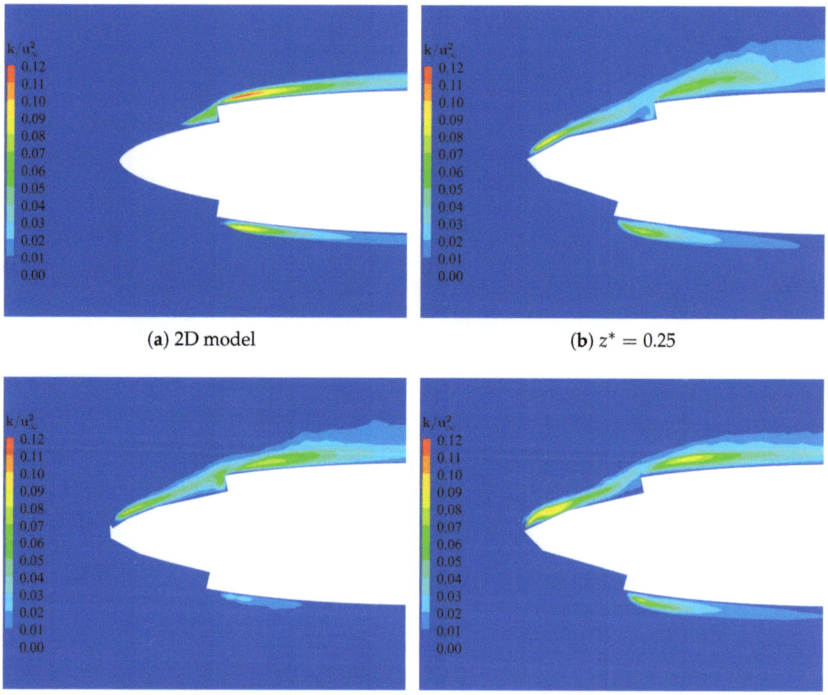

(a) 2D model (b) $z^* = 0.25$

(c) $z^* = 0.50$ (d) $z^* = 0.75$

Figure 13. Comparison between contour plots of the non-dimensional turbulent kinetic energy around the 2D airfoil (**a**) and the 3D blade (**b–d**) at 4° incidence. (**a**) Turbulent kinetic energy contour around the two-dimensional airfoil; (**b–d**) turbulent kinetic energy contour around the three-dimensional blade on a xy plane passing through 1/4, 1/2, and 3/4 of the span, respectively.

4.2. Clean and Damaged 2D Airfoil Comparison

As mentioned, to gradually increase the analysis complexity, we start by comparing the aerodynamics performance of the damaged 2D airfoil with the corresponding clean one. The same numerical model as presented in Section 3.1 is adopted for aerodynamic flow simulation of the damaged airfoil. Figure 14 compares the pressure coefficient trend distributions at various angles of attack. In particular, solid blue lines report variation of C_p for the non-eroded airfoil behavior and red lines for the damaged airfoils. The reader can observe that the delamination zone causes a strong discontinuity in the pressure coefficient distribution, especially in low-incidence angles. This is due to the sudden deceleration of the airflow with consequent pressure recovery in this region, a fact that tends to mitigate its intensity as the incidence angle increases since the flow at the leading edge reduces its kinetic energy content. However, immediately after the eroded region, a strong suction is recorded. Far from the damaged zone, similar wall pressure values are recovered between the clean and the damaged configurations.

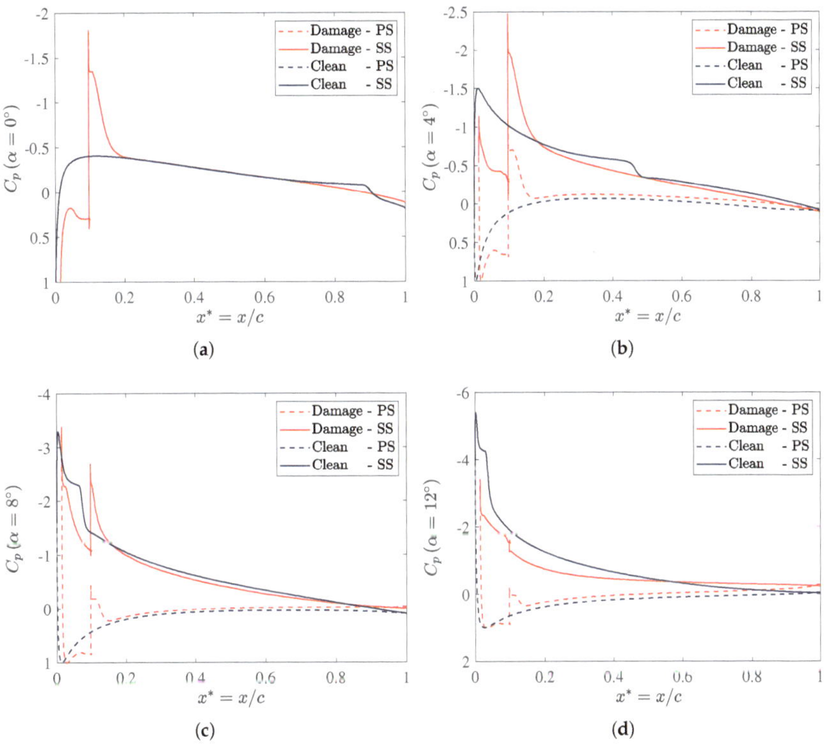

Figure 14. Pressure coefficients comparison between the clean airfoil and the damaged airfoil: the distribution on the Pressure Side (PS) is denoted with the dashed line, while the continuous line is used for the Suction Side (SS) trends. (**a**) C_p distribution at 0° incidence; (**b**) C_p distribution at 4° incidence; (**c**) C_p distribution at 8° incidence; (**d**) C_p distribution at 12° incidence.

Figure 15 provides a comparison for the friction coefficient distributions. The same pattern is adopted, and, in particular, the damaged and the clean airfoil results are compared at different angles of attack. Only suction side distributions are shown to enhance the clarity of the results. Looking at the clean airfoil friction coefficient, the latter presents a steep rise in the damaged zone around the leading edge, followed by a smooth decrease until the $C_f = 0$ condition is met. Flow detachment due to the low-Reynolds conditions happens near the trailing edge for the zero-angle of attack configuration and it moves toward the leading edge at higher incidence angles. The clean geometry smoothly regulates such a transition, and the separation location meets the leading edge at high incidence angles. Conversely, the leading edge erosion promotes an instantaneous recirculation of the boundary layer, and even for the zero-incidence case, the C_f coefficient experiences negative values near the leading edge.

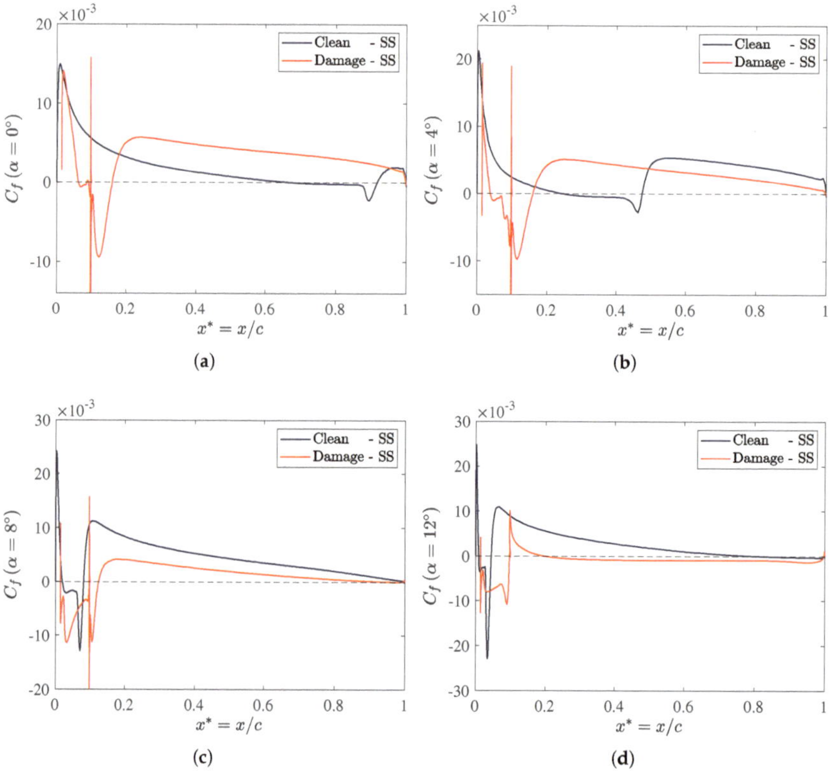

Figure 15. Friction coefficients comparison between the clean airfoil and the damaged airfoil. C_f distribution at (**a**) 0° incidence; (**b**) 4° incidence; (**c**) 8° incidence; (**d**) 12° incidence.

The difference of pressure and friction coefficients between the clean and the damaged case induces a significant reduction in the off-design eroded profile performance. In this regard, Figure 16 reports the polar diagrams comparison between the two geometries. The incidence values up to the stall condition are studied. To analyze the behavior of the airfoil beyond this condition, a non-stationary model should be set up, which is beyond the scope of this paper. Moreover, the most significant eroded areas are confined towards the blade tip, due to the higher rotational velocity. Airfoils in this area rarely experience very off-project angles of attack. Looking at Figure 15a, simulations predict the erosion to have a small impact on the lift coefficient at low incidences, while a more pronounced discrepancy is observed at higher angles of attack. Eroded airfoil drag polar plot, on the other hand, is systematically shifted upward by a value of 0.15–0.18 in the whole operation range. Thus, both C_l and C_d curves disclose a degradation of the aerodynamic performance of the damaged airfoil that generally increases with the angle of attack.

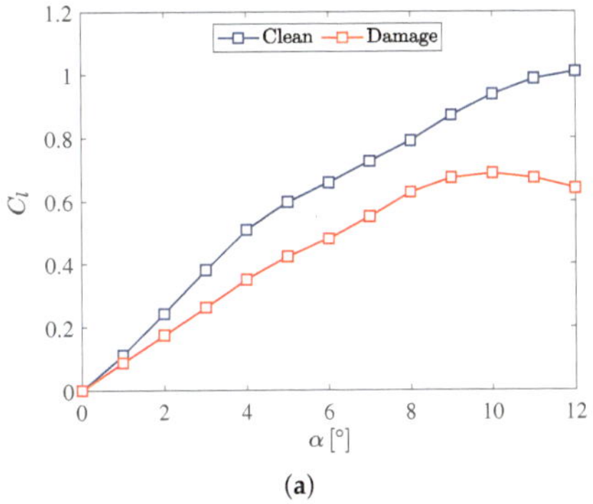

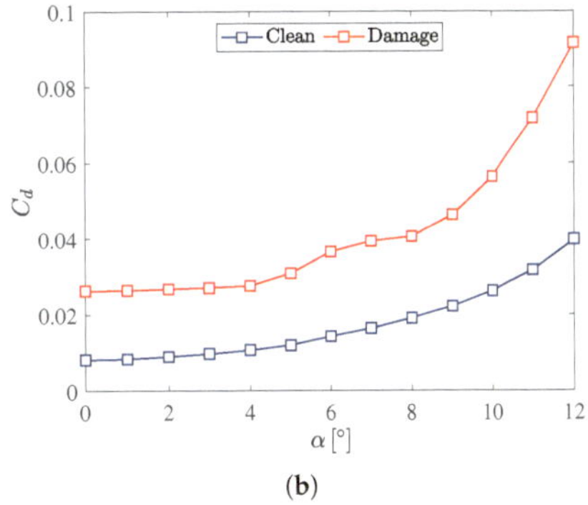

Figure 16. Comparison between polar diagrams of the clean airfoil and the damaged airfoil. (**a**) C_l polar plot comparison; (**b**) C_d polar plot comparison.

4.3. Damaged 3D Performance Against 2D Eroded Assessment

Having understood the essential aspects relating to the damage 2D airfoil performance compared to the clean one, the present section aims to reach the aerodynamic characteristics of the 3D blade with the eroded 2D profile.

Since wall quantities depend on non-dimensional chordwise, $x^* = x/c$ and spanwise, $z^* = z/b$, coordinates, wall properties are post-processed accordingly to spanwise-average operations. Thus, the three-dimensional spanwise-averaged coefficient distribution can be calculated from:

$$\overline{C_x}(x_i^*) = \frac{1}{N_z} \sum_{k=1}^{N_z} C_x(x_i^*, z_k^*) \tag{10}$$

Here, $C_x(x_i^*, z_k^*)$ denotes the N_z pressure (or friction) coefficient values corresponding to a certain x_i^* and z_k^* blade surface location, the latter discretely selected as in Figure 17. Signal variations are then quantitatively traced by computing the root-mean-square of the difference between the local pressure and friction coefficient distributions and the respective spanwise-average values, as

$$\sigma[C_x(x_i^*)] = \sqrt{\frac{1}{N_z} \sum_{k=1}^{N_z} \left(C_x(x_i^*, z_k^*) - \overline{C_x}(x_i^*)\right)^2} \tag{11}$$

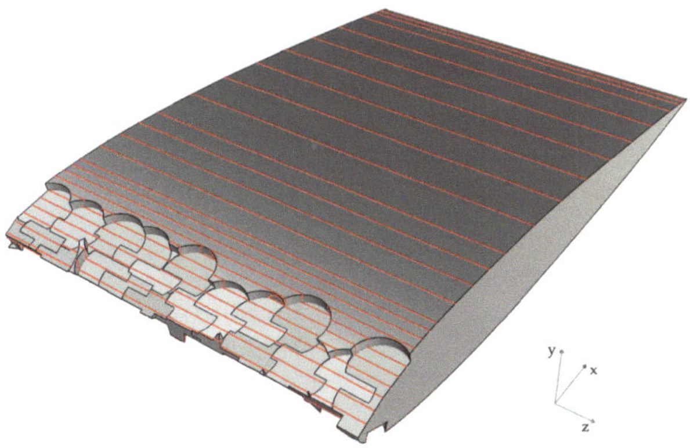

Figure 17. Geometrical representation of the lines where friction and pressure coefficient are evaluated.

Results of the analysis are reported in Figures 18 and 19. In particular, Figure 18 provides the three-dimensional spanwise-averaged pressure coefficient distribution, $\overline{C_p}(x^*)$, compared to two-dimensional damaged reference. Results refer to the 0° (Figure 18a) and 4° (Figure 18b) incidence angles, respectively. As can be observed, although the flow is strongly three-dimensional, its mean characteristics are very closely fit by the two-dimensional eroded model. In particular, far from the area where the delaminated and healthy portion meet, the pressure field trend provided by the 2D model adheres perfectly to the forecast provided by the three-dimensional model. Mild discrepancies are detected around the geometrical jump. Here, in particular, the $\overline{C_p}(x^*)$ distribution shows a smoother behavior in the eroded region with a less marked discontinuity than the 2D airfoil values. This fact is easily explained since the two-dimensional setup is arranged so that a localized discontinuity perfectly uncouples the delaminate and the clean portions of the airfoil. Conversely, the three-dimensional model offers a complex pattern to deal with a realistic eroded leading edge. The latter induces an early separation of the flow in some areas with consequent coverage of part of the area of swirling flows and thickened boundary layer.

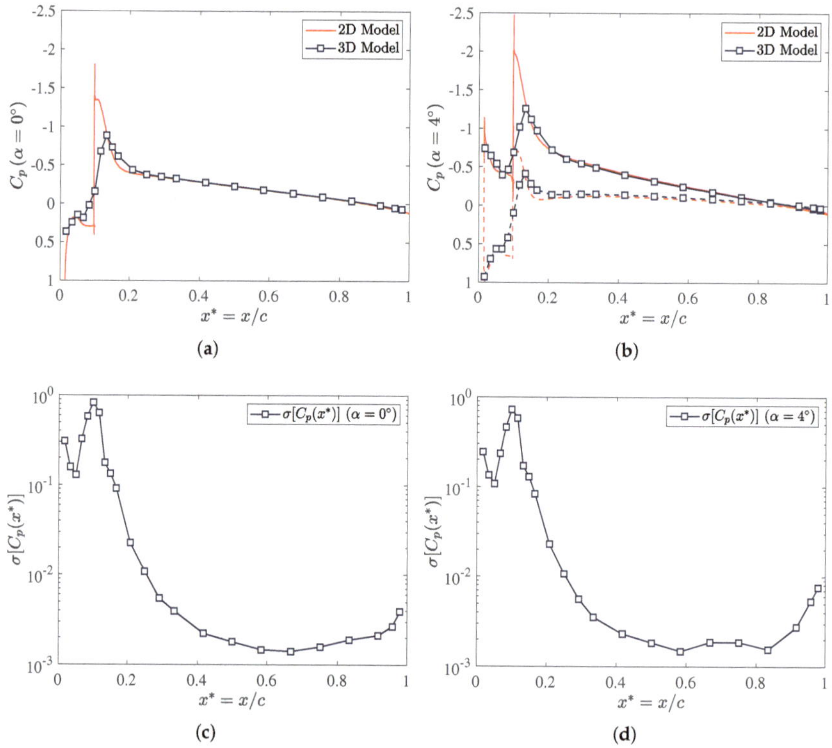

Figure 18. Pressure coefficients comparison between the 2D damaged airfoil and the 3D eroded blade. Distributions on the pressure side are denoted with the dashed line while solid lines are used for suction side trends. Dark blue squares denote the $\overline{C_p}(x^*)$ value over the three-dimensional blade. (**a**) Pressure coefficients comparison at 0° incidence; (**b**) Pressure coefficients comparison at 4° incidence; (**c**) Pressure coefficient standard deviation on the blade at 0° incidence; (**d**) Pressure coefficient standard deviation on the blade at 4° incidence.

Figure 18c,d, on the other hand, provide the trend of $\sigma[C_p(x^*)]$ along the blade. Here it is worth mentioning that such information can only be provided through a complete three-dimensional model. Nevertheless, the σ distribution makes it possible to verify a posteriori how significant the average data of the pressure coefficient is with respect to its true distribution. Thus, we observe that a maximum $\sigma[C_p(x^*)]$ deviation of the order of unity is detected, the peak being clustered around $x^* = 0.1$ where the geometrical discontinuity is nominally located. Conversely, far from the eroded area, the pressure coefficient fluctuations from the mean value are bounded in a small range, with an average of around 1×10^{-2}. Such a result allows us to state that the real three-dimensionality of the flow is clustered almost exclusively in the area where the eroded part is connected to the clean one, while elsewhere, the flow recovers its two-dimensional nature.

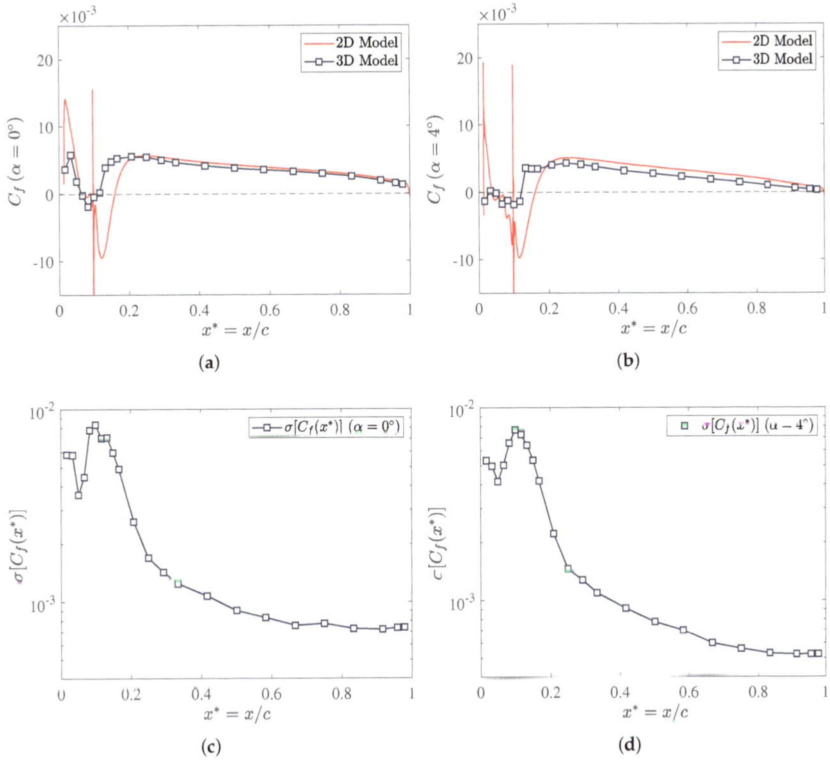

Figure 19. Friction coefficients comparison between the 2D damaged airfoil and the 3D eroded blade. Dark blue squares denote the $\overline{C_f}(x^*)$ value over the three-dimensional blade. (**a**) C_f distribution at 0° incidence; (**b**) C_f distribution at 4° incidence; (**c**) Friction coefficient standard deviation on the blade at 0° incidence; (**d**) Friction coefficient standard deviation on the blade at 4° incidence.

Having compared the pressure coefficient trends between the 2D and 3D models, we now move forward to the wall velocity gradients analysis by comparing the friction coefficient distributions and observing their average local fluctuations. Figure 19 reports the results of the analysis. Moreover, in this case it can be observed that the two-dimensional model well reproduces the friction coefficient's global trend. In particular, far from erosion, the 2D and 3D predictions are almost superimposable both for the case with 0° (Figure 19a) and 4° (Figure 19b) angle of attacks. The most significant discrepancy between the two modeling methodologies is observed in the eroded area, where the 3D model always predicts a slightly anticipated flow separation, whereas the 2D model concentrates it almost exclusively around the geometric discontinuity.

The analysis of the root-mean-square deviation distributions (Figure 19c,d) shows that the 3D friction values are well confined around their average, with a deviation that does not exceed the second decimal digit. Nevertheless, if pressure coefficient distributions are impressively superimposed between the two-dimensional and three-dimensional model, a 3D eroded pattern leads to more significant discrepancies in the boundary layer flow mechanics, with consequent less compatibility between the 2D and 3D friction coefficients.

Finally, Figure 20 compares the off-design behavior of the 2D and 3D arrangements by displaying the lift and drag coefficient distributions as a function of the angle of attack. As expected, the lift coefficient trends for the 2D and 3D configurations are almost identical (Figure 20a). This is because the 2D and 3D pressure fields are effectively equivalent at an extensive range of angles of attack, resulting in a similar lift coefficient between the two modeling techniques. On the other hand, the three-dimensional model's drag coefficient is systematically higher (by roughly 17%) than in the damaged 2D instance (Figure 20b). This is owing to the differing wall dynamics experienced by the 3D geometry near the leading edge. Although the friction coefficient is an essential factor in predicting blade aerodynamics, in any digital twin model that aims at evaluating a wind turbine's service life would have the primary goal of accurately estimating the lift coefficient. Such a parameter, in fact, is the primary contributor to yearly electricity production, and as such, it is the critical datum for estimating maintenance interventions. As a result of present investigations, a simple eroded 2D model, adequately calibrated, has been shown to thoroughly represent the overall aerodynamics of a wind turbine blade under eroded leading edge conditions at a much lower computational cost than a fully 3D setup. In particular, present analyses are performed on a workstation composed of 2 Intel® Xeon® X5650 CPUs and 24GB of RAM. With such an architecture, two-dimensional simulations took about 30 min to get the convergence of the flow quantities, while 3D cases take about 15 h. Thus, the computational cost of the 2D model is negligible compared to the 3D approach, making 2D modeling an underlying basis in defining a digital twin for predicting the service life of a wind turbine.

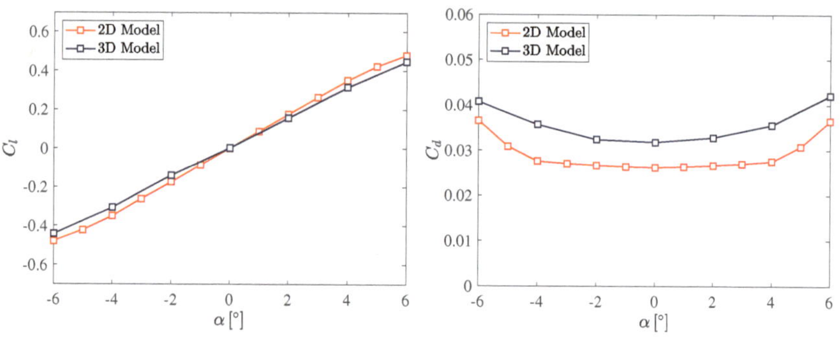

Figure 20. Comparison between polar diagrams of the 2D damaged airfoil and the 3D eroded blade. (**a**) Comparison of the lift coefficient distribution as a function of the angle of attack; (**b**) Comparison of the drag coefficient distribution as a function of the angle of attack.

5. Conclusions

The present work compares three-dimensional CFD RANS modeling of leading-edge eroded wind turbine blades with simplified two-dimensional setups.

First, the numerical models' behavior is evaluated to establish the optimal arrangement for grid resolution and turbulence model. Thus, a validation campaign with experimental data is reported for global and local quantity convergence, determining the optimal mesh refinement level and the most appropriate turbulence model for predicting the airfoil and the blade aerodynamic performance at low Reynolds numbers which are experienced in wind turbines operation. According to our analyses, a 500 k-element 2D assessment coupled with the TSST model provides a good balance between computational time and accuracy, while a 5 million-element 3D setup, always combined with the TSST model, ensures grid independency for the blade aerodynamics.

Qualitative (i.e., observation of fluid fields) and quantitative observations (i.e., analysis of wall flow properties) are then provided. From the first set of analyses, we found that, regardless of the strongly three-dimensionality behavior of the blade, 2D and 3D turbulent kinetic energy fields are closely similar. This encourages us to address deeper comparison levels by looking at pressure and friction coefficient distribution as well as the polar diagrams. Quantitative observations show that the 2D approach accurately fits surface pressure trends, as well as excellent agreements in lift coefficient in a wide range of angles of attack are recovered. On the other hand, a systematic underestimation of the drag coefficient of about 17% in the whole range of operations is observed. The outcomes of this research are crucial since they confirm that the system's inviscid characteristics (i.e., pressure distribution and lift coefficient) are not characterized by extremely three-dimensional effects and that a simplified two-dimensional model can forecast their trends accurately. Thus, a reduced-order model can well represent the more significant characteristics of the complex three-dimensional severely eroded blade, and do so at a much lower computational cost. In addition, since the lift coefficient is the fundamental parameter for forecasting the AEP of a wind power plant, simplified 2D models like the one proposed by the authors may serve as a foundation for simplified predictive models of the wind farm's working life accounting for blade erosion.

Future works will investigate other stages of the erosion process as well propose an algebraic model calibrated on CFD data to determine the operating life of wind turbines accounting for the erosion of the leading edge.

Author Contributions: Conceptualization, F.D.V.; methodology, M.C. and F.D.V.; software, M.C. and F.D.V.; validation, M.C., F.D.V.; formal analysis, M.C. and F.D.V.; investigation, M.C. and F.D.V.; resources, M.C., F.D.V., E.B., F.Z. and H.H.; data curation, M.C. and F.D.V.; writing—original draft preparation, F.D.V.; writing—review and editing, M.C., F.D.V., E.B., F.Z., A.H. and H.H.; visualization, M.C., F.D.V., E.B. and H.H.; supervision, F.D.V. and E.B.; project administration, F.D.V. All authors have read and agreed to the published version of the manuscript.

Funding: This research received no external funding.

Institutional Review Board Statement: Not applicable.

Informed Consent Statement: Not applicable.

Data Availability Statement: Not applicable.

Conflicts of Interest: The authors declare no conflict of interest.

Abbreviations

The following abbreviations are used in this manuscript:

AEP	Annual Energy Production
CFD	Computational Fluid Dynamics
LEE	Leading Edge Erosion
PS	Pressure Side
RANS	Reynolds-Averaged Navier–Stokes
SA	Spalart–Allmaras
SS	Suction Side
SST	Shear Stress Transport
TSST	Transition Shear Stress Transport

References

1. Wind Europe. *Wind Energy in Europe: 2021 Statistics and the Outlook for 2022–2026*; Technical Report; Wind Europe: Brussels, Belgium, 2022. Available online: https://windeurope.org/intelligence-platform/product/wind-energy-in-europe-2021-statistics-and-the-outlook-for-2022-2026/ (accessed on 11 August 2022).
2. IRENA. *Renewable Energy Statistics 2022*; Technical Report; The International Renewable Energy Agency: Abu Dhabi, UAE, 2022. Available online: https://irena.org/publications/2022/Apr/Renewable-Capacity-Statistics-2022 (accessed on 11 August 2022).
3. Butterfield, C.P.; Scott, G.; Musial, W. Comparison of wind tunnel airfoil performance data with wind turbine blade data. *J. Sol. Energy Eng.* **1992**, *114*, 119–124. [CrossRef]
4. Somers, D.M.; Tangler, J.L. Wind tunnel test of the S814 thick root airfoil. *J. Sol. Energy Eng.* **1996**, *118*, 217–221. [CrossRef]
5. Devinant, P.; Laverne, T.; Hureau, J. Experimental study of wind-turbine airfoil aerodynamics in high turbulence. *J. Wind Eng. Ind. Aerodyn.* **2002**, *90*, 689–707. [CrossRef]
6. Selig, M.S.; McGranahan, B.D. Wind tunnel aerodynamic tests of six airfoils for use on small wind turbines. *J. Sol. Energy Eng.* **2004**, *126*, 986–1001. [CrossRef]
7. Chamorro, L.P.; Porté-Agel, F. A wind-tunnel investigation of wind-turbine wakes: Boundary-layer turbulence effects. *Bound.-Layer Meteorol.* **2009**, *132*, 129–149. [CrossRef]
8. Chamorro, L.P.; Porté-Agel, F. Effects of thermal stability and incoming boundary-layer flow characteristics on wind-turbine wakes: A wind-tunnel study. *Bound.-Layer Meteorol.* **2010**, *136*, 515–533. [CrossRef]
9. Chamorro, L.P.; Porte-Agel, F. Turbulent flow inside and above a wind farm: A wind-tunnel study. *Energies* **2011**, *4*, 1916–1936. [CrossRef]
10. Adaramola, M.; Krogstad, P.Å. Experimental investigation of wake effects on wind turbine performance. *Renew. Energy* **2011**, *36*, 2078–2086. [CrossRef]
11. Sun, H.; Gao, X.; Yang, H. A review of full-scale wind-field measurements of the wind-turbine wake effect and a measurement of the wake-interaction effect. *Renew. Sustain. Energy Rev.* **2020**, *132*, 110042. [CrossRef]
12. Porté-Agel, F.; Bastankhah, M.; Shamsoddin, S. Wind-turbine and wind-farm flows: A review. *Bound.-Layer Meteorol.* **2020**, *174*, 1–59. [CrossRef]
13. Yang, S.; Chang, Y.; Arici, O. Incompressible Navier-Stokes computation of the NREL airfoils using a symmetric total variational diminishing scheme. *J. Solar Energy Eng.* **1994**, *116*, 174–182. [CrossRef]
14. Wolfe, W.; Ochs, S.; Wolfe, W.; Ochs, S. CFD calculations of S809 aerodynamic characteristics. In Proceedings of the 35th Aerospace Sciences Meeting and Exhibit, Reno, NV, USA, 6–9 January 1997; p. 973. [CrossRef]
15. Munduate, X.; Ferrer, E. CFD predictions of transition and distributed roughness over a wind turbine airfoil. In Proceedings of the 47th AIAA Aerospace Sciences Meeting Including the New Horizons forum and Aerospace Exposition, Orlando, FL, USA, 5–8 January 2009; p. 269. [CrossRef]
16. Sanderse, B.; Van der Pijl, S.; Koren, B. Review of computational fluid dynamics for wind turbine wake aerodynamics. *Wind Energy* **2011**, *14*, 799–819. [CrossRef]
17. Mittal, A.; Sreenivas, K.; Taylor, L.K.; Hereth, L.; Hilbert, C.B. Blade-resolved simulations of a model wind turbine: Effect of temporal convergence. *Wind Energy* **2016**, *19*, 1761–1783. [CrossRef]
18. Van der Laan, M.; Hansen, K.S.; Sørensen, N.N.; Réthoré, P.E. *Predicting Wind Farm Wake Interaction with RANS: An Investigation of the Coriolis Force*; Journal of Physics: Conference Series; IOP Publishing: Bristol, UK, 2015; Volume 625, p. 012026. [CrossRef]
19. Astolfi, D.; Castellani, F.; Terzi, L. A study of wind turbine wakes in complex terrain through RANS simulation and SCADA data. *J. Solar Energy Eng.* **2018**, *140*. [CrossRef]
20. Zhong, W.; Tang, H.; Wang, T.; Zhu, C. Accurate RANS simulation of wind turbine stall by turbulence coefficient calibration. *Appl. Sci.* **2018**, *8*, 1444. [CrossRef]
21. Mehta, D.; Van Zuijlen, A.; Koren, B.; Holierhoek, J.; Bijl, H. Large Eddy Simulation of wind farm aerodynamics: A review. *J. Wind Eng. Ind. Aerodyn.* **2014**, *133*, 1–17. [CrossRef]
22. Thé, J.; Yu, H. A critical review on the simulations of wind turbine aerodynamics focusing on hybrid RANS-LES methods. *Energy* **2017**, *138*, 257–289. [CrossRef]
23. Stevens, R.J.; Martínez-Tossas, L.A.; Meneveau, C. Comparison of wind farm large eddy simulations using actuator disk and actuator line models with wind tunnel experiments. *Renew. Energy* **2018**, *116*, 470–478. [CrossRef]
24. Solís-Gallego, I.; Argüelles Díaz, K.M.; Fernández Oro, J.M.; Velarde-Suárez, S. Wall-resolved LES modeling of a wind turbine airfoil at different angles of attack. *J. Marine Sci. Eng.* **2020**, *8*, 212. [CrossRef]
25. Vermeer, L.; Sørensen, J.N.; Crespo, A. Wind turbine wake aerodynamics. *Prog. Aerosp. Sci.* **2003**, *39*, 467–510. [CrossRef]
26. Enevoldsen, P.; Xydis, G. Examining the trends of 35 years growth of key wind turbine components. *Energy Sustain. Dev.* **2019**, *50*, 18–26. [CrossRef]
27. Keegan, M.H.; Nash, D.; Stack, M. On erosion issues associated with the leading edge of wind turbine blades. *J. Phys. D: Appl. Phys.* **2013**, *46*, 383001. [CrossRef]
28. Rempel, L. Rotor blade leading edge erosion-real life experiences. *Wind Syst. Mag.* **2012**, *11*, 22–24. Available online: http://windaction.s3.amazonaws.com/attachments/2310/1012_BladeFeature.pdf (accessed on 11 August 2022).
29. Dalili, N.; Edrisy, A.; Carriveau, R. A review of surface engineering issues critical to wind turbine performance. *Renew. Sustain. Energy Rev.* **2009**, *13*, 428–438. [CrossRef]

30. Ehrmann, R.S.; White, E.B.; Maniaci, D.C.; Chow, R.; Langel, C.M.; Van Dam, C.P. Realistic leading-edge roughness effects on airfoil performance. In Proceedings of the 31st AIAA Applied Aerodynamics Conference, San Diego, CA, USA, 24 June 2013; p. 2800. [CrossRef]
31. Sareen, A.; Sapre, C.A.; Selig, M.S. Effects of leading edge erosion on wind turbine blade performance. *Wind Energy* **2014**, *17*, 1531–1542. [CrossRef]
32. Gaudern, N. *A Practical Study of the Aerodynamic Impact of Wind Turbine Blade Leading Edge Erosion*; Journal of Physics: Conference Series; IOP Publishing: Bristol, UK, 2014; Volume 524, p. 012031. [CrossRef]
33. Langel, C.M.; Chow, R.; Hurley, O.F.; Van Dam, C.C.P.; Maniaci, D.C.; Ehrmann, R.S.; White, E.B. Analysis of the impact of leading edge surface degradation on wind turbine performance. In Proceedings of the 33rd Wind Energy Symposium, Kissimmee, FL, USA, 5–9 January 2015; p. 0489. [CrossRef]
34. Maniaci, D.C.; White, E.B.; Wilcox, B.; Langel, C.M.; van Dam, C.; Paquette, J.A. *Experimental Measurement and CFD Model Development of Thick Wind Turbine Airfoils with Leading Edge Erosion*; Journal of Physics: Conference Series; IOP Publishing: Bristol, UK, 2016; Volume 753, p. 022013. [CrossRef]
35. Han, W.; Kim, J.; Kim, B. Effects of contamination and erosion at the leading edge of blade tip airfoils on the annual energy production of wind turbines. *Renew. Energy* **2018**, *115*, 817–823. [CrossRef]
36. Elhadi Ibrahim, M.; Medraj, M. Water droplet erosion of wind turbine blades: Mechanics, testing, modeling and future perspectives. *Materials* **2019**, *13*, 157. [CrossRef]
37. Dashtkar, A.; Hadavinia, H.; Sahinkaya, M.N.; Williams, N.A.; Vahid, S.; Ismail, F.; Turner, M. Rain erosion-resistant coatings for wind turbine blades: A review. *Polym. Polym. Compos.* **2019**, *27*, 443–475. [CrossRef]
38. Shankar Verma, A.; Jiang, Z.; Ren, Z.; Hu, W.; Teuwen, J.J. Effects of onshore and offshore environmental parameters on the leading edge erosion of wind turbine blades: A comparative study. *J. Offshore Mech. Arct. Eng.* **2021**, *143*. [CrossRef]
39. Law, H.; Koutsos, V. Leading edge erosion of wind turbines: Effect of solid airborne particles and rain on operational wind farms. *Wind Energy* **2020**, *23*, 1955–1965. [CrossRef]
40. Castorrini, A.; Cappugi, L.; Bonfiglioli, A.; Campobasso, M. *Assessing Wind Turbine Energy Losses due to Blade Leading Edge Erosion Cavities with Parametric CAD and 3D CFD*; Journal of Physics: Conference Series; IOP Publishing: Bristol, UK, 2020; Volume 1618, p. 052015. [CrossRef]
41. Wang, Y.; Wang, L.; Duan, C.; Zheng, J.; Liu, Z.; Ma, G. CFD simulation on wind turbine blades with leading edge erosion. *J. Theor. Appl. Mech.* **2021**, *59*. [CrossRef]
42. Ortolani, A.; Castorrini, A.; Campobasso, M.S. *Multi-Scale Navier-Stokes Analysis of Geometrically Resolved Erosion of Wind Turbine Blade Leading Edges*; Journal of Physics: Conference Series; IOP Publishing: Bristol, UK, 2022; Volume 2265, p. 032102. [CrossRef]
43. Mishnaevsky, L., Jr.; Hasager, C.B.; Bak, C.; Tilg, A.M.; Bech, J.I.; Rad, S.D.; Fæster, S. Leading edge erosion of wind turbine blades: Understanding, prevention and protection. *Renew. Energy* **2021**, *169*, 953–969. [CrossRef]
44. Eisenberg, D.; Laustsen, S.; Stege, J. Wind turbine blade coating leading edge rain erosion model: Development and validation. *Wind Energy* **2018**, *21*, 942–951. [CrossRef]
45. ANSYS. ANSYS 19.2 FLUENT User's Guide; ANSYS: Canonsburg, PA, USA, 2018.
46. MATLAB. *9.7. 0.1190202 (R2019b)*; MathWorks Inc.: Natick, MA, USA, 2018. Available online: https://it.mathworks.com/products/new_products/release2019b.html (accessed on 11 August 2022).
47. Spalart, P.; Allmaras, S. A one-equation turbulence model for aerodynamic flows. In Proceedings of the 30th Aerospace Sciences Meeting and Exhibit, Reno, NV, USA, 6–9 January 1992; p. 439. [CrossRef]
48. Menter, F.R. Two-equation eddy-viscosity turbulence models for engineering applications. *AIAA J.* **1994**, *32*, 1598–1605. [CrossRef]
49. Menter, F.R.; Langtry, R.B.; Likki, S.; Suzen, Y.; Huang, P.; Völker, S. A correlation-based transition model using local variables—Part I: Model formulation. *J. Turbomach.* **2006**, *128*, 413–422. [CrossRef]
50. ANSYS FLUENT Theory Guide. Available online: https://www.afs.enea.it/project/neptunius/docs/fluent/html/th/main_pre.htm (accessed on 11 August 2022).
51. Campobasso, M.S.; Castorrini, A.; Cappugi, L.; Bonfiglioli, A. Experimentally validated three-dimensional computational aerodynamics of wind turbine blade sections featuring leading edge erosion cavities. *Wind Energy* **2022**, *25*, 168–189. [CrossRef]
52. Sapre, C. Turbine Blade Erosion and the Use of Wind Protection Tape. 2012. Available online: https://www.ideals.illinois.edu/items/31357 (accessed on 11 August 2022).
53. Avanzi, F.; De Vanna, F.; Ruan, Y.; Benini, E. Design-Assisted of Pitching Aerofoils through Enhanced Identification of Coherent Flow Structures. *Designs* **2021**, *5*, 11. [CrossRef]
54. De Vanna, F.; Bof, D.; Benini, E. Multi-Objective RANS Aerodynamic Optimization of a Hypersonic Intake Ramp at Mach 5. *Energies* **2022**, *15*, 2811. [CrossRef]
55. De Vanna, F.; Picano, F.; Benini, E.; Quinn, M.K. Large-Eddy Simulations of the Unsteady Behavior of a Hypersonic Intake at Mach 5. *AIAA J.* **2021**, *59*, 3859–3872. [CrossRef]
56. Drela, M. XFOIL: An analysis and design system for low Reynolds number airfoils. In *Low Reynolds Number Aerodynamics*; Springer: Berlin, Germany, 1989; pp. 1–12. [CrossRef]
57. Gregory, N.; O'Reilly, C. Low-Speed Aerodynamic Characteristics of NACA 0012 Aerofoil Section, Including the Effects of Upper-Surface Roughness Simulating Hoar Frost. ARC, R&M-3726. 1970. Available online: https://reports.aerade.cranfield.ac.uk/handle/1826.2/3003 (accessed on 11 August 2022).

fluids

Article

Comparison of FLACS and BASiL Model for Ro-Pax Ferry LNG Bunkering Leak Analysis

Boon How Lim [1,2] and Eddie Y. K. Ng [1,*]

1 College of Engineering, School of Mechanical and Aerospace Engineering, Nanyang Technological University, 50 Nanyang Ave, Singapore 639798, Singapore
2 Sembcorp Marine, 80 Tuas South Boulevard, Singapore 637051, Singapore
* Correspondence: mykng@ntu.edu.sg

Abstract: Performing liquefied natural gas (LNG) bunkering involves the risk of accidental leakage. When released from containment, LNG rapidly vaporizes into flammable natural gas and could lead to flash fire and explosion. Hence, LNG bunkering needs to take place in an area without an ignition source called a safety zone. This study compares the safety zone estimated by the Bunkering Area Safety Information for LNG (BASiL) model with that of the computational fluid dynamic (CFD) software FLACS, for Ro-Pax ferry bunkering. Horizontal leaks covering different wind speeds in eight wind directions were compared between the two models. Additionally, a grid refinement study was performed systematically to quantify the discretization error uncertainty in the CFD. Of 24 leak cases, FLACS and the BASiL model results agreed on 18 cases. In three cases validation was inconclusive due to the CFD error uncertainty. The BASiL model underestimated the safety zone distance in three cases compared with FLACS. Future work would be to perform a higher grid refinement study to confirm inconclusive comparison and examine ways to reduce gas dispersion spread for the worst result.

Keywords: LNG leak; BASiL model; FLACS; Ro-Pax ferry

Citation: Lim, B.H.; Ng, E.Y.K. Comparison of FLACS and BASiL Model for Ro-Pax Ferry LNG Bunkering Leak Analysis. *Fluids* **2022**, 7, 272. https://doi.org/10.3390/fluids7080272

Academic Editors: Ernesto Benini, Francesco De Vanna and Mehrdad Massoudi

Received: 11 July 2022
Accepted: 4 August 2022
Published: 8 August 2022

1. Introduction

Liquefied natural gas (LNG) will become an important energy source as the world aims to lower its greenhouse gas emissions [1]. According to the DNV Energy Transition Outlook 2020, the marine industry. especially shipping, will switch from predominantly oil fuel to low carbon fuels and LNG [2]. When performing LNG bunkering, ship owners have a responsibility to ensure safety when an accidental LNG leak occurs. The specification for the bunkering of liquefied natural gas-fueled vessels (ISO 20519) requires risk assessment of accidental LNG leaks, especially for flammable gas spreading [3]. LNG rapidly vaporizes into natural gas (NG) when spilled, creating a dispersing flammable cloud. This flammable cloud can lead to flash fire or explosion if it encounters an ignition source. Therefore, it is important to set up a safety zone that does not contain any ignition sources in which the LNG can disperse. However, safety zone evaluation is challenging because factors such as leak rate, leak frequency, wind speed, wind direction, and the surrounding geometry will influence the gas dispersion [4].

Walter and colleagues have summarized the two main approaches developed to study the dispersion of LNG leaks [5]. The two approaches are the integral model and the Navier–Stokes model. The integral model uses a simplified conservation equation with a mathematical complexity of one dimension. Well-known commercial software which uses the integral model includes PHAST [6] and ALOHA [7]. The Navier–Stokes model covers the complete representation of fundamental fluid dynamics over three dimensions, solving the time-dependent conservation equation of momentum, mass, energy, and species. While the Navier–Stokes model offers higher accuracy, it is time-consuming and often the integral model is used for preliminary risk assessment. The Navier–Stokes model is used in computational fluid dynamic (CFD) software. Please see the CFD papers by Mattia and colleagues on safety zone evaluation for LNG pool leaks within a harbor [8,9]. The software PHAST and Fire Dynamics Simulator (FDS) were utilized to predict the extent of LNG dispersion during LNG leak scenarios. Three types of bunkering operation, LNG unloading operation, Shore to Ship, and Truck to Ship were considered for leak simulation. The heat flux from the ground was found to be the most impactful parameter on the size of the flammable plume size. The study also highlighted how uneven terrain aids in flammable gas mixing efficiency, thereby limiting the flammable area to the proximity of the ship.

Marko and colleagues compared the LNG pool dispersion results from PHAST, FDS, and Ansys Fluent at the port of Koper, Slovenia [10]. The results show that all three models produce comparable results for the gas dispersion distance on flat terrain. It was concluded that PHAST is not reliable for LNG leaks that occur near lower structures such as piers, houses, and moored ships. This is because the integral model does not consider the heights of geometry and will not capture the stratification of a vaporization gas cloud. The presence of the ship significantly influences the dispersion of the gas cloud shape and length. This paper concluded that CFD should be considered for gas dispersion in ports with a defined gas evaporation rate obtained from experimental tests or simulated separately and validated for a specific spill case.

The Society for Gas as Marine Fuel (SGMF) has developed an approach named Bunkering Area Safety Information for LNG (BASiL) to estimate the safety zone for bunkering activity [11]. BASiL contains a database of 1.4 million combinations of eight input parameters: location (latitude and longitude); amount of LNG transferred and the duration of the transfer; LNG supply temperature and pressure; LNG composition; transfer pressure; transfer system elevation from the ground; hose/transfer system diameter; and emergency shut down (ESD) method. The model interpolates simultaneously within the most appropriate points in the database to derive the safety zone for LNG bunkering. BASiL neither falls under the integral model nor the Navier–Stokes model category and is akin to a data analysis approach. Given the novelty of the BASiL approach to predict LNG dispersion, it will be worthwhile to validate it with CFD software. This paper compares the safety zone predicted by BASiL and the Flame Acceleration Simulator (FLACS) for a Ro-Pax ferry bunkering operation [12]. FLACS is developed by the company Gexcon and has been validated extensively with LNG spill experiments such as Burro [13], Maplin Sands [14], and Falcon [15].

2. Materials and Methods

The leak scenario used for comparison is the LNG bunkering of a Ro-Pax ferry in Hong Kong. The Ro-Pax ferry fueling was performed with a 4-inch hose, using a pump from two LNG road tankers filled at 2.5 bars and pumped to 6 bars. The ferry's manifold is located in the hull, 10 m below the main deck. The hose has a minimum height above the ground of 0.1 m. The emergency shut down (ESD) system is semi-automatic, acting after 120 s. Considering a 6 mm leak hole in a 4-inch hose, the BASiL model predicted a horizontal jet radius of 27 m from the leak source [11]. Figure 1 is an illustration of the bunkering setup and the semi-circle in pink represents the safety zone.

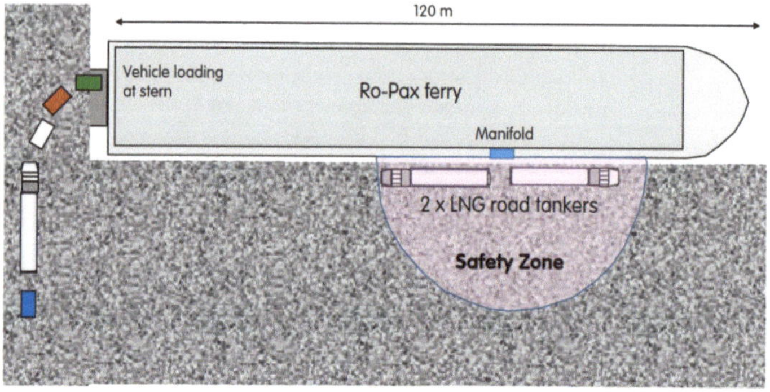

Figure 1. BASiL safety zone for Ro-Pax ferry LNG bunkering in Hong Kong (extracted from SGMF manual).

The CFD simulation was performed using a Dell Precision T7820, equipped with dual Intel Xeon Gold 6150 2.7 GHz processor and 128 GB (8 × 16 GB) 2666 MHz DDR4 RAM. CFD software FLACS version 20.1 simulated the leak scenario of the Ro-Pax ferry for comparison. FLACS used the finite volume method on a structured cartesian grid to solve Favre averaged transport equations for mass, momentum, enthalpy, turbulent kinetic energy, the rate of dissipation of turbulent kinetic energy, and mass-fraction of fuel. The porosity concept was used to capture the congestion effect of geometries within the cartesian grid which affect gas dispersion. The governing equation for gas transport equation is as follows:

$$\frac{\partial}{\partial t}(\beta_v \rho Y_{fuel}) + \frac{\partial}{\partial x_j}\left(\beta_j \rho u_j Y_{fuel}\right) = \frac{\partial}{\partial x_j}\left(\beta_j \frac{\mu_{eff}}{\sigma_{fuel}} \frac{\partial Y_{fuel}}{\partial x_j}\right) + R_{Fuel} \tag{1}$$

R_{fuel} is the fuel reaction rate (zero since LNG is not reacting with air). σ_{fuel} is the Prandtl/Schmidt number pre-defined as 0.7. β_v is the volume porosity of the mesh. β_j is the area porosity of the mesh in the j direction. ρ is the air density. Y_{fuel} is the gas concentration. u_j is the velocity of the wind in the j direction. x_j is the axis in the j direction. μ_{eff} is the effective viscosity.

The effective viscosity, μ_{eff}, is as follows:

$$\mu_{eff} = \mu + \rho C_\mu \frac{k^2}{\varepsilon} \tag{2}$$

μ is the gas viscosity and C_μ is pre-defined as 0.09. k is turbulent kinetic energy and ε is the dissipation rate of turbulent kinetic energy. The turbulence effect on gas dispersion is based on the k-ε turbulence model.

The transport equation for the dissipation rate of turbulent kinetic energy is as follows:

$$\frac{\partial}{\partial t}(\beta_v \rho \varepsilon) + \frac{\partial}{\partial x_j}\left(\beta_j \rho u_j \varepsilon\right) = \frac{\partial}{\partial x_j}\left(\beta_j \frac{\mu_{eff}}{\sigma_\varepsilon} \frac{\partial \varepsilon}{\partial x_j}\right) + \beta_v P_\varepsilon - C_{2\varepsilon} \beta_v \rho \frac{\varepsilon^2}{k} \tag{3}$$

$C_{2\varepsilon}$ is pre-defined as 1.92 and P_ε is the production of dissipation.

The transport equation for turbulent kinetic energy is as follows:

$$\frac{\partial}{\partial t}(\beta_v \rho k) + \frac{\partial}{\partial x_j}\left(\beta_j \rho u_j k\right) = \frac{\partial}{\partial x_j}\left(\beta_j \frac{\mu_{eff}}{\sigma_k} \frac{\partial k}{\partial x_j}\right) + \beta_v P_K - \beta_v \rho \varepsilon \tag{4}$$

P_K is the summation of the production of turbulent kinetic energy due to shear stress and buoyancy. σ_k is pre-defined as 1.

The SIMPLE (Semi-Implicit Method for Pressure-Linked Equation) algorithm is applied for pressure and velocity correction [16]. The FLACS numerical model employs a second-order hybrid scheme (second order upwind and second order central difference, with delimiters for certain conditions) for diffusive/convective flux equations. The first order backward Euler scheme is applied for the time-stepping scheme. The convergence criterion is mass residual and shall be less than 10^{-4}.

For the wind boundary condition, the wind velocity at an elevation z is as follows [17]:

$$U(z) = \frac{u^*}{\kappa}\left[\ln\left(\frac{(z - z_d) + z_0}{z_0}\right) - \psi_u(z)\right] \tag{5}$$

z_0 is atmospheric roughness length and z_d is canopy height (0 because the simulation takes place on flat ground). z_0 and z_d are user-defined. κ is the von Kármán constant given as 0.41.

u^* is the friction velocity and is as follows:

$$u* = \frac{U_0 \kappa}{\ln\left(\frac{(z_{ref} - z_d) + z_0}{z_0}\right) - \psi_u(z_{ref})} \tag{6}$$

U_0 is the wind speed at reference height z_{ref}. U_0 and z_{ref} are user-defined.

In this paper, Pasquill class D (neutral) is applied and the function $\psi_u(z)$ is given zero value.

The turbulence parameter profiles of the wind boundary at elevation, z, are as follows:

$$k(z) - \begin{cases} 6(u*)^2 & \text{if } z \leq 0.1h \\ 6(u*)^2(1 - \frac{z}{h})^{1.75} & \text{if } z > 0.1h \end{cases} \tag{7}$$

$$\varepsilon(z) = \begin{cases} \frac{(u*)^3}{\kappa z}(1.24 + 4.3\frac{z}{h})^2 & \text{if } z \leq 0.1h \\ \frac{(u*)^3}{\kappa z}(1.24 + 4.3\frac{z}{L})(1 - 0.85\frac{z}{h})^{3/2} & \text{if } z > 0.1h \end{cases} \tag{8}$$

u^* is the friction velocity. k is turbulent kinetic energy and ε is the dissipation rate of turbulent kinetic energy. κ is the von Kármán constant given as 0.41. h is the height of the atmospheric mixing layer [18]. L is the Monin–Obukhov length.

For inflow/parallel boundaries, Equations (5)–(8) are applied to set the boundary condition. For the outflow boundaries, the momentum and continuity equations are solved. For more information on FLACS, please refer to the FLACS-CFD 20.1 User Manual, 2020 [12].

The ship model was based on Adinda Windu Karsa, a Ro-Pax ferry working in the Indo-Java Sea [19]. The ferry measures 114.8 m by 22 m and the freeboard height is approximately 13.2 m above the sea surface. Figure 2 is a picture of the Ro-Pax ferry model used in the CFD simulation.

Figure 2. Model of Ro-Pax ferry, Adinda Windu Karsa, used in CFD simulation.

The LNG truck tankers model was based on the Mercedes Benz Actros truck series. The two trucks were modeled back-to-back, each connected to a trailer loaded with two 20-foot tanks. A pump was located between the two road tankers to transfer the LNG into the Ro-Pax ferry. The road tankers, pump, and Ro-pax vessel positioning were identical to Figure 1. Figure 3 shows the horizontal plan view of the model, the modeled wind directions (red arrow), and the LNG leak direction (blue arrow). The leak direction was chosen to be east because the leak would not be impinged and give a conservative result. The breach leaked horizontally at a height of 0.4 m above the ground.

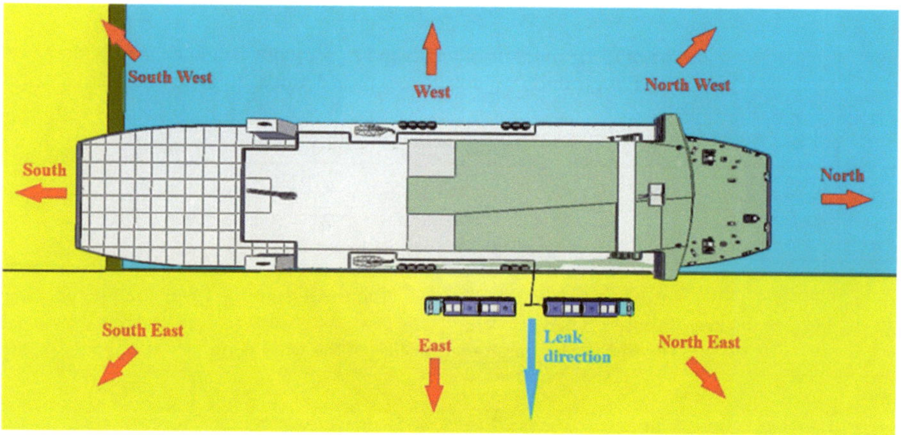

Figure 3. Plan view of the model, respective directions, and leak direction.

A structural non-uniform grid was applied by following FLACS best practice guidelines. The grid around the leak was locally refined to 0.257 m. The neighboring cells increased gradually at a rate of approximately 10% up to the domain boundaries. The ferry and trucks occupied a space of approximately 30 m by 115 m by 22 m. To allow the wind boundary to be established, the FLACS manual advises that the distance between object and domain be at least half the object dimension. Therefore, a grid domain covering 200 m by 300 m by 50 m in the x, y, and z-direction respectively was selected. The Ro-Pax ferry was placed approximately at the center of the simulated volume. Figure 4 shows the coarse grid setup.

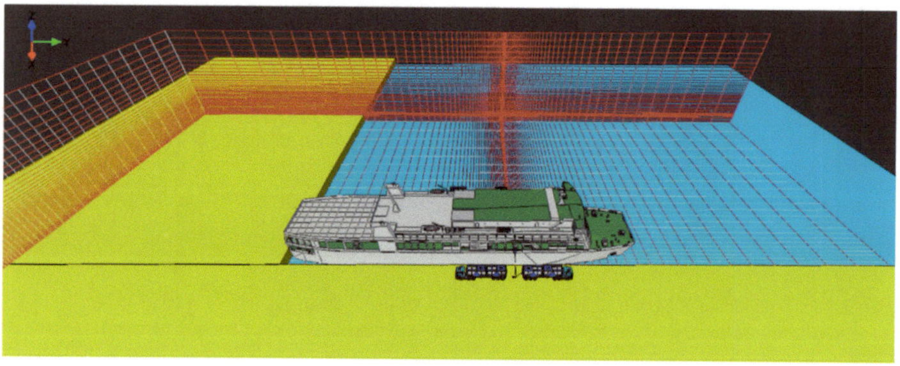

Figure 4. Coarse grid setup of the model.

A grid-independent study for four wind directions (south, west, north, and east) was conducted to test the coarse grid's suitability for the simulation. Table 1 shows the comparison of the coarse grid (406,980), 90% grid size (534,000), and 85% grid size (631,470) simulation result. The criterion for comparison is the furthest lower flammable limit (LFL) distance obtained from the different grid simulations. Please refer to Section 3 for a better explanation of LFL distance. The percentage difference ranges from 1.98 to 10.2%, demonstrating that the coarse grid size is adequate for the simulation.

Table 1. Comparison of coarse grid, 90% grid size, and 85% grid size simulation result.

Wind Speed (m/s)	Direction	Mesh Type	LFL Distance (m)	Difference Percentage from Coarse Grid (%)
8.1 (mean)	South	Coarse	23.03	N/A
		90% mesh size	21.07	−8.51
		85% mesh size	25.38	10.20
	West	Coarse	11.57	N/A
		90% mesh size	11.87	2.59
		85% mesh size	12.27	6.05
	North	Coarse	25.93	N/A
		90% mesh size	27.17	4.78
		85% mesh size	27.22	4.97
	East	Coarse	16.65	N/A
		90% mesh size	17.88	7.39
		85% mesh size	16.98	1.98

To measure the discretization error of the FLACS simulation, a grid refinement study based on Roache's work was performed [20]. The Roache grid convergence index (GCI) estimated the discretization error by comparing two discrete solutions from the coarse and refined grid. The GCI formula is calculated by:

$$\mathrm{GCI} = \frac{F_S}{\mathrm{r}^{\mathrm{P}} - 1}\left|\frac{f_2 - f_1}{f_1}\right| \tag{9}$$

F_S is the factor of safety set as 3, and r is the grid refinement factor set as 2 in this paper. P is the FLACS order of accuracy set as 2, f_2 is the coarse grid solution, and f_1 is the fine grid solution. The refined grid has 3,255,840 controlled volumes and is 8 times the coarse grid value (406,980). Figure 5 shows the refined grid setup.

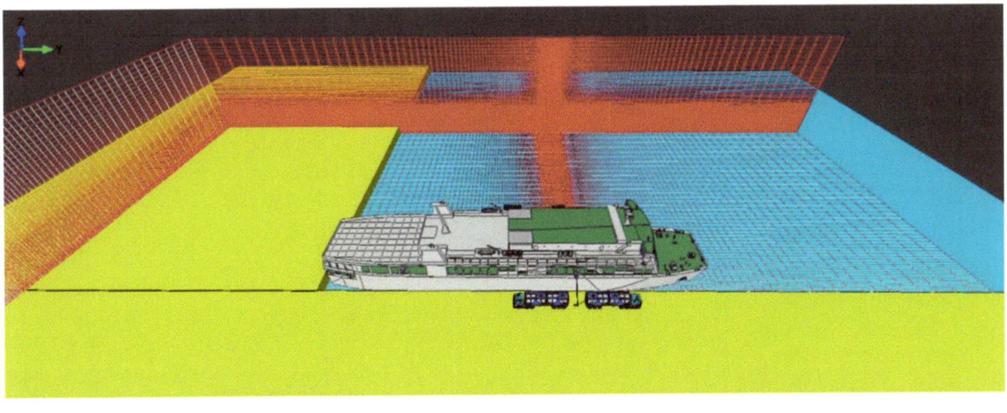

Figure 5. Refined grid setup of the model.

Based on Hong Kong January to May 2021 weather statistics, the initial conditions were set as an average temperature of 22.25 °C and 1 bar ambient pressure [21]. Ground roughness was set at 0.03 m because the terrain has few obstacles. The chosen solver was transient mode. Courant number (CFLC) based on sound velocity was set at 5. Courant number (CFLV) based on flow velocity was set at 0.5. The Hong Kong observatory wind record for January to May 2021 is presented as a wind rose in Figure 6. The prevailing wind direction is East North East (ENE). ENE mean wind speed is 8.1 m/s (elevation of 32 m above sea level) with a standard deviation of 2.38 m/s. Wind speed higher than the mean is associated with rainy weather which prevents LNG bunkering from taking place. Nubli and the team have concluded that lower wind speed leads to bigger gas dispersion [22]. To be conservative, only wind speeds at the mean speed and below were considered for the CFD simulation. Three wind speeds, 8.1 m/s (mean), 5.7 m/s (minus 1 standard deviation), and 3.3 m/s (minus 2 standard deviation) were selected for the simulation. These wind speeds cover the 5th to the 50th percentile wind speed distribution. The ENE wind profile was used for all the wind directions because no specific Hong Kong site was used in this study. Eight wind directions blowing north, northeast, east, southeast, south, southwest, west, and northwest were considered in the study. Pasquill stability class was set as neutral (D).

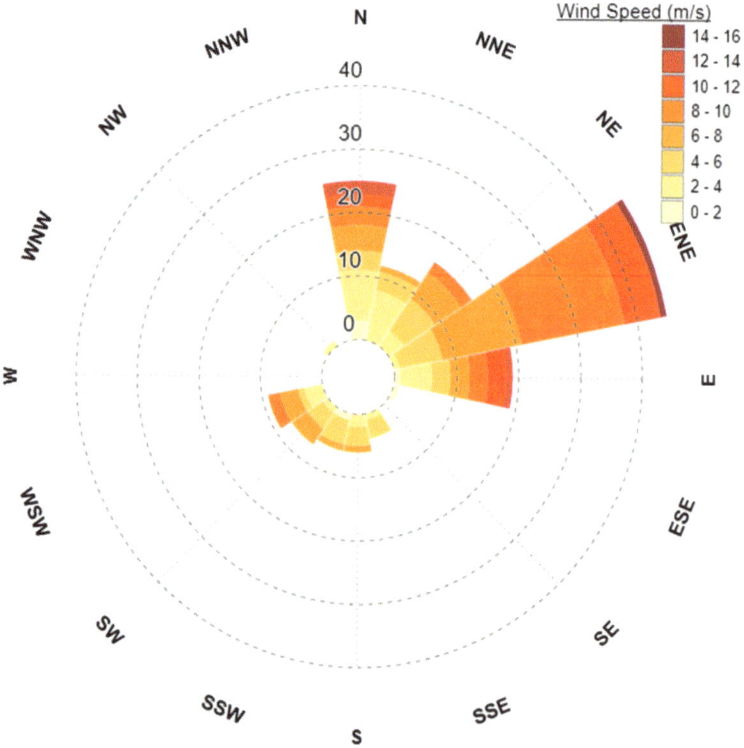

Figure 6. Windrose of Hong Kong weather from January to May 2021.

The initial leak condition was set as a 6 mm orifice releasing at 6 bar pressure. The leak duration was set as 120 s. An assumption was made that all the leaked LNG vaporized after leaving the containment and no LNG pool was formed. This assumption is supported by the recent leak experiment conducted by INERIS [23]. The experiment result showed that non-impinged pressurized LNG leaks from a breach less than 10 mm fully vaporize upon leaving containment. The Australian LNG source of 99.8% methane concentration was set as the gas composition [24]. Given the extremely high methane concentration, LNG is considered pure methane in the simulation. Considering pure methane would give a conservative result because heavier alkane disperses at a slower rate.

3. Results and Discussion

3.1. FLACS Coarse Grid Results

Natural gas (NG) is flammable between 5% to 15% methane concentration in the air. The 5% concentration is called the lower flammable limit (LFL), and the 15% concentration is termed the upper flammable limit (UFL). The LFL distance covered by each coarse grid simulation is presented in Table 2. The LFL distance is the furthest distance from the leak source where the methane concentration has diluted to 5% in the air. All simulation was performed in a transient state and the numerical data was stored at a frequency of one second. The FLACS post-processing tool, Flowvis was used to visualize the LFL distance for the 120 s leaks. A total of 24 cases were simulated with FLACS and five cases (bolded in red) exceeded the safety zone (27 m) advised by the BASiL model. To be conservative, an assumption of 30% discretization error was applied to the results. All amplified results exceeding 27 m were repeated with the refined grid to reduce discretization error. Table 2 column "GCI study" shows 12 cases to be re-simulated with the refined grid.

Table 2. LFL distance of leaks from coarse grid simulation.

Wind Speed (m/s)	Direction	LFL Distance (m)	130% LFL Distance (m)	GCI Study
8.1 (mean)	South	23.03	29.94	Yes
	South-West	17.05	22.17	No
	West	11.57	15.04	No
	North-West	20.32	26.42	No
	North	25.93	33.71	Yes
	North-East	25.17	32.72	Yes
	East	16.65	21.65	No
	South-East	11.28	14.66	No
5.7 (1 standard deviation)	South	**26.83**	34.88	Yes
	South-West	21.71	28.22	Yes
	West	20.65	26.85	No
	North-West	17.47	22.71	No
	North	21.07	27.39	Yes
	North-East	28.42	36.95	Yes
	East	20.65	26.85	No
	South-East	14.28	18.56	No
3.3 (2 standard deviation)	South	31.73	41.25	Yes
	South-West	17.81	23.15	No
	West	29.95	38.94	Yes
	North-West	20.27	26.35	No
	North	29.37	38.18	Yes
	North-East	27.92	36.30	Yes
	East	26.15	34.00	Yes
	South-East	20.73	26.95	No

3.2. FLACS Refined Grid Results

Table 3 provides the GCI value, LFL distance of the refined grid results, LFL distance range, and BASiL model percentile ranking. Using Equation (1), the GCI value was calculated from the coarse and refined grid LFL distance. The GCI value estimates the discretization error percentage to quantify the upper and lower LFL uncertainty boundaries. For example, a GCI value of 0.09 means the true LFL distance is between 91% and 109%

of the refined grid LFL distance. The percentile ranking gauges the accuracy of the BASiL model and is calculated as follows:

$$\text{Percentile} = 100 \times \frac{27 - \text{lower LFL distance}}{\text{Upper LFL distance} - \text{lower LFL distance}} \quad (10)$$

Table 3. LFL distance of leaks from refined grid simulation.

Wind Speed (m/s)	Direction	GCI Value	LFL Distance (m)	LFL Distance Range (m)	BASiL Model Percentile Ranking
8.1 (mean)	South	0.098	25.53	23.03–28.03	79.40
	North	0.017	26.37	25.93–26.81	121.59
	North-East	0.090	27.67	25.17–30.17	36.60
5.7 (1 standard deviation)	South	0.063	28.63	26.83–30.43	4.72
	South-West	0.126	19.28	16.85–21.71	208.85
	North	0.225	27.17	21.07–33.27	48.61
	North-East	0.009	28.17	27.92–28.42	N/A
3.3 (2 standard deviation)	South	0.114	35.83	31.73–39.93	N/A
	West	0.191	25.15	20.35–29.95	69.3
	North	0.240	38.62	29.37–47.87	N/A
	North-East	0.282	21.77	15.62–27.92	92.52
	East	0.048	24.95	23.75–26.15	135.42

If the lower LFL distance boundary is more than 27 m, no percentile is given because the BASiL model value does not fall within the boundaries. The highest GCI value is 0.282, roughly 2% less than the assumed discretization error for using the coarse grid. Hence, in cases that were not checked with the refined grid, it is certain that the LFL distance did not exceed 27 m.

Using 50 percentile ranking as the passing criteria, there are three cases in which the BASiL model may have underpredicted the safety zone which are bolded red in Table 3. There are three cases in which the BASiL model underpredicted which are listed as N/A in the percentile ranking column. Figure 7 shows the gas dispersion profiles (minimum 5% gas concentration in air) of the underpredicted cases, 5.7 m/s northeast wind, 3.3 m/s south wind, and 3.3 m/s north wind sequentially. The blue cloud shows the location where the methane concentration is at least LFL level. The paths where the dispersion exceeded 27 m in radius are generally in the wind direction except for the northeast wind case. Figure 8 shows the wind streamlines on the XY plane at 1 m above the ground. The different color depicts the wind speed in different regions. The leak area is blocked from the incoming wind and causes the nearby region air to flow toward the manifold. The black lines outline the streamlined formation. The leaked plume disperses in the streamlined direction and the result is shown in Figure 7a.

None of the leaks disperse beyond 3 m height and this observation is explained by the stratification effect of a cryogenic gas leak. Methane gas has negative buoyancy relative to air at a temperature below −117 degrees Celsius [5]. Therefore, the flammable gas spreads more in the horizontal axes than in the vertical direction.

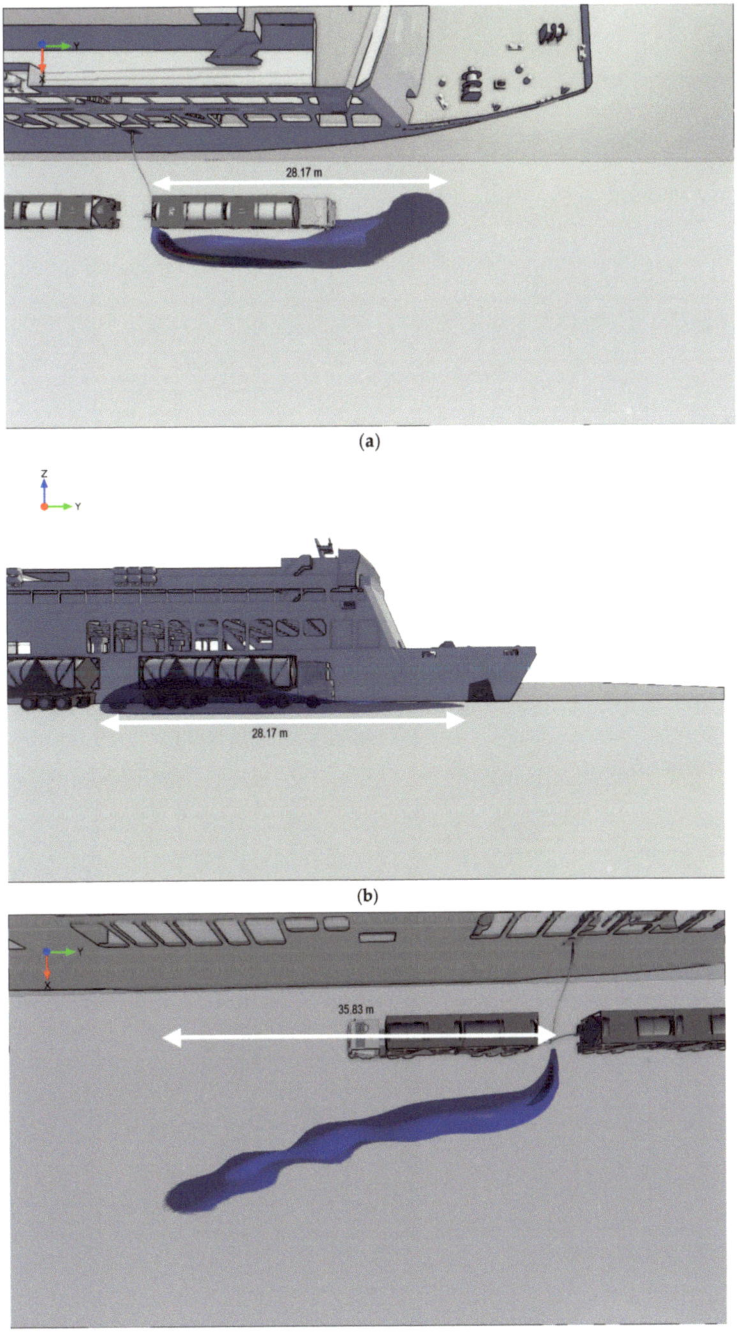

(**a**)

(**b**)

(**c**)

Figure 7. *Cont.*

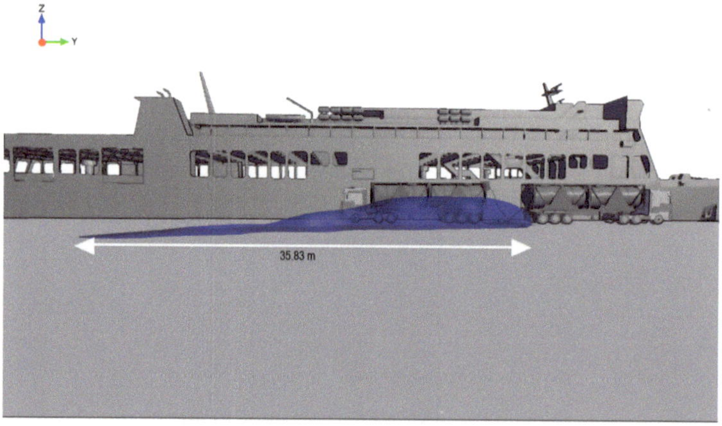

(d)

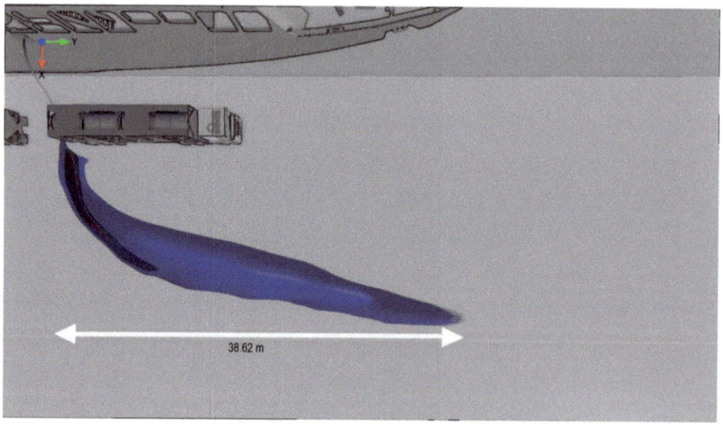

(e)

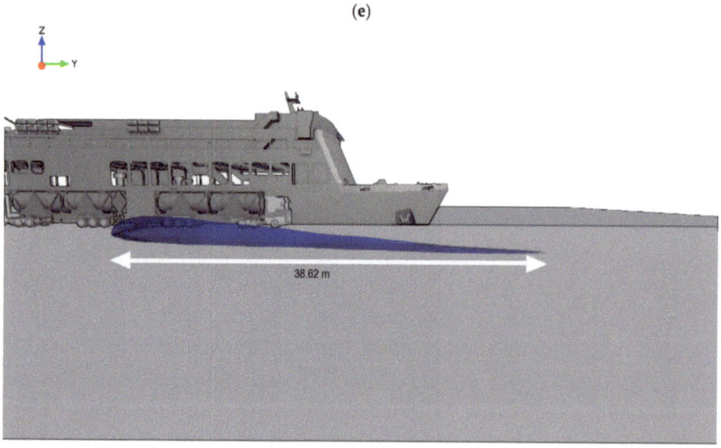

(f)

Figure 7. Snapshots of LNG dispersion profiles that exceeded 27 m: (**a**) Plan view of 5.7 m/s northeast wind, (**b**) Section view of 5.7 m/s northeast wind (**c**) Plan view of 3.3 m/s south wind, (**d**) Section view of 3.3 m/s south wind, (**e**) Plan view of 3.3 m/s north wind, (**f**) Section view of 3.3 m/s north wind.

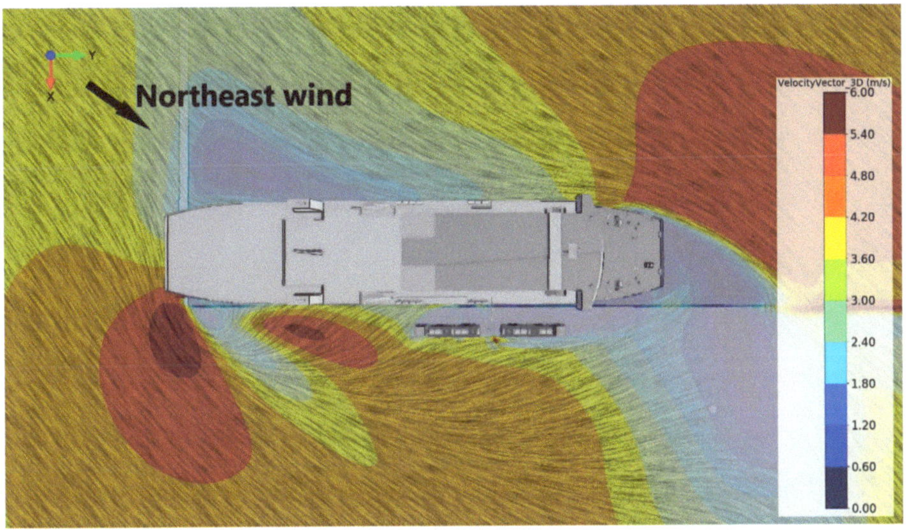

Figure 8. Wind streamlines with a 5.7 m/s northeast wind.

Table 4 summarizes the validation of the FLACS result with the BASiL model. Of 24 FLACS simulations, 18 cases did not exceed the 27 m safety zone advised by the BASiL model. Three cases were inconclusive as the BASiL model value falls in the lower 50th percentile of the LFL uncertainty boundaries. From the GCI analysis, three cases of gas dispersion were certain to have exceeded the 27 m. The cases in which BASiL had underestimated the safety zone are for wind speeds below the mean. This is unsurprising as BASiL is based on data interpolation to advise the safety zone. Assuming the 1.4 million-combination database has more information on parameter values centering near the normal values, it may be unsuitable to capture the 3.3 m/s wind speed cases which are outliers (5% percentile) in the wind distribution curve. The average LFL distance of the 12 cases in Table 4 is 27.4 m which is close to the 27 m advised by BASiL. This finding demonstrates the interpolation methodology nature adopted by BASiL.

Table 4. Summary of FLACS results compared with BASiL model.

Wind Speed (m/s)	Direction	BASiL Model Safety Zone Exceeded
8.1 (mean)	South	No
	South-West	No
	West	No
	North-West	No
	North	No
	North-East	No
	East	No
	South-East	Uncertain
5.7 (1 standard deviation)	South	Uncertain
	South-West	No
	West	No
	North-West	No
	North	Uncertain
	North-East	Yes
	East	No
	South-East	No

Table 4. *Cont.*

Wind Speed (m/s)	Direction	BASiL Model Safety Zone Exceeded
3.3 (2 standard deviation)	South	Yes
	South-West	No
	West	No
	North-West	No
	North	Yes
	North-East	No
	East	No
	South-East	No

4. Conclusions

FLACS was used to simulate 24 LNG horizontal leaks covering wind speed distribution up to the 5th percentile for a Ro-Pax ferry bunkering. GCI analysis was performed to reduce the CFD discretization error. The BASiL model safety zone estimation did not exceed FLACS findings except when wind speed was below the mean. Three validation cases have inconclusive findings and need further investigation. Future work would be to perform these three cases with four times grid refinement based on this paper's coarse grid setup. With three discrete solutions using a constant grid refinement of 2, the GCI safety factor can be reduced from 3 to 1.25 [20]. This would reduce the discretization error uncertainty and give a conclusive answer as to whether the BASiL model has underestimated the safety zone. However, this refinement will have approximately 26 million controlled volumes and require significant computational resources. The average computing time to perform the coarse grid simulation (406,980 controlled volumes) is 12.5 h and for the fine grid simulation (3,255,840 controlled volumes) is 200 h. It will need at least one month of computing time to perform simulations for 26 million controlled volumes with a commercial computer.

Other future work would be to examine ways to reduce the LFL distance for the leak cases, especially for wind speed conditions at and below the 5% percentile distribution. One potential solution is putting up a barrier to redirect the gas cloud. As observed in the northeast wind simulation, wind blockage can redirect the gas plume. This phenomenon may have the potential to aid in gas mixing and significantly influence LNG dispersion. The approach to studying port arrangement impact on LNG dispersion can be adopted from similar work done by Mariotti and team [25]. By reducing the LFL distance for the worst dispersion case, the ship owner can use the free space for simultaneous operation, thereby reducing the time spent in port.

Author Contributions: Conceptualization: B.H.L.; Methodology: B.H.L.; Formal analysis and investigation: B.H.L.; Writing—original draft preparation: B.H.L.; Writing—review and editing: B.H.L. and E.Y.K.N.; Funding acquisition: B.H.L. and E.Y.K.N.; Resources: B.H.L. and E.Y.K.N.; Supervision: E.Y.K.N. All authors have read and agreed to the published version of the manuscript.

Funding: This work was supported by the Industrial Postgraduate Programme (#001799-00001) initiated by the Singapore Economic Development Board (EDB).

Institutional Review Board Statement: Not applicable.

Informed Consent Statement: Not applicable.

Data Availability Statement: The data presented in this study are available on request from the corresponding author. The data are not publicly available due to the academic license agreement.

Acknowledgments: The authors would like to express their appreciation to Gexcon Bjørn Lilleberg for his advise on using FLACS.

Conflicts of Interest: The authors declare no conflict of interest.

References

1. DNV GL. *Mark Irvine Energy Transition Outlook 2019—A Global and Regional Forecast to 2050*; DNV GL: Høvik, Norway, 2019.
2. DNV GL. *Mark Irvine Energy Transition Outlook 2020 Executive Summary*; DNV GL: Høvik, Norway, 2020.
3. *ISO 20519:2017*; Ships and Marine Technology—Specification for Bunkering of Liquefied Natural Gas Fuelled Vessels. BSI Standards: Hong Kong, China, 2017; ISBN 978-0-580-89936-2.
4. Jeong, B.; Lee, B.S.; Zhou, P.; Ha, S. Man Evaluation of Safety Exclusion Zone for LNG Bunkering Station on LNG-Fuelled Ships. *J. Mar. Eng. Technol.* **2017**, *16*, 121–144. [CrossRef]
5. Ikealumba, W.C.; Wu, H. Some Recent Advances in Liquefied Natural Gas (LNG) Production, Spill, Dispersion, and Safety. *Energy Fuels* **2014**, *28*, 3556–3586. [CrossRef]
6. Process Hazard Analysis Software—PHAST. Available online: https://www.dnv.com/software/services/phast/index.html (accessed on 9 July 2022).
7. ALOHA Software. Available online: https://www.epa.gov/cameo/aloha-software (accessed on 9 July 2022).
8. Carboni, M.; Pio, G.; Vianello, C.; Mocellin, P.; Maschio, G.; Salzano, E. Numerical Simulation of LNG Dispersion in Harbours: A Comparison of Flammable and Visible Cloud. *Chem. Eng. Trans.* **2022**, *90*, 355–360. [CrossRef]
9. Carboni, M.; Pio, G.; Mocellin, P.; Vianello, C.; Maschio, G.; Salzano, E. Accidental Release in the Bunkering of LNG: Phenomenological Aspects and Safety Zone. *Ocean Eng.* **2022**, *252*, 111163. [CrossRef]
10. Gerbec, M.; Vidmar, P.; Pio, G.; Salzano, E. A Comparison of Dispersion Models for the LNG Dispersion at Port of Koper, Slovenia. *Saf. Sci.* **2021**, *144*, 105467. [CrossRef]
11. Society for Gas as a Marine Fuel Recommendation of Controlled Zones during LNG Bunkering. 2018. Available online: https://www.sgmf.info/ (accessed on 7 July 2022).
12. FLACS-CFD 20.1 User's Manual. Available online: https://support.flacs.com/support/login (accessed on 11 July 2022).
13. Koopman, R.P.; Cederwall, R.T.; Ermak, D.L.; Goldwire, H.C.; Hogan, W.J.; McClure, J.W.; McRae, T.G.; Morgan, D.L.; Rodean, H.C.; Shinn, J.H. Analysis of Burro Series 40-M3 Lng Spill Experiments. *J. Hazard. Mater.* **1982**, *6*, 43–83. [CrossRef]
14. Puttock, J.S.; Blackmore, D.R. Field Experiments on Dense Gas Dispersion. *J. Hazard. Mater.* **1982**, *6*, 13–41. [CrossRef]
15. Chan, S.T. Numerical Simulation of LNG Vapor Dispersion from a Fenced Storage Area. *J. Hazard. Mater.* **1992**, *30*, 195–224. [CrossRef]
16. Suhas, V. *Patankar Numerical Heat Transfer and Fluid Flow*, 1st ed.; CRC Press: Boca Raton, FL, USA, 1980.
17. Monin, A.S.; Obukhov, A.M. Basic Laws of Turbulent Mixing in the Surface Layer of the Atmosphere. *Tr. Akad. Nauk SSSR Geofiz.* **1959**, *24*, 163–187.
18. Golder, D. Relations among Stability Parameters in the Surface Layer. *Bound.-Layer Meteorol.* **1972**, *3*, 47–58. [CrossRef]
19. Adinda Windu Karsa. Available online: https://www.marinetraffic.com/en/ais/details/ships/shipid:2969507/mmsi:525003465/imo:9713789/vessel:ADINDA_WINDU_KARSA (accessed on 24 July 2021).
20. Roache, P.J. *Verification and Validation in Computational Science and Engineering/by Patrick J. Roache*; Hermosa: Albuquerque, NM, USA, 1998; ISBN 0913478083.
21. Hong Kong Observatory. Available online: https://www.hko.gov.hk/en/cis/climat.htm (accessed on 24 July 2021).
22. Nubli, H.; Sohn, J.M. CFD-Based Simulation of Accidental Fuel Release from LNG-Fuelled Ships. *Ships Offshore Struct.* **2020**, *17*, 339–358. [CrossRef]
23. Joubert, L.; Leroy, G.; Betteridge, S.; Vyazmina, E.; Bernard, L. Presentation of the Experimental JIP SPARCLING: Inside and beyond a Pressurised LNG Release. In Proceedings of the 13th International Symposium on Hazards, Prevention, and Mitigation of Industrial Explosions, Braunschweig, Germany, 27 July–31 August 2020; pp. 2–11.
24. Wood, D.A. Predicting Saturated Vapor Pressure of LNG from Density and Temperature Data with a View to Improving Tank Pressure Management. *Petroleum* **2020**, *7*, 91–101. [CrossRef]
25. Mariotti, A.; Galletti, C.; Brunazzi, E.; Salvetti, M.V. Steady Flow Regimes and Mixing Performance in Arrow-Shaped Micro-Mixers. *Phys. Rev. Fluids* **2019**, *4*, 034201. [CrossRef]

fluids

MDPI

Article

Evaluation of the Thermofluidic Performance of Climatic Chambers: Numerical and Experimental Studies

Bahareh Ramezani [1], António Tadeu [1,2,*], Tiago Jesus [1], Michael Brett [1,3] and Joel Mendes [4]

1 Itecons—Institute of Research and Technological Development in Construction, Energy, Environment and Sustainability, 3030-289 Coimbra, Portugal; bahareh.ramezani@itecons.uc.pt (B.R.); tiago.jesus@itecons.uc.pt (T.J.); michael.brett@itecons.uc.pt (M.B.)
2 ADAI—LAETA, Department of Civil Engineering, Faculty of Sciences and Technology, University of Coimbra, 3030-289 Coimbra, Portugal
3 ADAI—LAETA, Department of Mechanical Engineering, Faculty of Sciences and Technology, University of Coimbra, 3030-289 Coimbra, Portugal
4 ARALAB—Câmaras Climáticas e Ambientes Controlados, 2635-047 Rio de Mouro, Portugal; joel.mendes@aralab.pt
* Correspondence: tadeu@itecons.uc.pt

Abstract: Climatic chambers are highly important in research and industrial applications and are used to examine manufactured samples, specimens, or components in controlled environment conditions. Despite the growing industrial demand for climatic chambers, only a few published studies have specifically concentrated on performance analysis and functional improvements through numerical and experimental studies. In this study, a 3D computational fluid dynamics (CFD) model of a climatic chamber was developed using Ansys Fluent to simulate the fluid flow, heat, and mass transfer to obtain the velocity, temperature, and relative humidity fields in the interior box of a 1200 L climatic chamber. The results were then validated with experimental data from a prototype. Finally, the heat losses of the surrounding components of the chamber were calculated, and the relationship between the inside temperature and the overall thermal loss was modelled. This validated numerical model provides the possibility of optimising the performance of climate chambers by reducing the thermal loss from the walls and modifying the air flow pattern inside the chamber.

Keywords: climatic chamber; modelling; CFD; validation; optimisation

Citation: Ramezani, B.; Tadeu, A.; Jesus, T.; Brett, M.; Mendes, J. Evaluation of the Thermofluidic Performance of Climatic Chambers: Numerical and Experimental Studies. *Fluids* **2021**, *6*, 433. https://doi.org/10.3390/fluids6120433

Academic Editors: Ernesto Benini, Francesco De Vanna and Laura A. Miller

Received: 27 August 2021
Accepted: 25 November 2021
Published: 30 November 2021

1. Introduction

Climatic test chambers are widely used in research and product development in industrial applications to conduct tests on manufactured samples, specimens, or components under controlled environmental conditions. Despite the increasing interest in the use of climatic chambers in various research fields and industries, only a few valid scientific works have focused on the analysis of their thermofluidic performance, either experimentally or numerically. Liang et al. [1] provided a mathematical model to investigate the transient thermal performance of a full-sized vehicle climate chamber and suggested solutions to improve heating and cooling rates. Dostál et al. [2] presented a control algorithm using a predictive control method model for a climatic chamber, and García-Contreras et al. [3] proposed a theoretical and experimental methodology to obtain the thermal loads for a climatic vehicle test chamber. The performance of a temperature and humidity chamber was evaluated by Mensah et al. [4] experimentally. In another study, three-dimensional CFD thermofluidic modelling was carried out for a test chamber by Candanedo et al. [5], and the results were validated experimentally. A numerical and experimental study was performed by Lecoq et al. [6] on the evaporation rate in a cold chamber, and in a recent study, the application of CFD modelling to predict the air circulation in an evaporator of a large-scale heat pump was studied by Rogié et al. [7] using Ansys Fluent.

Yuan Liu et al. [8] were able to obtain good agreement between theoretical and numerical simulations, which coupled heat and moisture, but the data were not verified experimentally. In another study, Vojkuvkova et al. [9] used CFD modelling to investigate the condensation and freezing of water vapour in a mixing chamber. Their results showed that the most condensation occurs on the upper parts of walls. In another study, CFD modelling was applied by Shojaee Nasirabadi et al. [10] to simulate the moisture transfer throughout an electronic enclosure. The results show that the local environment inside the enclosure responds faster to the temperature changes than it does to the relative humidity variations. They also found good agreement between the numerical simulation results and experimental results from the literature.

This work, however, aims to fill a gap in the climatic chamber equipment modelling, which typically is only concerned with modelling the temperature and velocity fields. It provides a 3-D thermofluidic CFD model to analyse the fluid flow, heat, and mass transfer behaviour of a new climatic chamber prototype designed by the industrial company Aralab, with the collaboration of the Itecons research centre. This model is based on a 300 L inner volume chamber that was numerically and experimentally studied by Silva et al. [11].

This new climatic chamber has an inner box with 1200 L capacity, which is used to keep test specimens, and a smaller box in which the evaporator, electric heater resistance, and ventilation equipment is located (Figure 1). The technical information for the chamber is presented in Table 1.

Figure 1. Climatic chamber prototype under development: external and internal view.

Table 1. Specifications of the climatic chamber.

Temperature range	10–90 °C
Temperature precision	In time ≤ ±0.5 °C In space ≤ ±1.5 °C
Humidity range	10% RH to 98% RH
Humidity precision	In time ≤ ±1.0% RH In space ≤ ±2.5% RH
Heating	Tubular stainless steel electric heaters (details in Section 3.2)

Table 1. *Cont.*

Cooling	Air-cooled hermetic scroll compressor group with enforced ventilation
Humidity control system	Humidity: thermostatic bath with dew point control Drying: thermostatic bath with dew point control and additional dry coil
Humidity sensors	Psychrometric, capacitive (or both simultaneously)
Ventilation	Two fans (details in Section 3.2)

Compared with the previous model of this equipment and others readily available in the market, this new climatic chamber was equipped with variable compressors and variable ventilators, an optimised reduction of the rock—wool wall insulation thickness was also applied, and an eco-friendly refrigerant was adopted. These changes lead to a reduction in energy consumption and acoustic noise emission. However, they make it critically necessary to study their effect on the homogeneity of the temperature and the relative humidity inside the test chamber throughout its use, and to ascertain the need to modify or create optimised air flow routes.

This paper first describes the 3D CFD model of the climate chamber that was developed using Ansys Fluent to simulate the fluid flow, heat, and mass transfer. Then, the thermofluidic performance of the chamber is presented, based on the simulation results, and the heat loss from the surrounding components of the chamber is calculated to arrive at a fixed internal temperature. Finally, the numerical model was validated using experimental data, comparing the results for fluid flow velocity, temperature, and relative humidity.

2. Geometry

The 3D model of the climatic chamber was developed using the software SolidWorks and Ansys SpaceClaim 19.2. The key components of the chamber that were modelled included the top, bottom, and back perforated grids of the testing box, thermal insulation walls, door component together with a six-layer glass window, evaporator, heat resistances, piping, water storage (humidity system), and two fans. The climatic chamber under study had the following external dimensions: L = 1.25 m, H = 1.24 m, and W = 1.84 m (see Figure 2).

Additionally, the temperature was controlled using the heating electric resistances and the evaporator located near the fans, and the relative humidity was controlled via a humidity system, which heats up a reservoir of water using an electric resistance or cools down the fluid, leading to condensation of vapour into the reservoir, depending on whether the relative humidity is to be increased or reduced, respectively.

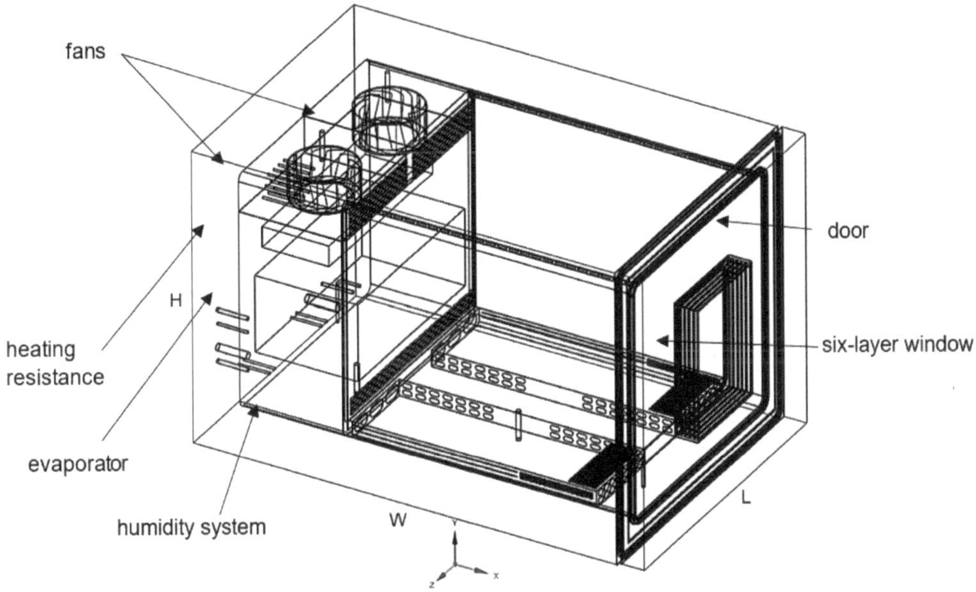

Figure 2. The geometry of the climatic chamber.

3. Model Setup

3.1. Mesh

The Ansys Fluent finite volume method (FVM) [12,13] was used to solve the fundamental equations governing heat transfer and fluid flow. This was performed using an Ansys Mesher and a nonstructured tetrahedral mesh. In the meshing process, a sweep method was used for meshing the sweepable components. The mesh was then converted into polyhedral elements in Ansys Fluent to increase precision, convergence, and robustness. Figure 3 illustrates the mesh structure used to model the solid parts and the fluid.

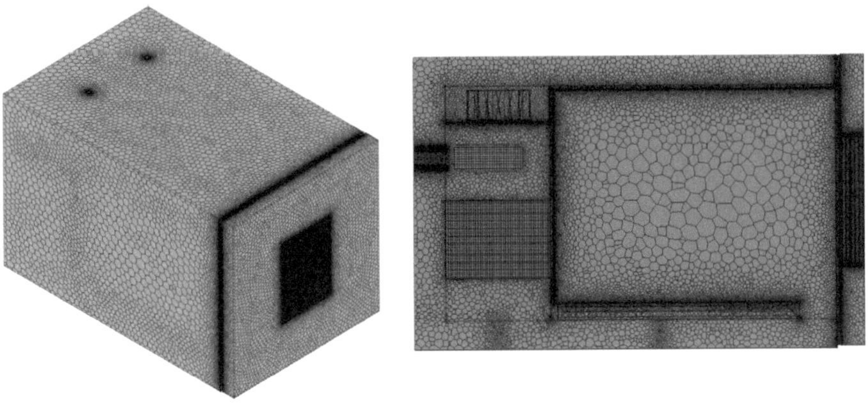

(**a**) (**b**)

Figure 3. Polyhedral mesh—3D view: (**a**) and cross-sectional view (**b**).

The mesh was refined in the zones that required more accuracy, such as the components with a more detailed geometry (Figure 4).

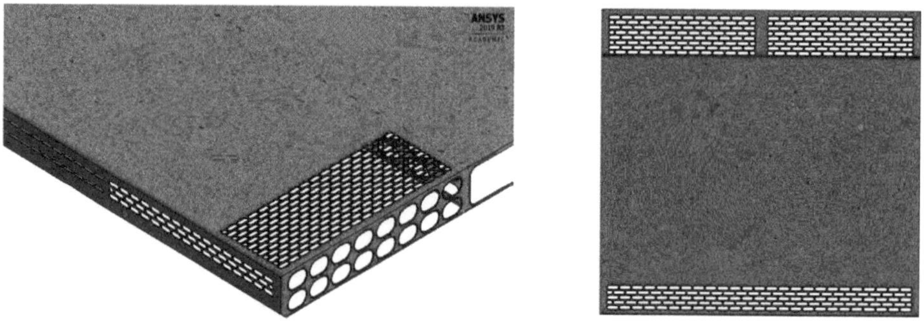

(a) (b)

Figure 4. Mesh refinement in bottom: (**a**) and back (**b**) perforated grids.

The mesh was optimised and partitioned in Ansys Fluent using the Cuthill–McKee inverse method [14] and METIS algorithm, respectively [15,16]. The final number of elements of the model was 33.14 million. The grid independence study was performed to ensure that the solution did not change with further refinement. With this in mind, while the number of elements increased, convergence of the monitored variables was checked. By increasing the number of elements to 48.19 million, the average temperature of the air inside changed by approximately 0.03%, which shows the high degree of independence of the results with respect to the mesh size. To ensure successful computations of the interior turbulent flows of the climatic chamber, mesh generation must be well defined in the regions where the mean flow undergoes rapid changes and the shear layers have large strain rates. By computing the first cells' height near the wall, based on the average fluid velocity, length scale, viscosity, and density, the target $y^+ \leq 1$ near the walls of the climatic chamber was achieved by imposing discretised inflation layers on 10 cells with a constant growth rate of 1.2, and the obtained y^+ was equal to 0.98.

In this study, Ansys Fluent was used to solve the mass, momentum, and energy conservation equations for the entire domain using the finite volume method (FVM) [17]. The fundamental equations of mass, momentum, and energy conservation can be found in Cengel et al. [18]. The fluid flow was considered to be a turbulent regime in the inlet during nominal operation conditions. Turbulent flow is governed by the Navier–Stokes momentum equations, the continuity equations, and in compressible flow, the energy and state equations. In this study, the turbulent flow was modelled using the Reynolds-averaged Navier-Stokes (RANS) equations and four different turbulent models: $k-\omega$ SST, $k-\varepsilon$ standard, $k-\varepsilon$ RNG, and $k-\varepsilon$ realisable. The governing equations can be found in detail in [14,19]. Reynolds-averaged Navier–Stokes (RANS) is the most widely used class of turbulence models in applied research due to their fast computation time and excellent convergence characteristics and because they can be used in combination with steady-state simulations [20]. To investigate the humidity distribution inside the chamber, the species transport model in Ansys Fluent was applied in the steady-state model. A certain amount of the mass fraction of water was introduced into the air circulation, depending on the relative humidity set point. As the main aim of the study is to assess the temperature and humidity distribution in the testing area of the chamber, the phase change due to evaporation and condensation is ignored in this model. The mass diffusion in turbulent flows is computed using the following equation:

$$\vec{J} = -\left(\rho D_m + \frac{\mu_t}{Sc_t}\right)\nabla Y - D_T \frac{\nabla T}{T} \quad (1)$$

where D_m is the mass diffusion coefficient for species, Y is the local mass fraction of species, μ_t is the turbulent viscosity, Sc_t is the turbulent Schmidt number, and D_T is the thermal diffusion coefficient.

Furthermore, the transport of enthalpy due to mass diffusion, which has a significant effect on the enthalpy field, was calculated in Ansys Fluent. More information regarding the species transport model can be found in the Ansys Fluent manual [17].

3.2. Boundary Conditions

3.2.1. Exterior Wall

The convection heat transferring through the external wall to the air surrounding the equipment can be calculated using Newton's law of cooling:

$$Q_{L,h} = \overline{h}_e A(T_s - T_a) \tag{2}$$

where $\overline{h}_e$ is the external convection coefficient, A is the area of the wall, T_s is the surface temperature, and T_a is the ambient temperature. This value is obtained from the average Nusselt number, which is given by the following correlation:

$$\overline{Nu_L} = \frac{\overline{h}_e L}{k_f} \tag{3}$$

where k_f is the thermal conductivity and L is the characteristic length of the geometry, which is equal to the lateral/horizontal size of the lateral/horizontal plate. For plates of various shapes (e.g., rounded corners), this characteristic length can be defined as [21]:

$$L \equiv \frac{A_s}{P} \tag{4}$$

where A_s and P are the plate surface area (one side) and perimeter, respectively.

For the horizontal plate, the flow patterns and heat transfer depend strongly on whether the surface is cold or hot and on whether it is facing up or down [22]. Therefore, the average Nusselt numbers for the top and bottom surfaces are obtained from the following equations. For the upper surface of the hot plate (Equation (4)) and the lower surface of the cold plate (Equation (5)), the average Nusselt numbers are:

$$\overline{Nu_L} = 0.54Ra_L^{1/4},\ \left(10^4 \leq Ra_L \leq 10^7,\ \mathrm{Pr} \geq 0.7\right) \tag{5}$$

$$\overline{Nu_L} = 0.15Ra_L^{1/3}, (10^7 \leq Ra_L \leq 10^{11},\ all\mathrm{Pr}) \tag{6}$$

and for the lower surface of the hot plate and the upper surface of the cold plate, the average Nusselt number is:

$$\overline{Nu_L} = 0.25Ra_L^{1/5},\ \left(10^4 \leq Ra_L \leq 10^9,\ \mathrm{Pr} \geq 0.7\right) \tag{7}$$

where, Pr is the Prandtl number and Ra_L is the Rayleigh number based on the plate length, which is given by the following equation:

$$Ra_L = \frac{g\beta_f(T_s - T_a)L^3}{v_f \alpha_f} \tag{8}$$

where g is the gravitational acceleration, β_f represents the fluid thermal expansion, v_f is the kinematic viscosity, and α_f is the thermal diffusivity.

For the lateral walls, the average Nusselt number can be calculated from the correlation, which has been recommended by Churchill and Chu [22] and may be applied over the entire range of Ra_L:

$$\overline{Nu_L} = \left\{ 0.825 + \frac{0.387 Ra_L^{1/6}}{\left[1 + (0.492/\text{Pr})^{9/16}\right]^{8/27}} \right\}^2 \quad (9)$$

For the inclined plates, it was suggested by Rich [23] that the convection coefficient could be determined from the vertical plate correlation if g is replaced by $g\cos\theta$ in computing the plate Rayleigh number. However, this approach is only satisfactory for the top and bottom surfaces of cold and hot plates, respectively. No recommendations are made for the opposite surfaces. The heat transfer calculations were performed according to an external air temperature of 22 °C.

3.2.2. Radial Fan

The climatic chamber uses two radial fans with variable speed, which rotate in opposite directions to ensure homogeneous distribution of temperature and moisture inside the chamber. The characteristic curve of the fan is presented in Figure 5. The diameter of the fan's rotor is 28 cm with a maximum flow rate of 1332 m^3/h and a maximum axial velocity of $u_{max} = 6$ m/s. The modelling was performed using the internal boundary condition and by introducing the uniform pressure jump variables for both fans, as defined by the manufacturer.

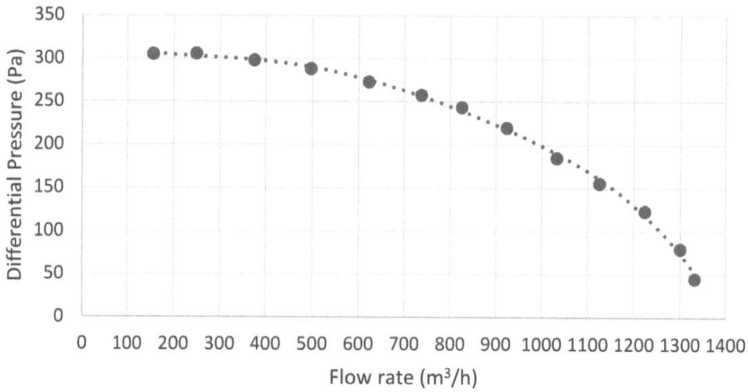

Figure 5. Radial fan characteristic curve.

The fan was considered to be adiabatic and with no entropy losses and, therefore, without any viscous dissipation. For simplification, the 3D rotating fans in the CFD model were ignored, and 1D flow through the pressure jump was considered. The pressure differential of the fan in Pascal as a function of the axial average velocity ($\overline{u}$) in m/s was obtained as follows:

$$\Delta p = -1.04\overline{u}^4 + 11.554\overline{u}^3 - 51.608\overline{u}^2 + 78.077\overline{u} + 271.67$$

3.2.3. Electric Resistance

The electrical resistance that supplies thermal energy was modelled as a uniformly distributed heat source in the location of the actual electric resistance in the chamber. The electrical supply of this chamber consists of 6 resistances, each with an electric power of $q = 3000$ W. Therefore, a total thermal power of $q = 18{,}000$ W was considered for the input electrical thermal power of the heat resistance, assuming no heat loss and 100% electrical-to-thermal-conversion efficiency.

3.2.4. Evaporator

The evaporator developed for this chamber was of the tube-fin heat exchanger type. It is important to model the evaporator to consider the pressure loss in the air flow. Modelling the evaporator with all the fins and tubes in Ansys Fluent is complex and inefficient. Hence, the pressure loss in the evaporator was computed and modelled as an anisotropic porous medium [24]. The pressure loss through the anisotropic porous medium depends on the magnitude of the velocity and is given by the equation:

$$\Delta p = \frac{1}{2}\rho_f C_2 v^2 \tag{10}$$

where, ρ_f is the density of the fluid, C_2 is the inertial resistance coefficient, and v is the velocity norm (perpendicular to the inflow). To obtain C_2 in Equation (10), the constant terms of Equation (11) for the core friction in the evaporator were calculated and equalised to C_2.

$$\Delta p = \frac{1}{2}\rho_f f \frac{L_c}{r_h} v^2 \tag{11}$$

where, f is the friction factor, L_c is the core length, and r_h is the hydraulic radius. According to the manufacturer, the value of 0.23 was considered for the air-side friction factor. It is important to note that to consider the effects of baffles on the fluid flow, the other pressure loss coefficients in the direction perpendicular to the fins were set to the large values. This provided a C_2 value of 40.35 m^{-1}, which was used as the simulation input data.

3.3. Material Properties

The air properties were considered to vary with temperature (T in Kelvin). The solids in this study are stainless steel, used in the box structure and pipes; rubber, glass, and copper for heat resistances and evaporator tubes; and rock–wool as the insulation to prevent heat loss from the chamber walls and door.

The air properties, including the density, dynamic viscosity, thermal conductivity, and specific heat, were assumed to vary with the temperature and were obtained from [11], based on interpolated data from Keenan [25] and Touloukian [26], and can be expressed as follows:

$$\rho_f = -2.542308 \times 10^{-3} T + 1.905009 \tag{12}$$

$$\mu_f = -2.745500 \times 10^{-11} T^2 + 6.273113 \times 10^{-8} T + 2.224504 \times 10^{-6} \tag{13}$$

$$k_f = 7.052556 \times 10^{-5} T + 4.555258 \times 10^{-3} \tag{14}$$

$$C_{p,f} = 5.929597 \times 10^{-4} T^2 - 3.653727 \times 10^{-1} T + 1.062852 \times 10^3 \tag{15}$$

The solid material properties are obtained from data provided by the manufacturer, as indicated in Table 2.

Table 2. Properties of the solid materials.

Material	**Density (kg/m³)**	**Thermal Conductivity (W/(m·K))**	**Emissivity (–)**
Stainless steel	8055 [c]	15.1 [c]	0.95 (exterior wall) [a]; 0.1 (inner wall) [a]
Rock–wool	120 [b]	$k = \mathbf{0.0001789}T + \mathbf{0.036261}$ [e]	(–)
Polyurethane	44 [b]	$k = \mathbf{0.0001089}T + \mathbf{0.020132}$ [e]	(–)
Rubber	1100 [d]	0.2 [d]	(–)
Glass	2500 [c]	1 [c]	0.837 [c]
Copper	8978 [b]	387.6 [b]	(–)

Source: [a] [11] [b] manufacturer's data; [c] Bergman et al. [21]; [d] [27]; [e] [28].

For the thermal insulation, the thermal conductivity was considered to be dependent on the temperature and can be achieved by the equations listed in Table 2 (where T is the temperature of the insulation in Kelvin).

3.4. Solver

The air behaviour was considered as an incompressible fluid, and a pressure-based solver was used, requiring less memory and allowing for flexibility in the solution procedure. In order to evaluate diffusive fluxes and velocity derivatives, the gradient of solution variables is required. The Green–Gauss cell-based approach was used for this, while the robustness of the SIMPLE algorithm led to it being used for the pressure–velocity coupling. The PRESTO! method was used for calculating cell-face pressure, which is highly recommended for flows involving steep pressure gradients or in strongly curved domains. The first-order upwind method was used for the remaining interpolation arrangements for robustness and easy convergence [12].

3.5. Convergence

At convergence, as mentioned in the Ansys customer training material, the following should be satisfied:

- The solution no longer changes with subsequent iterations.
- Overall mass, momentum, energy, and scalar balances are achieved.
- All equations (momentum, energy, etc.) are obeyed in all cells to a specified tolerance.

The convergence conditions were monitored for all turbulence models and were found to be adequate. Figure 6 illustrates the convergence of all scaled residuals associated with the mass, momentum, and energy conservation for the k−ω SST model and k−ε RNG model.

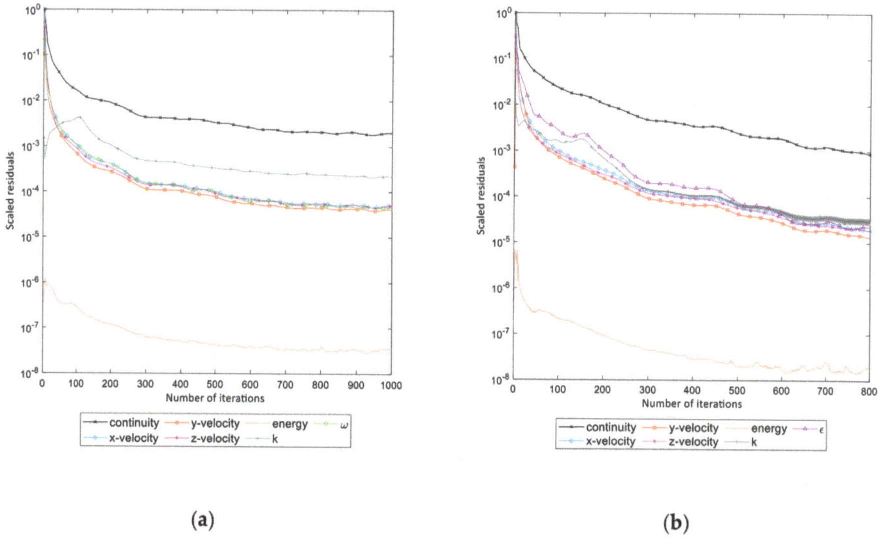

Figure 6. Convergence of scaled residuals: (**a**) k−ω SST model; (**b**) k−ε RNG model.

Additionally, three representative local velocities at six points (inside the box) including v1 (−0.13, 1.59, 0), v2 (−0.19, 1.66, 0.23), v3 (0, 1.46, 0.35), v4 (0, 1.4, −0.15), v5 (−0.43, 0.82, 0), and v6 (0.81, 1.46, −0.4) were monitored to ensure convergence (Figure 7).

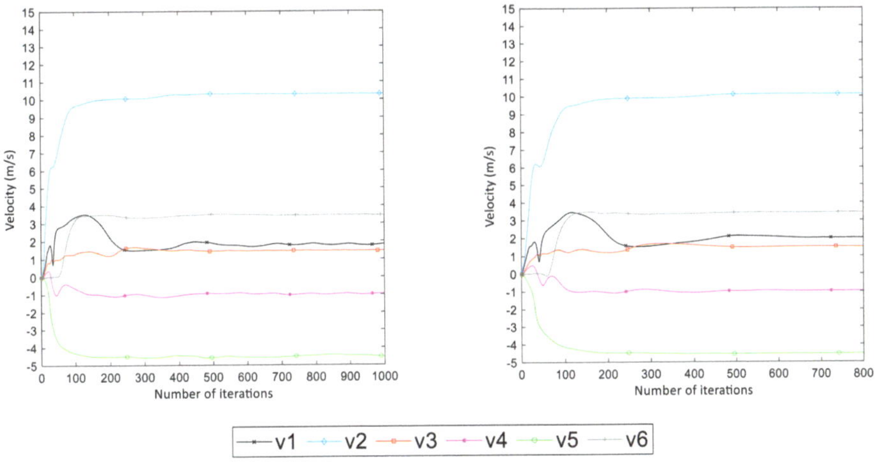

(a) (b)

Figure 7. Convergence of monitored variables: (**a**) k−ω SST; (**b**) k−ε RNG model.

4. Experimental Tests

A series of experiments were performed by recording the velocities and temperatures inside the climatic chamber. A unidirectional TA410 anemometer, with an accuracy of ±5%, and a calibrated type-K thermocouple, with an accuracy of ±0.3 °C, were installed to measure the velocity and temperature at 27 points inside the chamber. The positions of the monitoring points are shown in Figure 8.

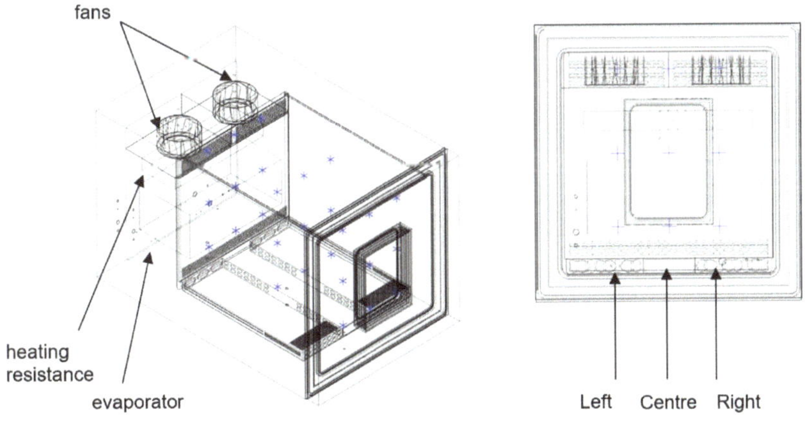

Figure 8. Positions of the monitoring points.

At this stage, the set point temperature of the equipment was set to 25 °C and was kept constant during the measurement. The velocity was measured for the three planes, including the left, centre, and right sections, and for three different distances from the door and the ceiling. Nine points were used in each plane. Figure 9 shows a comparison between the average results for each plane obtained from experimental measurements and those obtained from CFD modelling four turbulence models. The largest difference between the experimental and numerical analyses was observed for the velocity field over the top section of the chamber, where the larger velocity gradient is generated through the flow leaving the fans. This can be justified by the simplified fan model, which assumes a 1D flow through the pressure jump considered and ignores the 3D rotating fans in the CFD model. It is important to note that the difference in the precision of positioning the measuring devices relative to the monitoring points in the numerical model can make some differences. This also justifies the asymmetry obtained from experimental results for the left and right measuring columns. However, this difference is smaller for the outputs from the k−ε RNG model. Overall, the best experimental agreement was observed for the k−ε RNG model, with an average absolute error of 0.67 m/s.

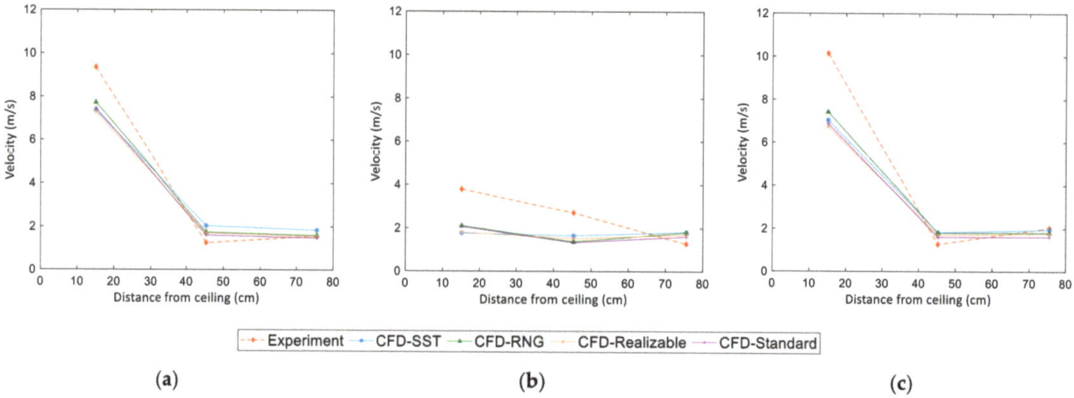

Figure 9. Comparison between numerical modelling and experimental outputs: (**a**) left column, (**b**) centre column, (**c**) right column.

The internal temperature field and humidity were also assessed at 23 points inside the test chamber. The temperature was measured with type-K thermocouples, and the relative humidity was measured using HIH-4000-004 sensors, which have an accuracy of ±3.5%. To monitor and acquire data from the sensors, a Keysight 34970A datalogger with 34901A multiplexers was used. The experimental results were compared with a numerical modelling output (for the four turbulence models). Figure 10 illustrates the comparison of experimental and numerical values for the k−ω SST model and k−ε RNG model for different measured points inside the chamber for different temperature and relative humidity set points: T = 10 °C, RH = 90%; T = 30 °C, RH = 65%; and T = 80 °C, RH = 60%. The results show that there is a small difference between experimental and numerical values, the maximum absolute error being 0.59 and 0.54 °C for temperature and 2.2% and 2.3% for relative humidity, for the k−ω SST and k−ε RNG models, respectively.

Figure 11 illustrates the measured average temperature and relative humidity for the bottom, middle, and top sections inside the chamber for different temperature and relative humidity set points: T = 10 °C, RH = 90%; T = 30 °C, RH = 65%; and T = 80 °C, RH = 60%. The negligible differences between the recorded values from different sections of the inner chamber show the homogeneous distribution of temperature and relative humidity inside the chamber.

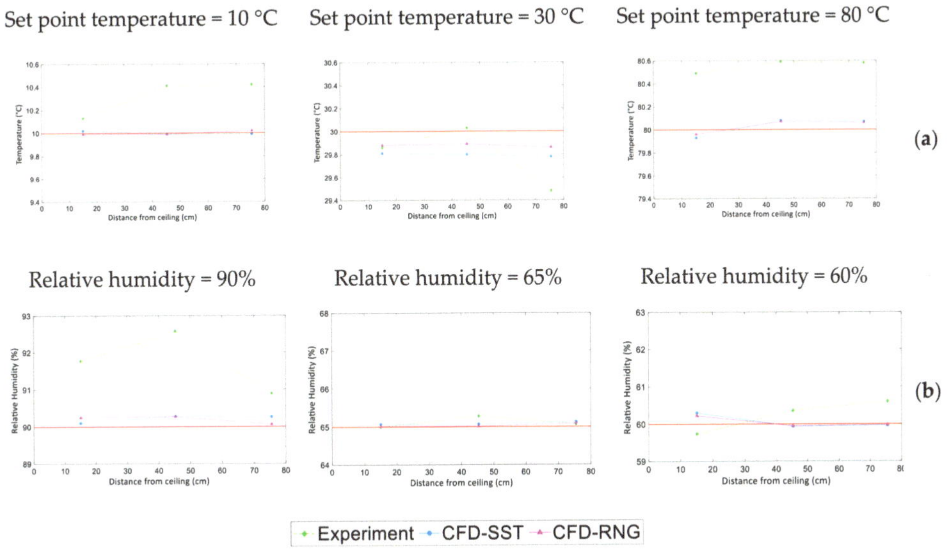

Figure 10. Comparison between numerical modelling and experimental outputs: (**a**) temperature; (**b**) relative humidity inside the chamber for different set points, experimental vs. numerical results.

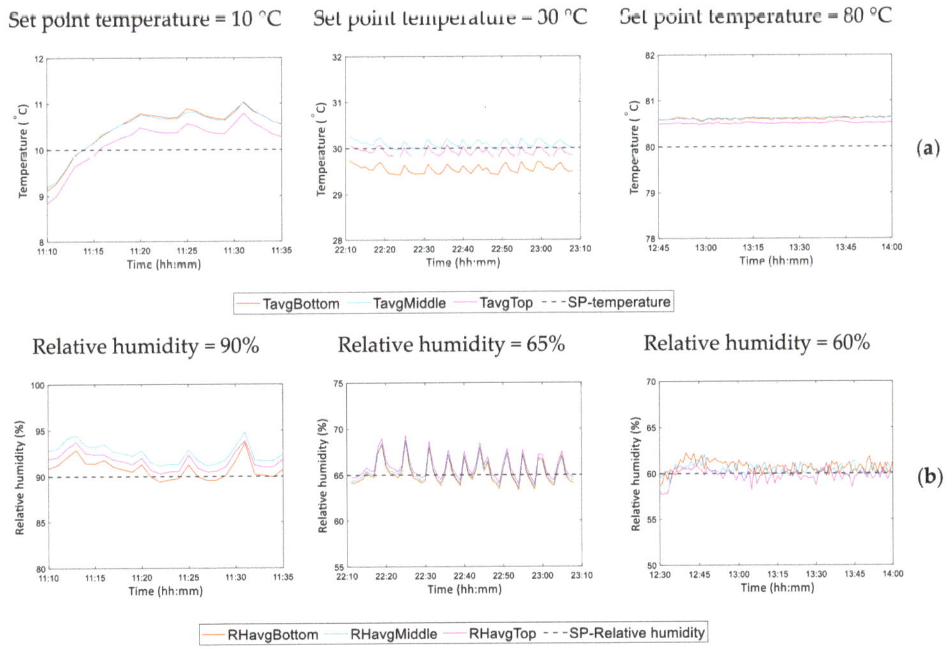

Figure 11. Measured (**a**) temperatures; (**b**) relative humidity inside the climatic chamber (SP-temperature: set point temperature; SP-relative humidity: set point relative humidity; TavgBottom: average temperature, bottom section of inside chamber; RHavgBottom: average relative humidity, bottom section of inside chamber; TavgMiddle: average temperature, middle section of inside chamber; RHavgMiddle: average relative humidity, middle section of inside chamber; TavgTop: average temperature, top section of inside chamber; RHavgBottom: average relative humidity, top section of inside chamber).

5. Results and Analysis

The thermofluidic performance of the climatic chamber was analysed by means of several simulations for the varying temperature of the air flow inside the inner chamber. The results for the velocity and temperature field are presented in the next sections, and the thermal loss variation with the air temperature is obtained.

5.1. Velocity Field

Figure 12 shows the results from the simulated velocity field over the vertical midplane of the chamber for the $k-\omega$ SST model. The velocity field shows the presence of the high-velocity zones around the top corners of the chamber. This is related to the fans' rotating direction, which causes high momentum of the air flow as it enters the inner chamber, mostly from the top corners.

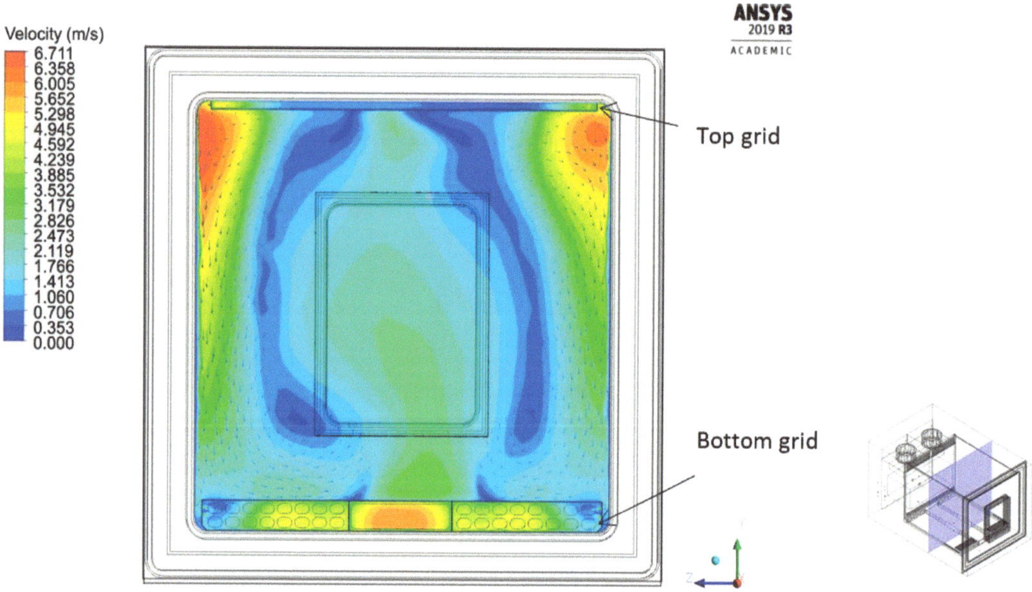

Figure 12. Front view of the velocity field in the vertical midplane (x/W = 0.4) obtained using the $k-\omega$ SST model The asymmetry in the fluid flow can be explained by the evaporator not being positioned symmetrically, and so it allows more fluid flow from one side than from the other. Figure 13 shows the details of the asymmetric position of the evaporator in the chamber and the velocity field viewed from above for the $k-\omega$ SST model.

Figure 14 shows the velocity field over the lateral plane, cutting half the left fan, using the $k-\omega$ SST model. As predicted, the high velocity zones occur around the inflow and outflow of the test zone. However, a homogeneous air flow persists over the central zone of the test box. Therefore, it can be expected that it would be the best location to keep the specimens during testing, given the uniform velocity field and hence uniform temperature distribution.

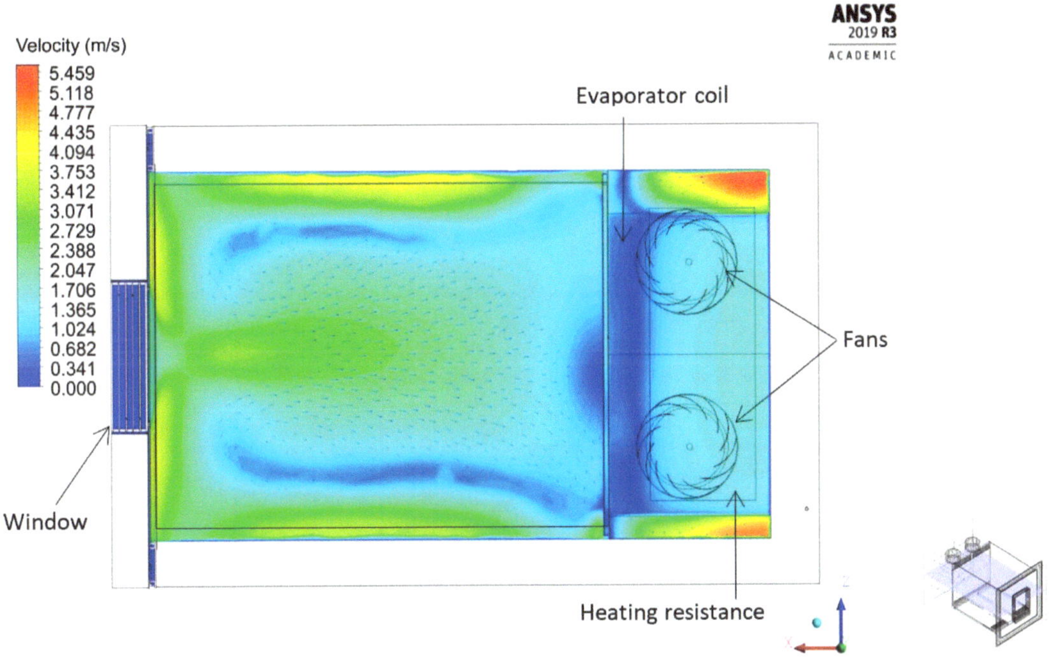

Figure 13. Top view of the velocity field on the horizontal midplane (y/H = 0.12) obtained using the k−ω SST model.

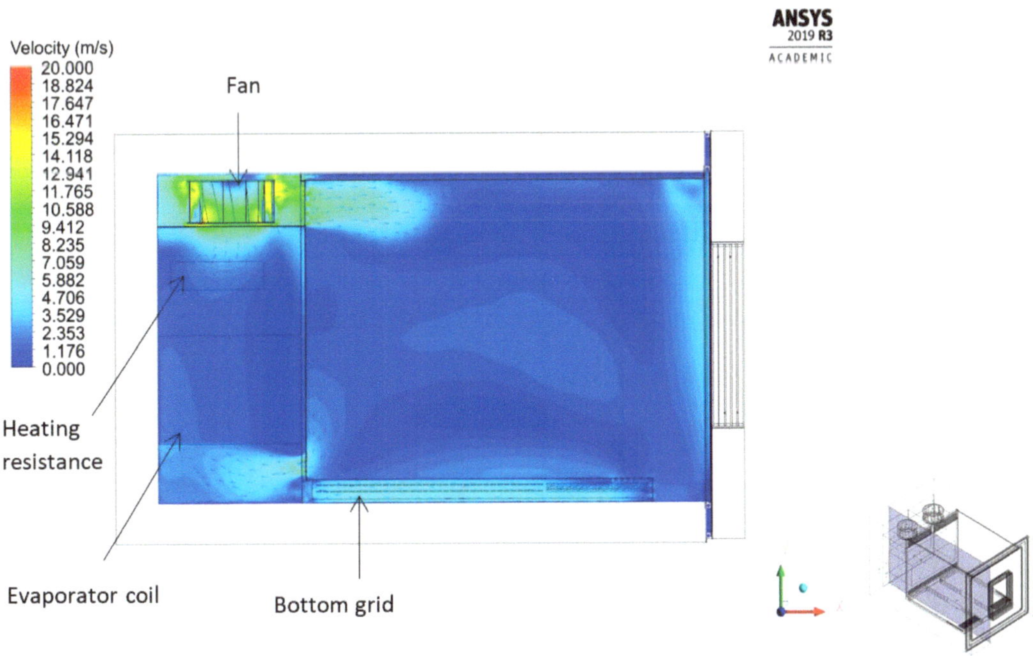

Figure 14. Lateral view of the velocity field in the vertical plane cutting half the left fan (z/L = 0.25) obtained using the k−ω SST model.

5.2. Humidity Field

Figure 15 illustrates the relative humidity distribution inside the chamber on the lateral midplane for temperature and relative humidity set points of 80 °C and 60%, respectively, for the k−ω SST model. It shows that the relative humidity is able to be diffused homogeneously throughout the chamber and stay in the desired set point. The highest concentration of humidity is observed in the top region of the chamber, above the top grid, and around the bottom area of the door, where the velocity field is around zero.

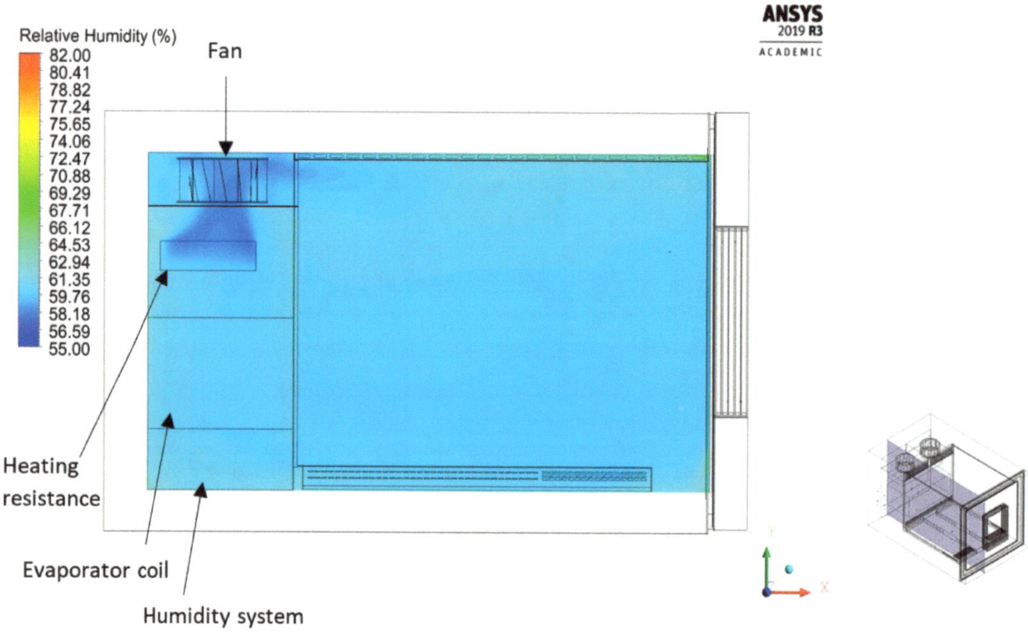

Figure 15. Relative humidity distribution on the lateral midplane (z/L = 0.5) obtained using the k−ω SST model.

The humidity field inside the chamber is shown from the top view on the horizontal midplane for the k−ω SST model in Figure 16. It shows an excellent humidity distribution in the middle region of the chamber with a standard deviation of 0.15% of relative humidity.

5.3. Temperature field

The temperature field was simulated for several fixed air temperatures inside the chamber (considering the humidity system is off). Figure 17 shows the temperature field for an inside temperature equal to 120 °C on the outside surface of the side wall of the chamber for the k−ω SST model, which represents the predicted thermal bridge at the door–wall junction.

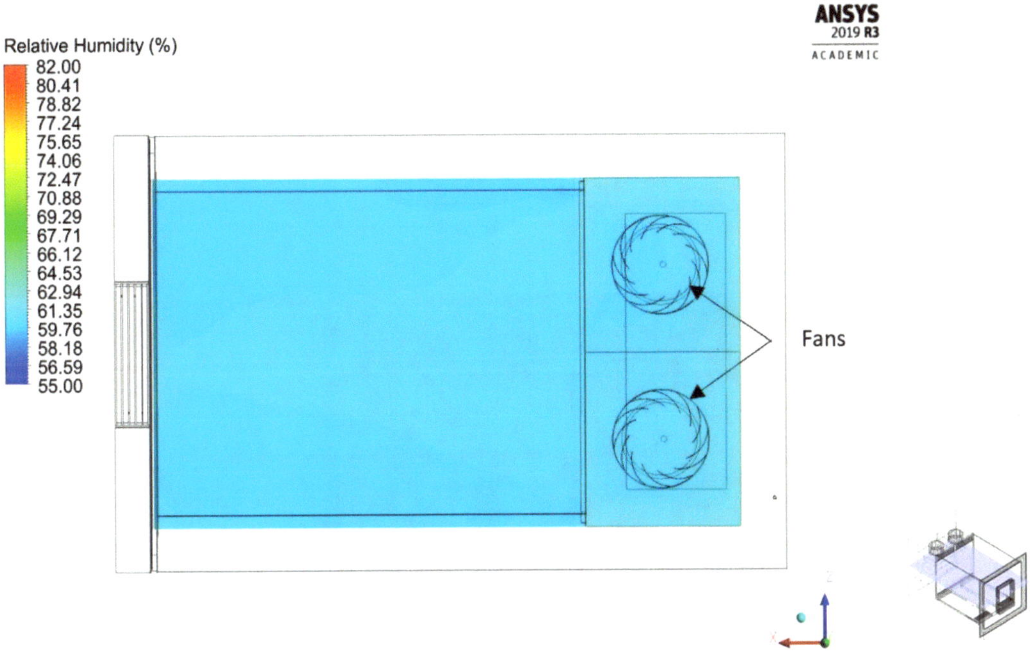

Figure 16. Top view of the relative humidity distribution on the horizontal midplane (y/H = 0.12) obtained using the k−ω SST model.

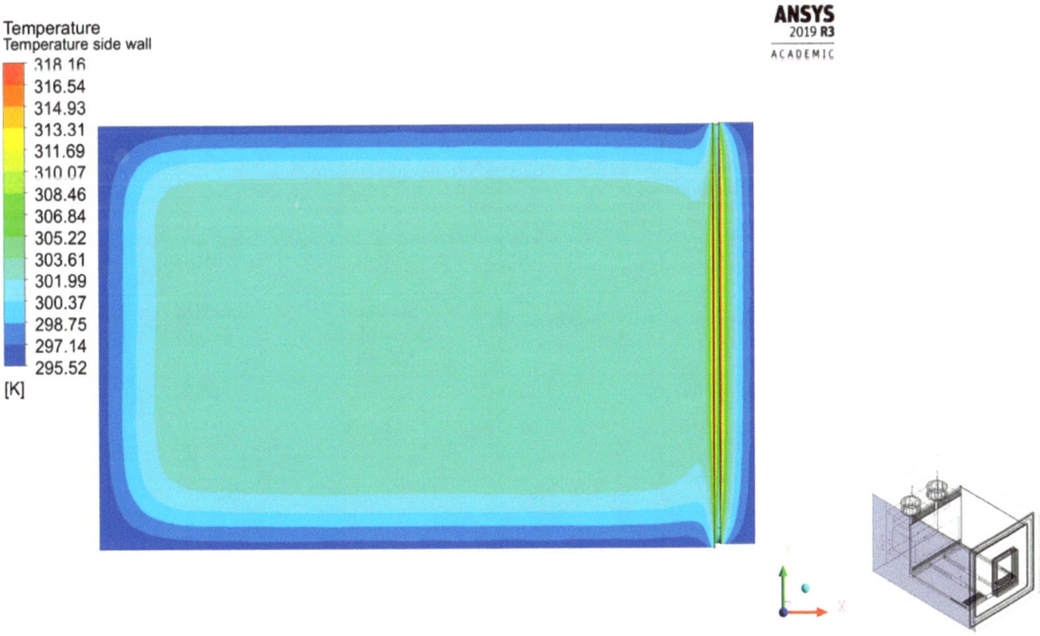

Figure 17. Temperature field on the side wall of the chamber obtained using the k−ω SST model.

The temperature distribution for an inside temperature equal to 120 °C on the vertical midplane of the chamber for the k−ω SST model is shown in Figure 18. From this result, it can be predicted that the temperature field is able to remain homogeneous inside the chamber during steady-state operation. It also shows how using the six-layer window is effective in reducing heat loss from the chamber.

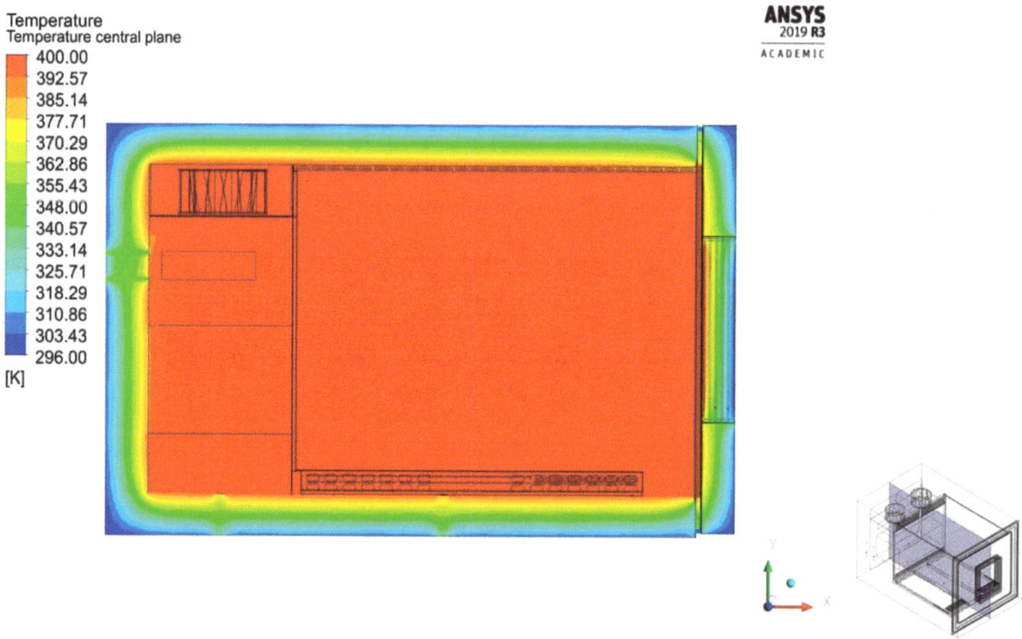

Figure 18. Temperature distribution on the lateral midplane (z/L = 0.5) obtained using the k−ω SST model.

The computations were repeated for all mentioned turbulence models. As shown in Table 3, negligible differences were found in the final outputs using the different turbulence models.

Table 3. Comparison of the average velocities and standard deviation of velocity, temperature, and relative humidity results inside the testing area obtained using the four different turbulence models.

Turbulence Model	Average Velocity	Standard Deviation of Velocity	Standard Deviation of Temperature	Standard Deviation of Humidity
k−ω SST	2.25 m/s	1.07 m/s	0.61 °C	0.72%
k−ε RNG	2.22 m/s	1.11 m/s	0.63 °C	0.76%
k−ε realisable	2.27 m/s	1.13 m/s	0.66 °C	0.74%
k−ε standard	2.14 m/s	1.14 m/s	0.67 °C	0.73%

5.4. *Thermal Losses*

The thermal losses from the chamber to the outside air were obtained for several fixed inner temperatures and for an external temperature of 22 °C. The results for the various components surrounding the inner chamber for a fixed inside temperature of 120 °C are shown in Table 4. The results show that 92% of the thermal loss of the chamber is through the insulation surrounding the external walls and door, which is mainly due to the large side area of the chamber (1258 m^2) compared with the other components (33.3 m^2). Heat loss through the thermal bridge created by the metal panel in the wall–door junction is responsible for 3.6% of the total heat loss. The smallest heat losses were recorded for the glass window and piping (including the fans' motors connecting shafts) with 2.7% and 1% of the total heat loss, respectively.

Table 4. Thermal losses.

Component	Heat Loss (W)
Walls	720.086
Window	20.981
Piping	8.858
Wall–door junction	28.246
Total	778.171

The total heat loss was also determined experimentally by recording the inside and outside temperature of the chamber as well as the energy consumption for a period of 2 days, and used the following equation (according to standard EN12897):

$$Q_{st} = E\left(\frac{120 - T_A}{T_i - T_A}\right) \tag{16}$$

where Q_{st} is permanent energy losses in (kWh/24 h), E is electrical energy consumption over a period of 24 h in (kWh/24 h), T_i is the interior temperature of the chamber over a period of 24 h in "°C", and T_A is the ambient temperature of the chamber for a 24 h period in "°C". The total heat losses for days 1 and 2 were calculated to be 665.15 and 791.32 W, respectively. Therefore, the average heat loss was 728.24 W, which is in good agreement with the numerical result (6% error).

To examine the effect of thermal insulation on the thermal performance of the chamber, the current thermal insulation was replaced by polyurethane insulation, and the total heat loss was calculated. The thermal and physical properties of polyurethane are listed in Table 2.

The simulations were repeated for different set point temperatures for both rock−wool and polyurethane. The total heat loss from the walls considering polyurethane as the thermal insulation was determined to be 728.44 W for a temperature set point of 120 °C, which shows a 6.4% decrease in the thermal loss from the walls compared with the rock−wool.

Figure 19 shows that the total heat loss from the chamber increases as a linear function with the air temperature inside the chamber. Therefore, as expected, the higher the set point temperature, the higher the predicted thermal loss.

5.5. Changing the Air Flow Pattern

Two different configurations were modelled numerically to optimise the air flow pattern inside the chamber: changing the fans' rotation direction and adding two deflectors behind the fans. Figure 20 shows the velocity field on the lateral and horizontal midplane of the chamber for both configurations. It shows that the inclusion of deflectors has a very small effect on the air flow field in the chamber. It also demonstrates that by reversing the fans' rotation direction, the velocity field shows an asymmetric pattern over the chamber, which can be explained by the asymmetric position of the evaporator. The standard deviations of velocity over the central midplane of the chamber were found to be equal to 1.19 and 1.75 m/s in the presence of deflectors and in inverting the fans' rotation, respectively.

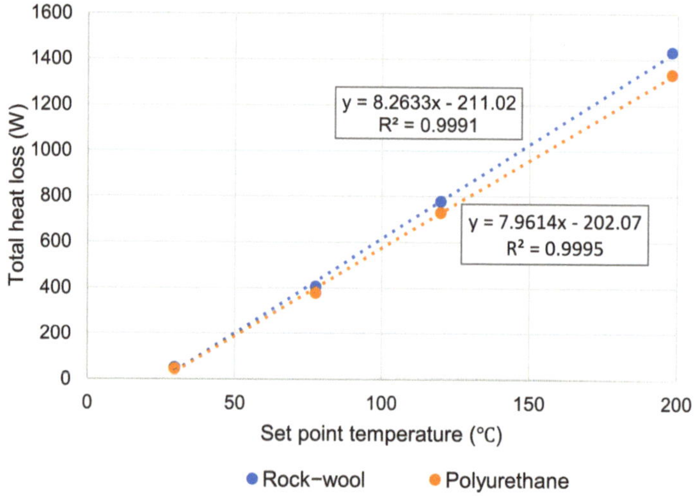

Figure 19. Variation of total heat loss of the chamber with the inside temperature (rock–wool vs. polyurethane).

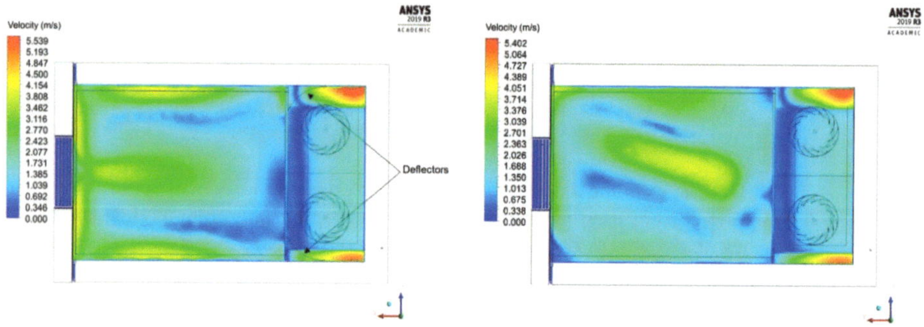

(**a**) (**b**)

Figure 20. Top view of the velocity field on the horizontal midplane (y/H = 0.12): (**a**) fans with added deflectors; (**b**) fans with inverse rotation.

The temperature and relative humidity fields on the lateral midplane of the chamber are presented in Figure 21 for both configurations, for temperature and humidity set points of 80 °C and 60%, respectively. Despite the asymmetric air flow circulation in the chamber, the better temperature distribution resulted from the fans with inverse rotation with a standard deviation of 0.43 °C. This behaviour is probably due to a higher air flow velocity, which helps create a uniform air distribution inside the testing area. With the deflectors added, the temperature and relative humidity distribution are almost identical to the current design, so attaching the deflectors is unnecessary. Even though reversing the rotation of the fans will produce a better temperature distribution, the current design offers a more uniform air flow within the chamber than either of the two other configurations.

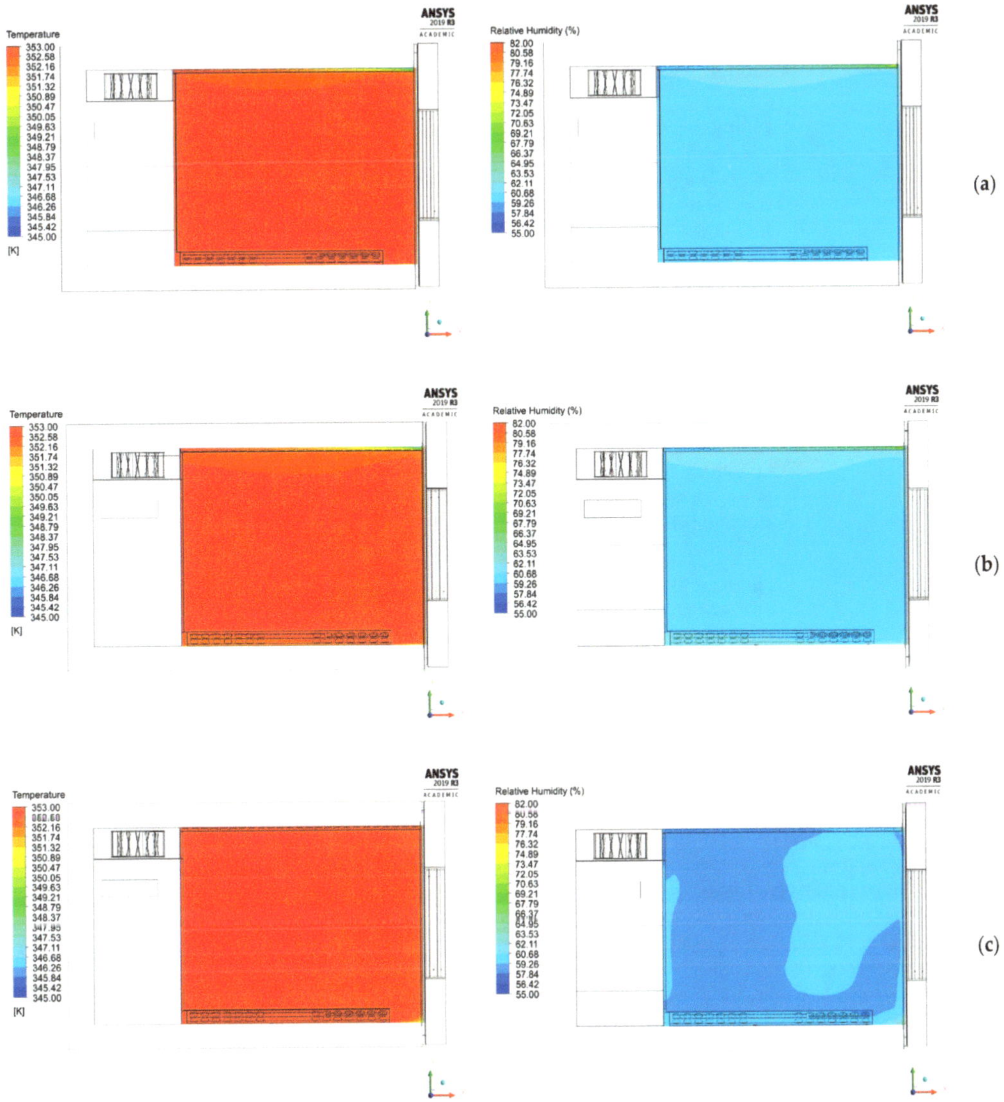

Figure 21. Temperature and relative humidity distribution in the testing area on the lateral midplane (z/L = 0.5): (**a**) current design, (**b**) fans with deflectors, (**c**) fans with inverse rotation.

6. Conclusions

In this study, a thermofluidic 3D model of a climate chamber was developed using Ansys Fluent and validated with experimental data. The numerical modelling was carried out using four different turbulence models: $k-\omega$ SST, $k-\varepsilon$ RNG, $k-\varepsilon$ realisable, and $k-\varepsilon$ standard. The numerical results showed good agreement with the experimental data, with an average absolute error of 0.67 m/s resulting from the $k-\varepsilon$ RNG model for the velocity. The CFD model demonstrated a homogeneous air flow inside the test zone of the chamber for all the turbulence models. Additionally, a uniform temperature distribution over the test zone with a standard deviation of 0.61 °C (for the $k-\omega$ SST model) was predicted. The relative humidity distribution inside the chamber was also modelled using a transport species approach. A homogeneous distribution of humidity was found inside the chamber with a standard deviation of 0.72% (for the k−ω SST model) inside the testing area. Furthermore, the thermal losses through the external components of the chamber, including the walls, front window, wall–door junction, and piping, were modelled, showing that the maximum heat loss was via the walls (720.08 W), followed by the thermal bridge created by the wall–door junction (28.24 W), which was in good agreement with the experimental data, with approximately 6% error. The model was simulated using polyurethane thermal insulation instead of the current rock−wool insulation as a further measure for improvement. The results showed a slight decrease (6.4%) in the total thermal loss from the chamber's walls. Finally, two strategies for modifying the air flow were proposed, by considering deflectors and inverting the fans' rotation. It was concluded that reversing the fans' rotation will result in a better general temperature distribution. Still, the current design appears to outperform both the proposed chamber configurations when it comes to a more uniform air flow within the chamber.

Author Contributions: Conceptualization, B.R., A.T., T.J., M.B. and J.M.; methodology, B.R., A.T., T.J., M.B.; software, B.R.; validation, T.J. and M.B.; formal analysis, B.R., A.T., T.J., M.B. and J.M.; investigation, B.R., M.B.; resources, B.R., A.T., T.J., M.B. and J.M.; data curation, B.R., A.T., T.J. and M.B.; writing—original draft preparation, B.R., M.B. and T.J.; writing—review and editing, B.R., A.T. and M.B.; visualization, T.J., M.B. and J.M.; supervision, A.T.; project administration, A.T.; funding acquisition, A.T. All authors have read and agreed to the published version of the manuscript.

Funding: This research is part of the Project TESTA+ More volume, the same technology (POCI-01-0247-FEDER-038356/LISBOA-01-0247-FEDER-038356) funded by the Operational Program for Competitiveness and Internationalization (POCI) of Portugal 2020, with the support of the European Regional Development Fund (FEDER).

Data Availability Statement: The data presented in this study are available on request from the corresponding author. The data are not publicly available due to companies' privacy policy.

Conflicts of Interest: The authors declare no conflict of interest.

References

1. Liang, Y.Y.; Hu, J.C.; Chen, J.P.; Shen, Y.G.; Du, J. A transient thermal model for full-size vehicle climate chamber. *Energy Build.* **2014**, *85*, 256–264. [CrossRef]
2. Dostál, J.; Ferkl, L. Model Predictive Control of Climatic Chamber with On-off Actuators. *IFAC Proc. Vol.* **2014**, *47*, 4423–4428. [CrossRef]
3. García-Contreras, R.; Gómez, A.; Fernández-Yáñez, P.; Armas, O. Estimation of thermal loads in a climatic chamber for vehicle testing. *Transp. Res. Part D Transp. Environ.* **2018**, *65*, 761–771. [CrossRef]
4. Mensah, K.; Byeong, M.Y.; Min, C.J.; Yoon, S. Study on the performance of a temperature and humidity chamber. In Proceedings of the Korean Society of Mechanical Engineers (KSME) Conference, Yeosu, South Korea, April 2016; pp. 351–352.
5. Candanedo, L.; Handfield, L.; Karava, P.; Bessoudo, M.; Tzempelikos, A.; Athienitis, A. Airflow and thermal simulation in a controlled test chamber with different heating configurations using CFD modeling. In Proceedings of the 2nd Canadian Solar Buildings Conference, Calgary, AB, Canada, 10–14 June 2007.
6. Lecoq, L.; Flick, D.; Plana-Fattori, A.; Laguerre, O. Characterization of evaporation rate, temperature, velocity and humidity fields in a cold chamber. In Proceedings of the 3rd IIR International Conference on Sustainability and the Cold Chain, London, UK, 23–25 June 2014; pp. 267–274.
7. Rogié, B.; Jensen, J.K.; Hansen, S.O.K.; Markussen, W.B. Analysis of Cold Air Recirculation in the Evaporators of Large-Scale Air-Source Heat Pumps Using CFD Simulations. *Fluids* **2020**, *5*, 186. [CrossRef]
8. Liu, Y.; Liu, S.; Zhang, H. Numerical Simulation of Heat and Moisture Transfer in Deep Air Buried. *Procedia Eng.* **2017**, *205*, 1927–1933. [CrossRef]
9. Vojkuvkova, P.; Sikula, O.; Weyr, J. Assessment of condensation of water vapor in the mixing chamber by CFD method. *EPJ Web Conf.* **2015**, *92*, 02112. [CrossRef]
10. Shojaee Nasirabadi, P.; Jabbari, M.; Hattel, J.H. CFD simulation and statistical analysis of moisture transfer into an electronic enclosure. *Appl. Math. Model.* **2017**, *44*, 246–260. [CrossRef]
11. Silva, R.; Brett, M.; Ferreira, A.D.; Serra, C.; Jesus, T.; Fino, M.; Tadeu, A.; Mendes, J.; Araújo, J.; Santos, R. Computational Fluid Dynamics Modelling and Experimental Validation of the Thermofluidic Performance of Climatic Chambers. *J. Therm. Sci. Eng. Appl.* **2020**, *12*, 021014. [CrossRef]
12. Ferziger, J.H.; Peric, M. *Computational Methods for Fluid Dynamics*, 3rd ed.; Springer: Berlin/Heidelberg, Germany, 2012.

13. Versteeg, H.K.; Malalasekera, W. *An Introduction to Computational Fluid Dynamics: The Finite Volume Method*, 2nd ed.; Pearson Education: Glasgow, UK, 2007.
14. Tu, J.; Yeoh, G.H.; Liu, C. *Computational Fluid Dynamics: A Practical Approach*, 2nd ed.; Butterworth Heinemann: Oxford, UK, 2012.
15. Karypis, G.; Kumar, V. *A Software Package for Partitioning Unstructured Graphs, Partitioning Meshes, and Computing Fill-Reducing Orderings of Sparse Matrices*; University of Minnesota, Department of Computer Science and Engineering, Army HPC Research Center: Minneapolis, MN, USA, 1998.
16. Karypis, G. METIS and ParMETIS. In *Encyclopedia of Parallel Computing*; Padua, D., Ed.; Springer: Boston, MA, USA, 2011; pp. 1117–1124. [CrossRef]
17. ANSYS, Inc. ANSYS FLUENT 12.0/12.1 Documentation. Available online: https://www.afs.enea.it/project/neptunius/docs/fluent/ (accessed on 26 August 2021).
18. Cengel, Y.A.; Cimbala, J.M. *Fluid Mechanics-Fundamentals and Applications*; McGraw-Hill Higher Education: Boston, MA, USA, 2006.
19. Menter, F.R.; Kuntz, M.; Langtry, R.B. Ten years of industrial experience with the SST turbulence model. *Heat Mass Transf.* **2003**, *4*, 625–632.
20. Durbin, P.A.; Reif, B.A. *Reynolds Averaged Navier-Stokes Equations. Statistical Theory and Modeling for Turbulent Flow*, 2nd ed.; John Wiley & Sons, Ltd.: London, UK, 2010; pp. 45–56. [CrossRef]
21. Bergman, T.L.; Lavine, A.S.; Incropera, F.P.; Dewitt, D.P. *Fundamentals of Heat and Mass Transfer*, 7th ed.; John Wiley & Sons: Hoboken, NJ, USA, 2011.
22. Churchill, S.W.; Chu, H.H.S. Correlating equations for laminar and turbulent free convection from a vertical pate. *Int. J. Heat Mass Transf.* **1975**, *18*, 1323–1329. [CrossRef]
23. Rich, B.R. An investigation of heat transfer from an inclined flat plate in free convection. *Trans. ASME* **1953**, *75*, 489–499.
24. Shah, R.K.; Sekulic, D.P. *Fundamentals of Heat Exchanger Design*; John Wiley & Sons: Hoboken, NJ, USA, 2003. [CrossRef]
25. Keenan, J.H.; Chao, J.; Kaye, J. *Gas Tables: International Version*, 2nd ed.; John Wiley & Sons: Hoboken, NJ, USA, 1983.
26. Touloukian, Y.S. *Thermophysical Properties of Matter: The TPRC Data Series*; A Comprehensive Compilation of Data; Purdue University, Thermophysical and Electronic Properties Information Center: Lafayette, IN, USA, 1970; Volume 1.
27. Engineering Toolbox. Available online: http://www.engineeringtoolbox.com (accessed on 16 March 2021).
28. Abdou, A.A.; Budaiwi, I.M. Comparison of thermal conductivity measurements of building insulation materials under various operation temperatures. *J. Build. Phys.* **2005**, *29*, 171–184. [CrossRef]

MDPI AG
Grosspeteranlage 5
4052 Basel
Switzerland
Tel.: +41 61 683 77 34

Fluids Editorial Office
E-mail: fluids@mdpi.com
www.mdpi.com/journal/fluids

MDPI

www.ingramcontent.com/pod-product-compliance
Lightning Source LLC
Chambersburg PA
CBHW041413210926
51850CB00013B/3259
* 9 7 8 3 7 2 5 8 3 5 7 6 8 *